Lipid Peroxidation

Lipid Peroxidation

Edited by **Donna Thompson**

New York

Published by Callisto Reference,
106 Park Avenue, Suite 200,
New York, NY 10016, USA
www.callistoreference.com

Lipid Peroxidation
Edited by Donna Thompson

International Standard Book Number: 978-1-63239-449-1 (Hardback)

Printed in the United States of America.

Contents

Preface

Lipid peroxidation has been an area of interest for scientists across the globe. This book focuses on current advances made in lipid peroxidation. The data compiled in this book has been contributed by researchers with extensive experience in various fields of study. We hope that the matter provided here is comprehensible to a wide audience, not only experts but also students interested in the above stated topics. This book includes topics related to the chemical mechanisms of lipid peroxidation and their biological implications. It also includes an analysis of the processes of lipid peroxidation. Furthermore, the book elucidates lipid peroxidation in vegetables, oils, plants and meats.

This book is a result of research of several months to collate the most relevant data in the field.

When I was approached with the idea of this book and the proposal to edit it, I was overwhelmed. It gave me an opportunity to reach out to all those who share a common interest with me in this field. I had 3 main parameters for editing this text:

1. Accuracy – The data and information provided in this book should be up-to-date and valuable to the readers.
2. Structure – The data must be presented in a structured format for easy understanding and better grasping of the readers.
3. Universal Approach – This book not only targets students but also experts and innovators in the field, thus my aim was to present topics which are of use to all.

Thus, it took me a couple of months to finish the editing of this book.

I would like to make a special mention of my publisher who considered me worthy of this opportunity and also supported me throughout the editing process. I would also like to thank the editing team at the back-end who extended their help whenever required.

Editor

Lipid Peroxidation: Chemical Mechanisms, Antioxidants, Biological Implications

Lipid Peroxidation: Chemical Mechanism, Biological Implications and Analytical Determination

Marisa Repetto, Jimena Semprine and Alberto Boveris

Additional information is available at the end of the chapter

1. Introduction

Currently, lipid peroxidation is considered as the main molecular mechanisms involved in the oxidative damage to cell structures and in the toxicity process that lead to cell death. First, lipid peroxidation was studied for food scientists as a mechanism for the damage to alimentary oils and fats, nevertheless other researchers considered that lipid peroxidation was the consequence of toxic metabolites (e.g. CCl4) that produced highly reactive species, disruption of the intracellular membranes and cellular damage (Dianzani & Barrera, 2008).

Lipid peroxidation is a complex process known to occur in both plants and animals. It involves the formation and propagation of lipid radicals, the uptake of oxygen, a rearrangement of the double bonds in unsaturated lipids and the eventual destruction of membrane lipids, with the production of a variety of breakdown products, including alcohols, ketones, alkanes, aldehydes and ethers (Dianzani & Barrera, 2008).

In pathological situations the reactive oxygen and nitrogen species are generated at higher than normal rates, and as a consequence, lipid peroxidation occurs with α-tocopherol deficiency. In addition to containing high concentrations of polyunsaturated fatty acids and transition metals, biological membranes of cells and organelles are constantly being subjected to various types of damage (Chance et al., 1979; Halliwell & Gutteridge, 1984). The mechanism of biological damage and the toxicity of these reactive species on biological systems are currently explained by the sequential stages of reversible oxidative stress and irreversible oxidative damage. Oxidative stress is understood as an imbalance situation with increased oxidants or decreased antioxidants (Sies, 1991a; Boveris et al., 2008). The concept implies the recognition of the physiological production of oxidants (oxidizing free-radicals and related species) and the existence of operative antioxidant defenses. The imbalance concept recognizes the physiological effectiveness of the antioxidant defenses in maintaining both oxidative stress and cellular damage at a minimum level in physiological conditions (Boveris et al., 2008).

Lipid peroxidation is a chain reaction initiated by the hydrogen abstraction or addition of an oxygen radical, resulting in the oxidative damage of polyunsaturated fatty acids (PUFA). Since polyunsaturated fatty acids are more sensitive than saturated ones, it is obvious that the activated methylene (RH) bridge represents a critical target site. The presence of a double bond adjacent to a methylene group makes the methylene C-H bond weaker and therefore the hydrogen in more susceptible to abstraction. This leaves an unpaired electron on the carbon, forming a carbon-centered radical, which is stabilized by a molecular rearrangement of the double bonds to form a conjugated diene which then combines with oxygen to form a peroxyl radical. The peroxyl radical is itself capable of abstracting a hydrogen atom from another polyunsaturated fatty acid and so of starting a chain reaction (Halliwell & Gutteridge, 1984) (Fig. 1).

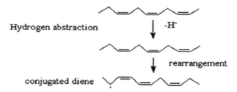

Figure 1. Initiation step of lipid peroxidation process.

Molecular oxygen rapidly adds to the carbon-centered radicals (R·) formed in this process, yielding lipid peroxyl radicals (ROO·). Decomposition of lipid peroxides is catalyzed by transition metal complexes yielding alcoxyl (RO·) or hydroxyl (HO·) radicals. These participate in chain reaction initiation that in turn abstract hydrogen and perpetuate the chain reaction of lipid peroxidation. The formation of peroxyl radicals leads to the production of organic hydroperoxides, which, in turn, can subtract hydrogen from another PUFA. This reaction is termed propagation, implying that one initiating hit can result in the conversion of numerous PUFA to lipid hydroperoxides. In sequence of their appearance, alkyl, peroxyl and alkoxyl radicals are involved. The resulting fatty acid radical is stabilized by rearrangement into a conjugated diene that retains the more stable products including hydroperoxides, alcohols, aldehydes and alkanes. Lipid hydroperoxide (ROOH) is the first, comparatively stable, product of the lipid peroxidation reaction (Halliwell & Gutteridge, 1984) (Fig. 2).

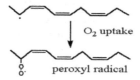

Figure 2. Initial phase of the propagation step of lipid peroxidation process indicating the oxygen uptake.

Reduced iron complexes (Fe^{2+}) react with lipid peroxides (ROOH) to give alkoxy radicals, whereas oxidized iron complexes (Fe^{3+}) react more slowly to produce peroxyl radicals. Both radicals can take part in the propagation of the chain reaction. The end products of these complex metal ion-catalyzed breakdowns of lipid hydroperoxides include the cytotoxic aldehydes and hydrocarbon gases such as ethane.

The free radical chain reaction propagates until two free radicals conjugate each other to terminate the chain. The reaction can also terminate in the presence of a chain-breaking anti-oxidant such as vitamin E (α-tocopherol) (Halliwell & Gutteridge, 1984).

In conditions in which lipid peroxidation is continuously initiated it gives non-radical products destroying two radicals at a time. In the presence of transition metal ions, ROOH can give rise to the generation of radicals capable of re-initiating lipid peroxidation by redox-cycling of these metal ions (Halliwell & Gutteridge, 1984).

Lipid peroxidation causes a decrease in membrane fluidity and in the barrier functions of the membranes. The many products of lipid peroxidation such as hydroperoxides or their aldehyde derivatives inhibit protein synthesis, blood macrophage actions and alter chemotactic signals and enzyme activity (Fridovich & Porter, 1981).

2. Biological implications of lipid peroxidation

The biological production of reactive oxygen species primarily superoxide anion (O_2^-) and hydrogen peroxide (H_2O_2) is capable of damaging molecules of biochemical classes including nucleic acids and aminoacids. Exposure of reactive oxygen to proteins produces denaturation, loss of function, cross-linking, aggregation and fragmentation of connective tissues as collagen (Chance et al., 1979). However, the most damaging effect is the induction of lipid peroxidation. The cell membrane which is composed of poly-unsaturated fatty acids is a primary target for reactive oxygen attack leading to cell membrane damage.

The lipid peroxidation of polyunsaturated fatty acids may be enzymatic and non-enzymatic. Enzymatic lipid peroxidation is catalyzed by the lipoxygenases family, a family of lipid peroxidation enzymes that oxygenates free and esterified PUFA generating as a consequence, peroxy radicals. Non enzymatic lipid peroxidation and formation of lipid-peroxides are initiated by the presence of molecular oxygen and is facilitated by Fe^{2+} ions (Repetto et al., 2010a).

Oxidative breakdown of biological phospholipids occurs in most cellular membranes including mitochondria, microsomes, peroxisomes and plasma membrane. The toxicity of lipid peroxidation products in mammals generally involves neurotoxicity, hepatotoxicity and nephrotoxicity (Boveris et al., 2008). The principal mechanism involves detoxification process in liver. Toxicity from lipid peroxidation affect the liver lipid metabolism where cytochrome P-450s is an efficient catalyst in the oxidative transformation of lipid derived aldehydes to carboxylic acids adding a new facet to the biological activity of lipid oxidation metabolites. Cytochrome P-450-mediated metabolism operates in parallel with other metabolic transformations of aldehydes; hence, the P450s could serve as reserve or compensatory mechanisms when other high capacity pathways of aldehyde elimination are compromised due to disease or toxicity. Finally, 4-hydroxynonenal (HNE), unsaturated aldehydes, such as acrolein, trans-2-hexenal, and crotonaldehyde, are also food constituents or environmental pollutants, P-450s may be significant in favoring lipid peroxidation that has significant downstream effects and possibly play a major role in cell signaling pathways. Oxidized lipids appear to have a signaling function in pathological situations, are pro-inflammatory agonists and contribute to neuronal death under conditions in which membrane lipid peroxidation occurs. For example, mitochondrial lipid cardiolipin makes up to 18% of the total phospholipids and 90% of the fatty acyl chains are unsaturated. Oxidation of cardiolipin may be one of the critical factors initiating apoptosis by liberating cytochrome c from the mitochondrial inner membrane and facilitating permeabilization of the outer membrane. The release of cytochrome c activates a proteolytic cascade that culminates in apoptotic cell death (Navarro & Boveris, 2009).

Previous results indicate that lipid peroxidation has a role in the pathogenesis of several pathologies as neurodegenerative (Dominguez et al., 2008; Famulari et al., 1996; Fiszman et al., 2003), inflammatory (Farooqui & Farooqui, 2011), infectious (Repetto et al., 1996), gastric (Repetto et al., 2003) and nutritional diseases (Repetto et al., 2010b).

Oxidative damage in liver is associated with hepatic lipid metabolism, and may be affecting the absorption and transport mechanisms of α-tocopherol in this organ. In the liver, the morphological damage is previous to the lipid peroxidation and the consumption of endogenous antioxidants. In kidney and heart, indeed, lipid peroxidation and oxidative damage preceded necrosis (Repetto et al., 2010b).

Lipid peroxidation is a chain reaction process characterized by repetitive hydrogen abstraction by HO· and RO·, and addition of O_2 to alkyl radicals (R·) resulting in the generation of ROO· and in the oxidative destruction of polyunsaturated fatty acids, in which the methylene group (=RH-) is the main target (Halliwell & Gutteridge, 1984).

The association between increased phospholipid oxidation, free-radical mediated reactions and pathological states was early recognized (Cadenas, 1989; Verstraeten et al., 1997; Liu et al., 2003). The contribution by Sies of the concept of oxidative stress followed (Sies, 1991a,1991b) with the implication that increased free-radical mediated reactions, basically by HO· and RO·, would produce phospholipid, protein, lipid, DNA, RNA or carbohydrate oxidation, whatever is close (Halliwell & Gutteridge, 1984). The increased oxidation of the cell biochemical constituents is associated with ultra structural changes in mitochondrial morphology with mitochondrial swelling and increased matrix volume (Boveris et al., 2008). In human liver, the morphological changes can affect the organ structure and function as it is the case for the bile canaliculi that are damaged in liver transplanted patients; a fact that is interpreted as consequence of the oxidative damage that is associated to ischemia-reperfusion (Cutrin et al., 1996). Interestingly, there are reports in rat liver experimental models, of increased peroxidation secondary to increased mitochondrial production of O_2^- and H_2O_2 (Fridovich, 1978; Navarro &Boveris, 2007; Navarro et al., 2009).

3. Chemical mechanisms for lipid peroxidation process

The spectrum of oxygen reactive species that are considered responsible for biological oxygen toxicity include the intermediates of the partial reduction of oxygen, superoxide radical (O_2^-), hydrogen peroxide (H_2O_2), and other reactive species as hydroxyl radicals (HO·), peroxyl radical (ROO·), nitric oxide (NO), peroxinitrite (ONOO-) and singlet oxygen ($^1O_2$).

The biological effects of excess levels of the spectrum of these species are quite similar, and that is the reason they are collectively called reactive oxygen species (ROS). The main free-radical mediated chain reactions in biological systems are summarized in Fig. 3. The Beckman-Radi-Freeman pathway and the Cadenas-Poderoso shunt have been added to the original consecutive reactions of the Fenton/Haber-Weiss pathway and lipid peroxidation process to incorporate NO and ONOO· to the biochemical free-radical mediated chain reaction (Moncada et al., 1991; Boveris et al., 2008) (Fig. 3).

In the last years the denominations "reactive oxygen species" (ROS) and "reactive nitrogen species" (RNS) had became very popular. The ROS denomination involves the three chemical species of the Fenton/Haber-Weiss pathway (O_2^-, H_2O_2 and HO·), the products of the partial reduction of oxygen. Similarly, the RNS denomination is loosely referring to the three chemical species of the Beckman-Radi-Freeman pathway (NO, ONOO·, and NO_2) (Moncada et al., 1991). The reference as a whole to either group, ROS and RNS, is usually made to explain or to refer to their biological activity, what reflects the fact that each group, ROS and RNS, are auto-propagated in biological systems from their promoters, O_2^- and NO. Nevertheless, the advantage and facility in referring to the biological effects implies the ignorance of the biochemistry of the process.

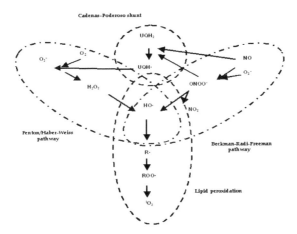

Figure 3. The free-radical mediated chain reaction in biochemistry. O_2^-, superoxide radical; H_2O_2, hydrogen peroxide, HO·, hydroxyl radical; NO, nitric oxide; ONOO·, peroxinitrite; ·NO2, nitrogen dioxide; UQH2, ubiquinol; UQH·, ubisemiquinone; R·, alkyl radical; ROO·, peroxyl radical; $^1O_2$, singlet oxygen.

The individual steps of the free-radical mediated chain reaction of biological systems (Fig. 3) are in majority non-enzymatic second order reactions with fast reaction rates, about 10^7 M^{-1} s^{-1}. The exceptions are the enzymatic dismutation of O_2^- (10^{10} M^{-1} s^{-1}, catalyzed by the antioxidant enzyme superoxide dismutase, SOD), the first order reaction of decomposition of ONOO⁻, and the relatively lower rate (10^5 M^{-1} s^{-1}) of the homolysis of H_2O_2 catalyzed by Fe^{2+} (Boveris et al., 2008).

Concerning the molecular mechanisms that produces lipid peroxidation in biological systems previous, it is accepted that lipid peroxidation may be a consequence of a) intermediates of the partial reduction of oxygen (homolysis of H_2O_2 and HO generation), b) direct autoxidation of lipids, c) intermediates of the nitric oxide metabolism, and d) modifications of lipid membrane surface structure (Fridovich & Porter, 1981; Boveris et al., 2008; Navarro & Boveris, 2009; Repetto et al., 2010a;).

The lipid peroxidation process is induced for the pro-oxidant effect of transition metals. A vast evidence supports the occurrence of reactions of metal ions with H_2O_2, and hydroperoxides in the cytosol and in biological membranes. The latter ones are the main target of oxidative damage. In other words, by one mechanism, transition metals produce lipid peroxidation by stimulation of the oxidative capacity of H_2O_2 by promoting free-radical mediated processes (Fridovich, 1978; Moncada et al., 1991; Verstraeten et al., 1997; Repetto et al., 2010a; Repetto & Boveris, 2012), and by another mechanism, they bind to negatively charged phospholipids which alters the physical properties of the bilayer and favors the initiation and propagation reactions of lipid peroxidation (Repetto et al., 2010a; Repetto & Boveris, 2012).

Lipid peroxidation is a chain reaction initiated by hydrogen abstraction or by addition of an oxygen radical, resulting in the oxidative damage of polyunsaturated fatty acids (PUFA). Since polyunsaturated fatty acids are more sensitive than saturated ones, it is obvious that the activated methylene (RH) bridge represents a critical target site. This initiation is usually performed by a radical of sufficient reactivity (Eq.1):

$$R_1H + R\cdot \quad \rightarrow \quad R_1\cdot + RH \tag{1}$$

Molecular oxygen rapidly adds to the carbon-centred radical (R·) formed in this process, yielding the lipid peroxyl radical (ROO) (Eq. 2):

$$R\cdot + O_2 \quad \rightarrow \quad ROO \tag{2}$$

The formation of peroxyl radicals leads to the production of organic hydroperoxides, which, in turn, can abstract hydrogen from another PUFA, analogous to reaction 1:

$$R_1H + ROO\cdot \quad \rightarrow \quad R_1\cdot + ROOH \tag{3}$$

This reaction is termed propagation, implying that one initiating hit results in the conversion of numerous PUFA to lipid hydroperoxides.

In the sequence of their appearance, alkyl, peroxyl, and alkoxyl radicals are generated in the free radical chain reaction.

The alkyl radical is stabilized by rearrangement into a conjugated diene that is a relatively stable product.

Lipid hydroperoxide (ROOH) is the first stable product of the lipid peroxidation reaction. Under conditions where lipid peroxidation is continuously initiated, radical anhilation or termination occurs with the destroying of two radicals at once:

$$ROO^{\cdot} \; + \; ROO^{\cdot} \; \rightarrow \; ROH + RO^{\cdot} + {}^1O_2 \tag{4}$$

In the presence of transition metal ions, ROOH gives raise to the generation of radicals capable of (re-)initiating the lipid peroxidation by redox-cycling of the metal ions (Repetto et al., 2010a; Repetto & Boveris, 2012):

$$ROOH + Me^{\,n+} \; \rightarrow \; RO^{\cdot} \; + Me^{\,(n-1)+} \tag{5}$$

$$ROOH + Me^{\,(n-1)+} \; \rightarrow \; ROO^{\cdot} \; + Me^{\,n+} \tag{6}$$

3.1. Autoxidation of lipids: Non-enzymatic lipid peroxidation

Non-enzymatic lipid peroxidation is a free radical driven chain reaction in which one free radical induces the oxidation of lipids, mainly phospholipids containing polyunsaturated fatty acids. Autoxidation of lipids in biological systems is a direct process that occurs by homolysis of endogenous hydroperoxides by scission of ROOH and production of RO and ROO.

The polyunsaturated fatty acids such as linoleic and arachidonic acids, which are present as phosphoglyceride esters in lipid membranes, are particularly susceptible to autoxidation. Moreover, autoxidation in biological systems has been associated with such important pathological events as damage to cellular membranes in the process of aging and the action of certain toxic substance. The autoxidation of most organic substrates in homogeneous solution is a spontaneous free-radical chain process at oxygen partial pressures above 100 torr (Repetto et al., 2010a).

Lipid hydroperoxides, in presence or absence of catalytic metal ions, produce a large variety of products including short and long chain aldehydes and phospholipids and cholesterol ester aldehydes, which provide an equivalent hydrogen abstraction from an unsaturated fatty acid and formation of free radical. The secondary products can be used to assess the degree of lipid peroxidation in a system (Sies, 1991a) (Eq. 7 to 9).

Eq. 7 requires some comments. As written is thermodinamically non spontaneous since it involves the breaking of a C-H bond (435 kJ/mol). However, polyunsaturated fatty acids in solutions are readily autooxidized, likely catalized by transition metal ions. The R· radicals reaction with O_2 yielding ROO·.

$$RH \; \rightarrow \; R^{\cdot} \; + H^{\cdot} \tag{7}$$

$$R^{\cdot} \; + \; O_2 \; \rightarrow \; ROO^{\cdot} \tag{8}$$

$$ROO \cdot + RH \rightarrow ROOH + R \cdot \tag{9}$$

Transition metal ions Fe^{2+} and Cu^{+} stimulate lipid peroxidation by the reductive cleavage of endogenous lipid hydroperoxides (ROOH) of membrane phospholipids to the corresponding alkoxyl (RO·) and peroxyl (ROO·) radicals in a process that is known as ROOH-dependent lipid peroxidation (Eqs. 10 and 11):

$$Fe^{2+} + ROOH \rightarrow RO \cdot + OH^- + Fe^{3+} \tag{10}$$

$$Fe^{3+} + ROOH \rightarrow RO_2 \cdot + H^+ + Fe^{2+} \tag{11}$$

The mechanisms of these two reactions appear to involve the formation of Fe(II)-Fe(III) or Fe(II)-O₂-Fe(III) complexes with maximal rates of HO· radical formation at a ratio Fe(II)/Fe(III) of 1 (Repetto et al., 2010a; Repetto & Boveris, 2012).

Cu^{2+} and Cu^+ are known for their capacity to decompose organic hydroperoxides (ROOH) to form RO· and ROO· (Eqs. 12 and 13) (Sies, 1991a; Repetto et al., 2010a; Repetto & Boveris, 2012).

$$Cu^+ + ROOH \rightarrow RO \cdot + OH^- + Cu^{2+} \tag{12}$$

$$Cu^{2+} + ROOH \rightarrow RO_2 \cdot + H^+ + Cu^+ \tag{13}$$

3.2. Lipid peroxidation generated for intermediates of the partial reduction of oxygen

The physiological generation of the products of the partial reduction of oxygen, O_2^- and H_2O_2, constitute the biological basis of the process of lipid peroxidation in mammalian aerobic cells. From a molecular point of view hydroxyl radical (HO·) generation, formed from H_2O_2 and Fe^{2+} by the Fenton reaction, has been considered for a long time as the likely rate-limiting step for physiological lipid peroxidation (Verstraeten et al., 1997; Repetto & Boveris, 2012). The Fenton reaction and Fenton-like reactions (Eq. 14) are frequently used to explain the toxic effects of redox-active metals (Eq. 5), where $M^{(n)+}$ is usually a transition metal ion:

$$Fe^{2+} + H_2O_2 \rightarrow [Fe(II)H_2O_2] \rightarrow Fe^{3+} + HO \cdot + HO \tag{14}$$

Trace (nM) levels of cellular and circulating active transition metal ions seem enough for the catalysis of a slow Fenton reaction *in vivo*, at the physiological levels of H_2O_2 (0.1-1.0 μM) (Chance et al., 1979).

Reactive oxygen species mainly include O_2^- and H_2O_2, which are physiologically generated as by-products of mitochondrial electron transfer. The formation of O_2^- is originated from the auto-oxidation of the ubisemiquinone of complexes I and III and the production of H_2O_2 occurs by intramitochondrial Mn-SOD catalysis (Navarro & Boveris, 2004; Navarro et al., 2007, 2010). When the electron transfer process is blocked at complexes I and III, electrons pass directly to O_2 producing O_2^-. The reactive oxygen and nitrogen species, although kept in low steady-state concentrations by antioxidant systems, are able to react and damage biomolecules (Fig. 3). Mitochondria are considered the main intracellular source of oxidizing reactive oxygen species (Navarro Boveris, 2004; Navarro et al., 2005, 2009, 2010).

At low level of H_2O_2, Fe^{2+} induces lipid peroxide decomposition, generating peroxyl and alkoxyl radicals and favoring lipid peroxidation. These results indicate that the onset of the Fe^{3+} stimulatory effect on Fe^{2+}-dependent lipid peroxidation is due to reactive oxygen species production via Fe^{2+} oxidation with endogenous ROOH (Repetto & Boveris, 2012).

The Cu^+ ion is considered an effective catalyst for the Fenton reaction (Eq. 15) [3].

$$Cu^+ + H_2O_2 \rightarrow [Cu(I)\text{-}H_2O_2] \rightarrow Cu^{2+} + HO^- + HO^· \qquad (15)$$

The process of lipid peroxidation has been recognized as a free radical-mediated and physiologically occurring with the supporting evidence of in situ organ chemiluminescence (Boveris et al., 1980). The main initiation reaction is understood to be mediated by $HO^·$ or by a ferryl intermediate, both with the equivalent potential for hydrogen abstraction from an unsaturated fatty acid, with formation of an alkyl radical ($R^·$) (Repetto & Boveris, 2012) (Eq. 16):

$$HO^· + RH \rightarrow H_2O + R^· \qquad (16)$$

One effect of the reaction of hydroxyl radicals, their formation catalyzed by iron ions, with lipids is to make those lipids insoluble or fibrotic that can be considered causative of membrane disruption and oxidative damage associated in different pathologies.

3.3. Lipid peroxidation generated from intermediates of the nitric oxide metabolism

An area of interest that has currently increased over the past decades is the study of nitric oxide (NO) since the demonstration, in 1987, of its formation by the enzyme NO synthase in vascular endothelial cells. This NO radical accounts for the properties of the called endothelial derived relaxing factor, is the endogenous stimulator of the soluble guanylate cyclase and is a potent vasodilator *in vitro* (Moncada et al., 1991). Unsaturated fatty acids are susceptible to nitration reactions. The nitric oxide (NO)-derived species are diffusible across membranes, their concentration in the hydrophobic core of membranes and lipoproteins lead to react fast with fatty acids and lipid peroxyl radicals ($ROO^·$) during the lipid oxidadation process generating oxidized and nitrated products of free lipids (arachidonic acid, arachidonate oleate, linoleate) and esterified (cholesteryl linoleate). Lipid nitration process includes *in vivo* different molecular mechanisms: a) NO autooxidation to nitrite, which has oxidant and nitrating properties, b) electrophilic addition of NO relates species to unsaturated fatty acids, c) radical reactions between $ROO^·$ and NO, d) peroxynitrite ($ONOO^-$) derived free radicals mediate oxidation, nitrosation and nitration reactions. These species are considered currently as mediators of adaptive inflammatory responses.

NO is an endogenous mediator of many physiological functions through stimulation of the guanylate cyclase enzyme including the regulation of vascular relaxing, post-traslational protein changes, gene expression and inflammatory cell function (Moncada et al., 1991). Free and esterified fatty acids as arachidonic and linoleic acids are important components of lipoproteins and membranes that may be oxidized for different compounds. The NO and NO-derived radicals react with fatty acids generating oxidized and nitrated species as nitroalkenes and consequently, nitroalcohols. At low oxygen concentrations the most important biological NO derivatives is $ONOO^-$. The nitroalkylation process occurs *in vitro* and in vivo, is involved in redox processes and cell signaling through the reversible covalent bound and post-traslational modifications responsible for structure, function and subcellular distribution of proteins (Valdez et al., 2011) and regulating the pro-inflammatory effect of oxidant exposure (Nair et al., 2007).

A novel mechanism for hydroxyl radical production, which is not dependent on the presence of transition metals, has recently been proposed. This involves the production of peroxynitrite (Beckman et al., 1990, 1994; Rachmilewitz et al., 1993) which has proinflammatory effects *in vitro* (Moncada et al., 1991), from the reaction of NO with O_2^- (Eqs. 17 to 20):

$$NO + O_2^- \quad \rightarrow \quad ONOO^- \tag{17}$$

$$ONOO^- + H^+ \rightarrow ONOOH \tag{18}$$

$$ONOOH \quad \rightarrow \quad HO^- + NO_2 \tag{19}$$

$$2\,H^+ + O_2^- + O_2^- \rightarrow H_2O_2 + O_2^- \tag{20}$$

In pathological situations, macrophages and neutrophils, recruited to a site of injury, are activated to produce NO as part of the inflammatory response. Furthermore, SOD activity rapidly scavenges O_2^- and also prolongs the vaso-relaxant effects of NO (Murphy & Sies, 1991; Hogg et al., 1992; Rachmilewitz et al., 1993).

3.4. Modifications of lipid membrane structure

The presence of cholesterol in cell surface membranes influences their susceptibility to peroxidation, probably both by intercepting some of the radicals present and by affecting the internal structure of the membrane by interaction of its large hydrophobic ring structure with fatty acid-side-chains. As lipid peroxidation precedes in any membrane, several of the products produced have a detergent-like activity, specially released fatty acids or phospholipids with one of their fatty-acid side-chains removed. This will contribute to increased membrane disruption and further peroxidation.

The onset of lipid peroxidation within biological membranes is associated with changes in their physicochemical properties and with alteration of biological function of lipids and proteins. Polyunsaturated fatty acids and their metabolites play physiological roles: energy provision, membrane structure, fluidity, flexibility and selective permeability of cellular membranes, and cell signaling and regulation of gene expression (Catala, 2006). The hydroxyl radical generated as a consequence of the Fenton reaction, oxidizes the cellular components of biological membranes (Fig. 4).

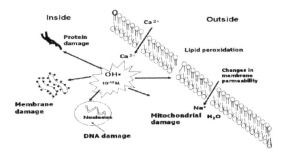

Figure 4. Lipid, DNA and protein oxidative damage from reactive hydroxyl radical.

The binding of positively charged species to a membrane (to the negatively-charged head-groups of phospholipids) can alter the susceptibility of the membrane to oxidative damage. This can be seen as either an enhancement or an inhibition of the rate of lipid peroxidation. Several metal ions such as Ca^{2+}, Co^{2+}, Cd^{2+}, Al^{3+}, Hg^{2+} and Pb^{2+} alter the rate of peroxidation in liposomes, erythrocytes and microsomal membranes, often stimulating the peroxidation induced by iron ions.

In the lipid peroxidation of the brain phosphatidylcholine-phosphatidylserine (PC-PS) liposomes (Repetto et al., 2010a) hydrogen abstraction occurred at the allilic carbons 9 and 10 of the oleic acid chain. Secondary initiation reactions are provided by hydrogen abstraction by $RO\cdot$ and $ROO\cdot$ (Eqs. 21 to 23) at the mentioned tertiary carbons:

$$RO\cdot + RH \rightarrow ROH + R \qquad (21)$$

$$RO_2\cdot + RH \rightarrow ROOH + R\cdot \qquad (22)$$

$$R\cdot + O_2 \rightarrow RO_2\cdot \qquad (23)$$

The $R\cdot$ and $ROO\cdot$ radicals (Eqs.21-23) are central to the free radical-mediated process of lipid peroxidation. The addition reaction of $R\cdot$ with O_2 to yield $ROO\cdot$ (Eq. 23) yield a product that is able to abstract hydrogen atoms and to regenerate $R\cdot$ for a new cycle of the free-radical chain-reaction [38]. The whole process, by repetition of reaction 23, consumes O_2 and produces malondialdehyde ($O=HC-CH_2-CH=O$), 4-hydroxynonenal and other dialdehydes as secondary and end products of lipid peroxidation. The process produces TBARS at an approximate ratio of 0.12 $TBARS/O_2$ and normally utilized as measurement of the rate and extent of lipid peroxidation (Junqueira et al., 2004).

There are two consequences of lipid peroxidation: structural damage to membranes and generation of secondary products. Membrane damage derives from the production of broken fatty acyl chains, lipid-lipid or lipid-protein cross-links, and endocyclization reactions to produce isoprostanes and neuroprostanes (Catala, 2006). This effect is severe for biological systems, produce damage of membrane function, enzymatic inactivation and toxic effects on cellular division and function.

4. Role of transition metal on lipid peroxidation process

Studies in the past two decades have shown that redox active metals undergo redox cycling reactions and possess the ability to produce reactive radicals such as superoxide anion radical and nitric oxide in biological systems. Disruption of metal ion homeostasis leads to oxidative stress, a state with increased formation of reactive oxygen species that overwhelms antioxidant protection and subsequently induces DNA damage, lipid peroxidation, protein modification and other effects, all symptomatic for numerous diseases, involving cancer, cardiovascular disease, diabetes, atherosclerosis, neurological disorders, and chronic inflammation.

The mechanism of lipid peroxidation in biological systems caused by free radicals has been the focus of scientific interest for many years (Chance et al., 1976; Fridovich & Porter, 1981; Fraga et al., 1988; Gonzalez-Flecha et al., 1991a, 1991b; Famulari et al., 1996; Fiszman et al., 2003; Junqueira et al., 2004; Catala, 2006; Boveris et al., 2008; Dianzani & Barrera, 2008; Dominguez et al., 2008; Repetto, 2008; Repetto et al., 2010b). Currently, it is known that the OH· radical, is formed mainly by the Haber-Weiss reaction, and it is responsible for the biological damage (Repetto et al., 2010a; Repetto et al., 2010b) (Eq.24):

$$O_2^- + H_2O_2 \rightarrow O_2 + HO\cdot + HO^- \tag{24}$$

However, this reaction would not proceed significantly *in vivo* because the rate constant for the reaction is lower than that of the dismutation reaction. Nevertheless, a modification of the Haber-Weiss reaction, the Fenton reaction and Fenton-like reactions, utilizes the redox cycling ability of iron to increase the rate of reaction, is more feasible *in vivo* (Chance et al., 1979; Boveris et al., 1980; Gonzalez Flecha et al., 1991b), and is frequently used to explain the toxic effects of redox-active metals where $M^{(n)+}$ is usually a transition metal ion.

As a transition metal that can exist in several valences and that can bind up to six ligands, iron is an important component of industrial catalysts in the chemical industry especially for redox reactions (Repetto et al., 2010a; Repetto & Boveris, 2012).

There are several reports on the role of transition metals in lipid peroxidation process associated with cellular toxicities, because once they enter our physiological systems, these metals play a role in oxidative adverse effects. Some transition metals including iron, chromium, lead, and cadmium generate lipid peroxidation *in vitro* e *in vivo*: fatty acids, cod liver oil, biological membranes, tissues and organs, suggesting that metals contribute to the oxidative effects of lipid peroxidation observed in various diseases (Repetto et al., 2010a; Repetto & Boveris, 2012).

The Fenton reaction occurs *in vivo* at a very low rate, and hence cannot account for any substantial production of OH· radicals in biology. On the other hand, when catalysed by transition metal ions, OH· radicals can be formed through reactions 25 and 26:

$$M^{(n)+} + O_2^- \rightarrow M^{(n-1)+} + O_2 \tag{25}$$

$$M^{(n-1)+} + H_2O_2 \rightarrow M^{(n)+} + HO\cdot + HO^- \tag{26}$$

The concentration of intracellular redox active transition metals is either low or negligible: free Fe^{2+} is 0.2-0.5 μM and the pool of free Cu^{2+} is about a single ion per cell. However, trace (nM) levels of cellular and circulating active transition metal ions seem enough for the catalysis of a slow Fenton reaction *in vivo* at the physiological levels of hydrogen peroxide (H_2O_2, 0.1-1.0 μM) (Repetto et al., 2010a; Repetto & Boveris, 2012).

It is well known that iron serves as a catalyst for the formation of the highly reactive hydroxyl radical via Fenton reaction. In addition to ferrous ion, many metal ions including Cu (I), Cr (II), and Co (II) were found to have the oxidative features of the Fenton reagent. Therefore, the mixtures of these metal compounds with H_2O_2 were named "Fenton like reagents". In actual in vivo systems, once organic peroxides (ROOH) are formed by the action of ROS, heat, and/or photo-irradiation, ROOH can be substituted for HO·, where ROOH reacts with metal ions to form alkoxyl radicals. Subsequently, a chain reaction of lipid peroxidation occurs.

The mechanisms for metal transition ions promoted lipid peroxidation are H_2O_2 decomposition and direct homolysis of endogenous hydroperoxides. The Fe^{2+}-H_2O_2-mediated lipid peroxidation takes place by a pseudo-second order process, and the Cu^{2+}-mediated process by a pseudo-first order reaction. Co^{2+} and Ni^{2+} alone, do not induce lipid peroxidation. Nevertheless, when they are combined with Fe^{2+}, Fe^{2+}-H_2O_2-mediated lipid peroxidation is stimulated in the presence of Ni^{2+} and is inhibited in the presence of Co^{2+} (Fig. 5) (Repetto et al., 2010a).

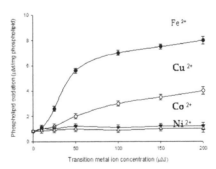

Figure 5. Phospholipid oxidation at different concentrations of transition metals.

There are many factors influencing lipid peroxidation products formation from lipids catalyzed by various metals. For example, the quantitative measurement of the reaction of Fe (II) and H_2O_2 has shown that a stoichiometric amount of hydroxyl radical is spin-trapped when ion concentration was less than 1 μM, suggesting that the strength of the Fenton system depended on the metal concentration. Since Fenton reported that a mixture of hydrogen peroxide and ferrous salts was an effective oxidant of a large variety of organic substrates in 1894 this reagent (the Fenton's reagent) has been used to investigate many subjects related to *in vitro* oxidation of organic substrates including lipids (Repetto et al., 2010a; Repetto & Boveris, 2012).

In the *in vitro* model of phosphatidylcholine/phosphatidyserine (60:40) liposomes and hydrogen peroxide (H_2O_2), Fe and Cu promote lipid peroxidation, interpreted as the consequence of the homolytic scission of H_2O_2 and of endogenous hydroperoxides (ROOH) and of the generation of hydroxyl ($HO^{•}$) and alcoxyl ($RO^{•}$) radicals (Cadenas, 1989) depending strictly on the participation of Fe and Cu as redox-reactive metals . However, Co^{2+} and Ni^{2+} alone, do not induce lipid peroxidation. Nevertheless, when they are combined with Fe^{2+}, Fe^{2+}-H_2O_2-mediated lipid peroxidation is stimulated in the presence of Ni^{2+} and inhibited in the presence of Co^{2+} (Repetto et al., 2010a; Repetto & Boveris, 2012).

Cr(III) occurs in nature and is an essential trace element utilized in the regulation of blood glucose levels. Cr(III) reacts with superoxide, subsequently Cr(II) yields hydroxyl radical via Fenton-like reaction with H_2O_2 to initiate lipid peroxidation.

Cadmium intoxication was shown to increase lipid peroxidation in rat liver, kidney and heart. However, the mechanisms of cadmium toxicity are not fully understood. Cadmium indirectly affects the generation of various radicals including superoxide and hydroxyl radical. The generation of hydrogen peroxide by cadmium ion may become a source of radicals in the Fenton system (Jomova & Valko, 2011).

5. Toxic effects of secondary products of lipid peroxidation

Many aldehydes are produced during the peroxidative decomposition of unsaturated fatty acids. Compared with free radicals, aldehydes are highly stable and diffuse out from the cell and attack targets far from the site of their production. About 32 aldehydes were identified as products of lipid peroxidation: a) saturated aldehydes (propanal, butanal, hexanal, octanal, being the decanal the most important); b) 2,3-trans-unsaturated-aldehydes (hexenal, octenal, nonenal, decenal and undecenal); c) a series of 4-hydroxylated,2,3-trans-unsaturated aldehydes: 4-hydroxyundecenal, being 4-hydroxinonenal (HNE) the most important quantitatively. Malonyldialdehyde (MDA) was considered for a long time as the most important lipid peroxidation metabolite. However, MDA is practically no toxic.

Recent studies have demonstrated that the most effective product of lipid peroxidation causing cellular damage is HNE. HNE produces different effects: acts as an intracellular signal able to modulate gene expression, cell proliferation, differentiation and apoptosis. The hydroxyl-group close to a carbonyl group present in HNE chemical structure is related to its high reactivity with different targets (thiol and amine groups). HNE is easily diffusible specie, but its biological effect depends on the molecule target and behavior as a signal to produce the damage.

Oxidative stress is a well known mechanism of cellular injury that occurs with increased lipoperoxidation of cell phospholipids and that has been implicated in various cell dysfunctions (Sies, 1991a,b; Catala, 2006). Aldehydes exhibit high reactivity with bio-molecules, such as proteins, DNA and phospholipids generating intra and intermolecular adducts.

The physiological concentrations of these products are low; however, higher concentrations correspond to pathological situations. Therefore, DNA damage caused by lipid peroxidation end products could provide promising markers for risk prediction and targets for preventive measures. DNA-reactive aldehydes can damage DNA either by reacting directly with DNA bases or by generating more reactive bifunctional intermediates, which form exocyclic DNA adducts. Of these, HNE and MDA, acrolein, and crotonaldehyde have been shown to modify DNA bases, yielding promutagenic lesions and to contribute to the mutagenic and carcinogenic effects associated with oxidative stress-induced lipid peroxidation and HNE and MDA implicated carcinogenesis.

The end-products of lipid peroxidation (HNE and MDA) cause protein damage by addition reactions with lysine amino groups, cysteine sulfhydryl groups, and histidine imidazole groups (Esterbauer et al., 1991; Esterbauer, 1996). Modifications of protein by aldehyde products of lipid peroxidation contribute to neurodegenerative disorders, activation of kinases (Uchida et al., 1999; Uchida, 2003) and inhibition of the nuclear transcription factor (Camandola et al., 2000).

6. Lipid peroxidation of subcellular fragments

6.1. Microsomes

Microsomes isolated from liver have been shown to catalyze an NADPH-dependent peroxidation of endogenous unsaturated fatty acids in the presence of ferric ions and metal chelators, such as ADP or pyrophosphates. Microsomal membranes are particularly susceptible to lipid peroxidation owing to the presence of high concentrations of polyunsaturated fatty acids (Poyer & McCay, 1971). The mechanism involved in the initiation of peroxidation in the NADPH-dependent microsomal system do not appear to involve neither superoxide nor hydrogen peroxide, since neither superoxide dismutase nor catalase cause inhibition of peroxidation. Nevertheless, reduced iron plays an important role in both the initiation and propagation of NADPH-dependent microsomal lipid peroxidation (Shires, 1975).

Microsomal membrane lipids, particularly the polyunsaturated fatty acids, undergo degradation during NADPH-dependent lipid peroxidation. The degradation of membrane lipids during lipid peroxidation has been observed to result in the production of singlet oxygen, which is detected as chemiluminescence (Boveris et al., 1980).

Nonenzymatic peroxidation of microsomal membranes also occurs and is probably mediated in part by endogenous hemoproteins and transition metals. High concentrations of transition metals (50 µM) promote auto-oxidation of phospholipids (Repetto et al., 2010a).

6.2. Mitochondria

It is currently accepted that mitochondrial complex I is particularly sensitive to inactivation by oxygen free radicals and reactive nitrogen species. This special characteristic is frequently referred as complex I syndrome, with the symptoms of reduced mitochondrial respiration with malate-glutamate and ADP and of reduced complex I activity. This complex I syndrome has been observed in aging (Navarro et al., 2005; Navarro & Boveris, 2004, 2008), in ischemia-reperfusion (Gonzalez-Flecha et al., 1993), in Parkinson's disease, and in other neurodegenerative diseases (Schapira et al., 1990a, 1990b; Sayre et al., 1999; Carreras et al., 2004; Schapira, 2008; Navarro et al., 2009), and in this study, with the addition of the increased rates of production of O_2^- and H_2O_2 by complex I mediated reactions, reactions with the free radicals intermediates of the lipid peroxidation process (mainly ROO·), and amine-aldehyde adduction reactions. It is now understood that the three processes above mentioned alter the native non-covalent polypeptide interactions of complex I and promote synergistically protein damage and inactivation by shifting the noncovalent bonding to covalent cross linking (Navarro et al., 2005). Complex I oxidative protein damage has also been considered the result of protein modification by reaction with malonaldehyde and 4-HO-nonenal (Sayre et al., 1999). It was hypothesized that protein damage in the subunits of complexes I and IV follows to free radical-mediated cross-linking and inactivation. The subunits that are normally held together by noncovalent forces are shifted to covalent cross-linking after reaction with the hydroperoxyl radicals (ROO·) and the stable aldehydes produced during the lipid peroxidation process.

The hypothesis that cumulative free radical-mediated protein damage is the chemical basis of respiratory complexes I and IV inactivation (Berlett & Stadtman, 1997) offers the experimental approach of the chronic use of vitamin E, as an antioxidant for the lipid phase of the inner mitochondrial membrane and for the prevention of the mitochondrial /damage associated with aging. The adduction reactions of malonaldehyde and 4-HO-nonenal with protein evolve to stable advanced lipid peroxidation products (Sayre et al., 1999) and protein carbonyls (Nair et al., 2007; Navarro et al., 2008). The molecular mechanism involved in the inactivation of complex I is likely accounted for by ROO· and ONOO⁻. Upon aging, frontal cortex and hippocampal mitochondria show a decreased rate of respiration, especially marked with NAD-dependent substrates, and decreased enzymatic activities of complexes I and IV associated with an increase in the content of oxidation products (TBARS and protein carbonyls) (Navarro et al., 2008) (Fig. 6).

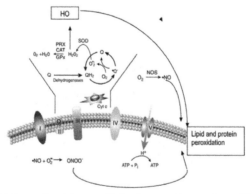

Figure 6. Lipid peroxidation and protein peroxidation by secondary products of lipid peroxidation in mitochondria.

7. Lipid peroxidation and human pathologies

The organism must confront and control the balance of both pro-oxidants and antioxidants continuously. The balance between these is tightly regulated and extremely important for maintaining vital cellular and biochemical functions. This balance often referred to as the redox potential, is specific for each organelle and biological site, and any interference of the balance in any direction might be deleterious for the cell and organism. Changing the balance towards an increase in the pro-oxidant over the capacity of the antioxidant is defined as oxidative stress and might lead to oxidative damage. Changing the balance towards an increase in the reducing power, or the antioxidant, might also cause damage and can be defined as reductive stress.

Oxidative stress and damage have been implicated in numerous disease processes, including inflammation, degenerative diseases, and tumor formation and involved in physiological phenomena, such as aging and embryonic development. The dual nature of these species with their beneficial and deleterious characteristics implies the complexities of their effects at a biological site.

Lipid peroxidation has been pointed out as a key chemical event in the oxidative stress associated with several inborn and acquired pathologies. Disruption of organelle and cell membranes together with calcium homeostasis alterations are the main supramolecular events linked to lipid peroxidation. However, it is not clear if lipid peroxidation process is a cause, triggering step of the clinical manifestations of the disease, or a consequence of toxic effects of lipid peroxidation products.

In pathological situations the reactive oxygen species are generated and as a consequence lipid peroxidation occurs with α-tocopherol deficiency. In addition to containing high concentrations of polyunsaturated fatty acids and transitional metals, red blood cells are constantly being subjected to various types of oxidative stress. Red blood cells however are protected by a variety of antioxidant systems which are capable of preventing most of the adverse effects under normal conditions. Among the antioxidant systems in the red cells, α-tocopherol possesses an important and unique role. α-tocopherol may protect the red cells from oxidative damage via a free radical scavenging mechanism and as a structural component of the cell membrane (Chitra & Shyamaladevi, 2011).

Levels of Met-Hb· are regarded as an index of intracellular damage to the red cell and it is increased when α-tocopherol is consumed and the rate of lipid peroxidation is increased. Scavenging of free radicals by α-tocopherol is the first and the most critical step in defending against oxidative damage to the red cells. When α-tocopherol is adequate, GSH and ascorbic acid may complement the antioxidant functions of α-tocopherol by providing reducing equivalents necessary for its recycling/regeneration.

On the other hand, when α-tocopherol is absent, GSH and ascorbic acid release transitional metals from the bound forms and/or maintain metal ions in a catalytic state. Free radical generation catalysed by transition metal ions in turn initiates oxidative damage to cell membranes. Membrane damage can lead to release of heme compounds from erythrocytes. The heme compounds released may further promote oxidative damage especially when reducing compounds are present (Boveris et al., 2008).

8. Lipid peroxidation and aging

Aging is a process directly related to systemic oxidative stress. Two components of the oxidative stress situation have been recognized in human aging: a decrease in availability of nutritional molecular antioxidants and an accumulation of products derived from the oxidation of biological structures. Oxidation of biomolecules is related to susceptibility to diseases, such as cancer and heart disease, as well as associated with the process of aging (Navarro et al., 2005; Navarro & Boveris, 2007, 2008).

The products derived from lipid peroxidation, measured in plasma by Junqueira et al., (2004) as fluorescent products, were higher in elderly than younger human subjects and even higher in disabled octogenarians and nonagenarians. This increase in lipid peroxidation products was directly correlated with age, and was associated with decreases in vitamin E and C.

9. Analytical determination of lipid peroxidation

Since the acceptation of the oxidative stress concept, scientists and physicians have been searching for a simple assay or a small group of determination that would result useful for the assessment of oxidative stress and lipid peroxidation in clinical situations. The determinations of marker metabolites are usually performed in blood, red blood cells or plasma. The markers for systemic oxidative stress are normally present in healthy humans and the assays for systemic oxidative stress are comparative, which makes necessary to have reference values from normal individuals.

At present, the plasma levels of oxidation products derived from free-radical mediated reactions and of antioxidants are used as indicators of systemic oxidative stress in humans and experimental animals. The more utilized determination of an oxidation product is MDA, which is determined with low specificity but with great efficiency by the simple and useful assay of TBARS with measurements made by spectrophotometry or spectrofluorometry. The normal plasma levels of TBARS are 2-3 μM (Junqueira et al., 2004).

Oxidative damage is characterized by increases in the levels of the oxidation products of macromolecules, such as thiobarbituric acid reactive substances (TBARS), and protein carbonyls. Many of these products can be found in biological fluids, as well as addition-derivatives of these reactive end-products. As a result of lipid peroxidation a great variety of aldehydes can be produced, including hexanal, malondialdehyde (MDA) and 4-hydroxynonenal (Catala, 2006).

Oxidation of an endogenous antioxidant reflects an oxidative stress that is evaluated by measuring the decrease in the total level of the antioxidant or the increase in the oxidative form. The only way not to be influenced by nutritional status is to measure the ratio between oxidized and reduced antioxidants present in blood. The published literature provides compelling evidence that a) MDA represents a side product of enzymatic PUFA-oxygenation and a secondary end product of no enzymatic (autoxidative) fatty peroxide formation and decomposition and b) sensitive analytical methods exist for the unambiguous isolation and direct quantification of MDA. Conceptually, these two facts indicate that MDA is an excellent index of lipid peroxidation. However, this conclusion is limited in practice by several important consideration: a) MDA yield as a result of lipid peroxidation varies with the nature of the PUFA peroxidised (specially its degree of instauration) and the peroxidation stimulus, b) only certain lipid oxidation products decompose yield MDA, c) MDA is only one of several end product of fatty peroxide formation and decomposition, d) the peroxidation environment influences both the formation of lipid-derived precursors and their decomposition to MDA, e) MDA itself is a reactive substance which can be oxidative and metabolically degraded, f) oxidative injury to no lipid biomolecules has the potential to generate MDA. With biological materials, it appears prudent to consider the TBARS test more than an empirical indicator of the potential occurrence of peroxidative lipid damage and not as a measure of lipid peroxidation (Repetto, 2008). The thiobarbituric acid test (TBARS) has been employed to a uniquely great degree over the last five decades to detect and quantify lipid peroxidation in a variety of chemical as well as biological material. Two underlying assumptions are implicit from the widespread use of the TBARS test to assess lipid peroxidation: a) an operative and quantitative relationship exists between lipid peroxidation and MDA, b) product formation during the TBARS test is diagnostic of the presence and amount of fatty peroxides.

Lipid peroxidation proceeds by a free-radical mediated chain reaction that includes initiation, propagation and termination reactions. The chain reaction is initiated by the abstraction of a hydrogen atom from a methylene group of an unsaturated fatty acid. Propagation is cycled through rounds of lipid peroxyl radical abstraction of the bis-methylene hydrogen atoms of a polyunsaturated fatty acyl chain to generate new radicals, after O_2 addition, resulting in the conversion of alkyl radical in hydroperoxyl radical. Termination involves the reaction of two hydroperoxyl radicals to form non-radical products. This reaction is particularly interesting since it is accompanied, although at low yield, by emission of light or chemiluminiscence. Some lipid peroxidation products are light-emitting species and their luminescence is used as an internal marker of oxidative stress (Chance et al., 1979; Boveris et al., 1980, Gonzalez-Flecha et al., 1991b; Sies, 1991a; Repetto, 2008). The measurement of light emission derived from $^1O_2$ and excited triplet carbonyl compounds, which are the most important chemiluminiscent species in the lipid peroxidation of biological systems, is directly related to the rate of lipid peroxidation and allows an indirect assay of the content of lipophilic antioxidants in the sample (Gonzalez-Flecha et al., 1991a). Lipophilic antioxidants react with lipid peroxyl radicals and lower antioxidant content is associated with higher chemiluminescence (Repetto, 2008).

The low-level chemiluminescence which accompanies the peroxidation of polyunsaturated fatty acids has been used as a tool in kinetic and mechanistic studies of biological samples to estimate the extent of the reactions and even to indicate tissue damage promoted by oxidants. Triplet carbonyls and singlet oxygen formed in the annihilation of intermediate peroxyl radicals (ROO·) have been identified as the chemiluminescence emitters.

Chemiluminescence is a very interesting way to evaluate an oxidative stress and lipid peroxidation in biological samples and living systems. The emission of light has been observed during stress in different experimental models. Chemiluminescence is very sensitive and thus can be applied to measure free radical production in human tissues.

Chemiluminescent systems may be classified in two classes based on the origin of the emitting molecule. In the first class, the emitter is a product of the chemical reaction (direct chemiluminescence). In the second class, there is energy transfer between an electronically excited product molecule and a second substance which then becomes the emitter (sensitized chemiluminescence) (Boveris et al., 1980; Gonzalez-Flecha et al., 1991b; Repetto, 2008).

The chemical mechanism responsible for spontaneous organ light emission is provided by the Russell's reaction in which two secondary or tertiary peroxyl radicals (ROO*) yield $^1O_2$ and excited carbonyl groups (=CO*) as products. In turn, two $^1O_2$, through dimol emission, lead to photoemission at 640 and 670 nm, whereas =CO* yields photons at the 460-470 nm band (Boveris et al., 1980). The main sources of the chemiluminescence detected in the direct and sensitized chemiluminescence is the dimol emission of $^1O_2$ (reaction 27) and the photon emission from excited carbonyl groups (reaction 28) (Boveris et al., 1980).

$$2\ ^1O_2 \quad \rightarrow \quad 2\ O_2 + h\upsilon\ (634\text{-}703\ nm) \tag{27}$$

$$RO^* \quad \rightarrow \quad RO + h\upsilon\ (380\text{-}460\ nm) \tag{28}$$

These reactions are accompanied by chemiluminescence whose intensity may serve as an indirect measure of peroxide free radical and α-tocopherol concentration in the sample.

Lipid peroxidation has been recognized as free radical-mediated and physiologically occurring (Navarro & Boveris, 2004, Navarro et al., 2010; Repetto & Boveris, 2012) with the supporting evidence of *in situ* organ chemiluminescence (Repetto, 2008). Spontaneous chemiluminescence of *in situ* organs directly reports the intracellular formation of singlet oxygen ($^1O_2$) (Boveris et al., 1980) and represents an issue of direct chemiluminescence. The generation of $^1O_2$ implies the collision of two peroxyl radicals (ROO·) with formation of excited species, $^1O_2$ itself and excited carbonyls, followed by photoemission. Light emission from *in situ* organs is a physiological phenomenon that provides a determination of the steady state concentration of singlet oxygen and indirectly of the rate of oxidative free radical reactions (Boveris et al., 1980). *In situ* liver chemiluminescence has been recognized as a reliable indicator of oxidative stress and damage in rat liver upon hydroperoxide infusion (Gonzalez-Flecha et al., 1991b), ischemia-reperfusion (Gonzalez-Flecha et al., 1993), and chronic and acute alcohol intoxication (Videla et al., 1983). The increases in photoemission observed were parallel to increased contents of indicators of lipid peroxidation (malonaldehyde and 4-HO-nonenal) but with a higher experimental/control ratio in organ chemiluminescence (Boveris et al., 1980).

Tert-butyl hydroperoxide initiated chemiluminescence is an example of sensitized chemiluminescence, and it has been used to enhance the chemiluminescence accompanying lipid peroxidation and the α-tocopherol content of tissues. This method has been successfully utilized to detect the existence of oxidative damage associated to experimental or pathological situations in tissue homogenates, subcellular fractions, and in human heart, liver and muscle biopsies (Gonzalez-Flecha et al., 1991b).

Tissue homogenates or blood samples are subjected to *in vitro* oxidative damage by supplementation with tert-butyl hydroperoxide. It reacts with hemoproteins and Fe^{2+} producing peroxyl and alcoxyl free radicals, which enter to the propagation phase of the lipid peroxidation radical chain reaction. The termination steps of the chain reaction generate compounds in an excited state: singlet oxygen and carbonyl groups. This assay is useful to evaluate the integral level of the non-enzymatic antioxidant defenses of a tissue (Gonzalez-Flecha et al., 1991a, 1993).

The increase of tert-butyl hydroperoxide-initiated chemiluminescence is indicative that α-tocopherol is the antioxidant consumed in erythrocytes and suggest that reactive oxygen species and lipid peroxidation catalyzed by reduced transition metals may be responsible for the onset of oxidative damage and the occurrence of systemic oxidative stress in patients suffering oxidative damage associated to neurological pathologies as Parkinson (Famulari et al., 1996, Dominguez et al., 2008), Alzheimer disease (Famulari et al., 1996; Repetto et al., 1999; Dominguez et al., 2008; Serra et al., 2009), and vascular dementia (Famulari et al., 1996, Dominguez et al., 2008; Serra et al., 2009); immunological diseases as HIV infection and AIDS (Repetto et al., 1996), hyperthyroidism and hypothyroidism (Abalovich et al., 2003). These methods were used to evaluate lipid peroxidation and oxidative damage in experimental models of oxidative stress in rats (Repetto et al., 2003, 2010; Ossani et al., 2007; Repetto & Ossani, 2008; Repetto & Boveris, 2010).

A common question of the researchers in the field is which the method of choice is. The answer is: none of them, and all of them. Each assay measures something different. Diene conjugation tells one about the early stages of peroxidation, as a direct measurement of lipid peroxides. In the absence of metal ions to decompose lipid peroxides there will be little formation of hydrocarbon gases, carbonyl compounds, or their fluorescent complexes, which does not necessarily mean therefore that nothing is happening. Even if peroxides do not decompose, the TBARS test can still detect them because of decomposition of peroxides. Changes in the mechanism of peroxide decomposition might alter the amount generated without any change in the overall rate of lipid peroxidation. Whatever method is chosen, one should think clearly what is being measured and how it relates to the overall lipid peroxidation process. Whatever possible, two or more different assay methods should be used.

10. Conclusion

Lipid peroxidation is a physiological process that takes place in all aerobic cells. Unsaturated fatty acids which are structural part of cell membranes are subjected to lipid peroxidation by a non enzymatic and free-radical mediated reaction chain. The molecular mechanisms of the lipid peroxidation process are known and it can be estimated that about 1 % of the total oxygen uptake of cells, organs and bodies in taken up by the reactions of lipid peroxidation. The initiation reactions are provided by the transition-metal catalyzed hemolytic scission of H_2O_2 and ROOH. In turn, H_2O_2 is mainly generated from the mitochondrial dismutation of superoxide radical (O_2^-). The products and by-products of lipid peroxidation are cytotoxic and lead in successive steps to oxidative stress, oxidative damage and apoptosis. In a long series of physiological and pathophysiological processes, including aging and neurodegenerative diseases, the rates of mitochondrial O_2^- and H_2O_2 are increased with a parallel increase in the rate of the lipid peroxidation process. It is expected that supplementation with adequate antioxidants, as for instance, α-tocopherol, will keep sensitive cells and organs in healthy conditions and increase lifespan.

Author details

Marisa Repetto, Jimena Semprine and Alberto Boveris
University of Buenos Aires, School of Pharmacy and Biochemistry,
General and Inorganic Chemistry,
Institute of Biochemistry and Molecular Medicine (IBIMOL-UBA-CONICET), Argentina

Acknowledgement

We thank to Dr. Jorge Serra for helping in the revision of this version.

11. References

Abalovich, M.; Llesuy, S.; Gutierrez, S. & Repetto, M. (2003) Peripheral markers of oxidative stress in Graves' disease. The effects of methimazole and 131 Iodine treatments. *Clinical Endocrinology.* Vol. 59, pp. 321-327, ISSN: 1365-2265

Beckman, J.; Beckman, T.; Chen, J.; Marshall, P. & Freeman, B. (1990) Apparent hydroxyl radical production from peroxynitrite. Implications for endothelial injury from nitric oxide oxide and superoxide. *Proceeding of the National Academy of Sciences of the United States.* Vol. 87, pp. 1620-1624, ISSN: 0027-8424

Beckman, J.; Chen, J.; Ischiropulos, H. & Crow, J. (1994) Oxidative chemistry of peroxynitrite. *Methods in Enzymology.* Vol. 233, pp. 229-240, ISSN: 0076-6879

Berlett, B.S. & Stadtman, E.R. (1997) Protein oxidation in aging, disease, and oxidative stress. *The Journal of Biological Chemistry.* Vol. 272, pp. 20313–20316, ISSN: 0021-9258

Boveris, A.; Cadenas, E.; Reiter, R.; Filipkowski, M.; Nakase, Y. & Chance, B. (1980) Organ chemiluminescence: noninvasive assay for oxidative radical reactions *Proceeding of the National Academy of Sciences of the United States.* Vol. 177, pp. 347-351, ISSN: 0027-8424

Boveris, A.; Fraga, C.; Varsavsky, A. & Koch, O. (1983) Increased chemiluminescence and superoxide production in the liver of chronically ethanol-treated rats. *Archives of Biochemistry and Biophysics.* Vol. 227, pp. 534-541, ISSN: 0003-9861

Boveris, A. & Navarro, A. (2008) Brain mitochondrial dysfunction in aging. *Life,* Vol. 60, No.5, pp. 308-314, ISSN: 1521-6543

Boveris, A.; Repetto, M.G.; Bustamante, J.; Boveris, A.D. & Valdez, L.B. (2008). The concept of oxidative stress in pathology. In: Álvarez, S.; Evelson, P. (ed.), *Free Radical Pathophysiology,* pp. 1-17, Transworld Research Network: Kerala, India, ISBN: 978-81-7895-311-3

Cadenas, E. (1989) Biochemistry of oxygen toxicity. *Annual Review of Biochemistry.* Vol. 58, pp. 79-110, ISSN:0066-4154

Camandola, S.; Poli, G. & Mattson, M. (2000) The lipid peroxidation product 4-hydroxy-2,3-nonenal inhibits constitutive and inducible activity of nuclear factor-bin neurons. *Molecular Brain Research.* Vol. 85, pp. 53–60, ISSN: 0021-9258

Carreras, M.C.; Franco, M.C.; Peralta, J.G. & Poderoso, J.J. (2004) Nitric oxide, complex I, and the modulation of mitochondrial reactive species in biology and disease. *Molecular Aspects of Medicine.* Vol. 25, pp. 125–139, ISSN: 0098-2997

Catala, A. (2006) An overview of lipid peroxidation with emphasis in outer segments of photoreceptors and the chemiluminescence assay. *The International Journal of Biochemistry and Cell Biology.* Vol. 38, pp. 1482-1495, ISSN: 1357-2725

Chance, B.; Sies, H. & Boveris, A. (1979) Hydroperoxide metabolism in mammalian organs. *Physiological Reviews.* Vol. 59, pp. 527-605, ISSN:©0031-9333

Cutrin, JC.; Cantino, D.; Biasi, F.; Chiarpotto, E.; Salizzoni, M.; Andorno, E.; Massano, G.; Lanfranco, G.; Rizetto, M.; Boveris, A. & Poli, G. (1996) Reperfusion damage to the bile canaliculi in transplanted human liver. *Hepatology.* Vol. 24, pp. 1053-1057, ISSN: 1527-3350

Dianzani, M. & Barrera, G. (2008) Pathology and physiology of lipid peroxidation and its carbonyl products. In: Álvarez, S.; Evelson, P. (ed.), *Free Radical Pathophysiology,* pp. 19-38, Transworld Research Network: Kerala, India, ISBN: 978-81-7895-311-3

Domínguez, R.O.; Marschoff, E.R.; Guareschi, E.M.; Repetto, M.G.; Famulari, A.L.; Pagano, M.A. & Serra, J.A. (2008). Insulin, glucose and glycated haemoglobin in Alzheimer's and vascular dementia with and without superimposed Type II diabetes mellitus condition. *Journal of Neural Transmission,* Vol. 115, pp. 77-84, ISSN: 0300-9564.

Esterbauer, H.; Schaur, J. & Zollner, H. (1991) Chemistry and biochemistry of 4-hydroxynonenal, malondialdehyde and related aldehydes. *Free Radical in Biology & Medicine.* Vol. 11, pp. 81–128, ISSN: 0891-5849

Esterbauer, H. (1996) Estimation of peroxidative damage. A critical review. *Pathologie Biologie.* Vol. 44, pp. 25–28, ISSN: 0031-3009

Famulari, A.; Marschoff, E.; Llesuy, S.; Kohan, S.; Serra, J.; Domínguez, R.; Repetto, M.G.; Reides, C. & Lustig, E.S. de (1996). Antioxidant enzymatic blood profiles associated with risk factors in Alzheimer's and vascular diseases. A predictive assay to differentiate demented subjects and controls. *Journal of the Neurological Sciences*, Vol. 141, pp. 69-78, ISSN: 0022-510X

Farooqui, T. & Farooqui, A. (2011) Lipid-mediated oxidative stress and inflammation in the pathogenesis of Parkinson's disease. *Parkinson's disease*. DOI: 10.4061/2011/247467

Fiszman, M.; D'Eigidio, M.; Ricart, K.; Repetto, M.G.; Llesuy, S.; Borodinsky, L.; Trigo, R.; Riedstra, S.; Costa, P.; Saizar, R.; Villa, A. & Sica, R. (2003). Evidences of oxidative stress in Familial Amyloidotic Polyneuropathy Type 1. *Archives of Neurology*, Vol. 60, pp. 593-597, ISSN 0003-9942

Fraga, C.; Leibovitz, B. & Tappel, A. (1988). Lipid peroxidation measured as thiobarbituric acid-reactive substances in tissue slices: characterization and comparison with homogenates and microsomes. *Free Radicals in Biology and Medicine*, Vol. 4, pp. 155-161, ISSN: 0891-5849

Fridovich, I. (1978) Superoxide radicals, superoxide dismutases and the aerobic lifestyle. *Photochemistry and Photobiology*. Vol. 28, pp. 733-741, ISSN: 1010-6030

Fridovich, S. & Porter, N. (1981) Oxidation of arachidonic acid in micelles by superoxide and hydrogen peroxide. *The Journal of Biological Chemistry*. Vol. 256, pp. 260-265, ISSN: 0021-9258

Gatto, E.; Carreras, M.C.; Pargament, G.; Reides, C.; Repetto, M.G.; Llesuy, S.; Fernández Pardal, M. & Poderoso, J. (1996). Neutrophil function nitric oxide and blood oxidative stress in Parkinson's disease. *Movement Disorders*, Vol. 11, pp. 261-267, ISSN: 0885-3185

Gatto, E.; Carreras, C.; Pargament, G.; Riobó, N.; Reides, C.; Repetto, M.; Fernández Pardal, N.; Llesuy, S. & Poderoso, J. (1997). Neutrophyl function nitric oxide and blood oxidative stress in Parkinson's Disease. *Focus Parkinson's Disease*, Vol. 9, pp. 12-14

Gonzalez Flecha, B., Repetto, M.; Evelson, P. & Boveris, A. (1991a) Inhibition of microsomal lipid peroxidation by α -tocopherol and α -tocopherol acetate. *Xenobiotica*. 21: 1013–1022, ISSN: 0049-8254

González Flecha, B.; Llesuy, S. & Boveris, A. (1991b). Hydroperoxide-initiated chemiluminescence: assay for oxidative stress in biopsies of heart, liver and muscle. Free Radicals in Biology and Medicine, Vol. 10, pp. 93-100, ISSN: 0891-5849

Gonzalez-Flecha, B.; Cutrin, J.C. & Boveris, A. (1993) Time course and mechanism of oxidative stress and tissue damage in rat liver subjected to in vivo ischemia-reperfusion. *Journal of Clinical Investigation*. Vol. 91, pp. 456–464, ISSN:⊕0021-9738

Halliwell, B. & Gutteridge, J.M.C. (1984). Oxygen toxicity, oxygen radicals, transition metals and disease. *Biochemical Journal*, Vol. 218, pp. 1-14, ISSN: 0264-6021

Hogg, N.; Darley-Usmar, V.; Wilson, M. & Moncada, S. (1992) Production of hydroxyl radicals from the simultaneous generation of superoxide and nitric oxide. *Biochemical Journal*. Vol. 281, pp. 419-424, ISSN: 0264-6021

Jomova, K. & Valko, M. (2011) Advances in metal-induced oxidative stress and human disease. *Toxicology.* Vol. 283, pp. 65-87, ISSN: 0300-483X.

Junqueira, V.; Barros, S.; Chan, S.; Rodríguez, L.; Giavarotti, L.; Abud, R. & Deucher, G. (2004) Aging and oxidative stress. *Molecular Aspects of Medicine.* Vol. 25, pp. 5–16, ISSN: 0098-2997

Liu, Q.; Raina, A.K.; Smith, M.A.; Sayre,. LM. & Perry, G. (2003) Hydroxynonenal, toxic carbonyls, and Alzheimer disease. *Molecular Aspects of Medicine.* Vol. 24, pp. 305–313, ISSN: 0098-2997

Moncada, S.; Palmer, R. & Higgs, E. (1991) Nitric oxide: Physiology, patophysiology and pharmacology. *Pharmaceutical Reviews.* Vol. 43, pp. 109-141, ISSN:⊕1918-5561

Murphy, M. & Sies, H. (1991) Reversible conversion of nitroxyl anion to oxide by superoxide dismutase. *Proceeding of the National Academy of Sciences of the United States.* Vol. 88, pp. 10860-10864, ISSN: 0027-8424

Nair, U.; Barstsch, H. & Nair, J. (2007) Lipid peroxidation-induced DNA damage in cancer-prone inflammatory diseases: a review of published adduct types and levels in humans. *Free Radical in Biology & Medicine.* Vol. 43, pp. 1109-1120, ISSN: 0891-5849

Navarro, A. & Boveris, A. (2004). Rat brain and liver mitochondria develop oxidative stress and lose enzymatic activities on aging. *American Journal of Physiology - Regulatory, Integrative and Comparative Physiology*, Vol. 287, pp. 1244-1249, ISSN: 0363-6119

Navarro, A.; Gomez, C.; Sanchez-Pino, MJ.; Gonzalez, H.;, Bandez, MJ.; Boveris, AD.; Boveris, A. (2005) Vitamin E at high doses improves survival, neurological performance, and brain mitochondrial function in aging male mice. *American Journal of Physiology - Regulatory, Integrative and Comparative Physiology*, Vol. 289, pp. 1392–1399, ISSN: 0363-6119

Navarro, A.; Boveris, A. (2007) The mitochondrial energy transduction system and the aging process. *American Journal of Physiology - Regulatory, Integrative and Comparative Physiology*, Vol. 292, pp. 670-686, ISSN: 0363-6119

Navarro, A.; Lopez-Cepero, JM.; Bandez, MJ.; Sanchez-Pino, MJ.; Gomez, C.; Cadenas, E.; Boveris, A. (2008) Hippocampal mitochondrial dysfunction in rat aging. *American Journal of Physiology - Regulatory, Integrative and Comparative Physiology*, Vol. 294, pp. 501-509, ISSN: 0363-6119

Navarro, A. & Boveris, A. (2009). Brain mitochondrial dysfunction and oxidative damage in Parkinson's disease. *Journal of Bioenergetics and Biomembranes,* Vol. 41, pp. 517-521, ISSN: 0145-479X

Navarro, A.; Boveris, A.; Bández, M.J.; Sánchez-Pino, M.J.; Gómez, C.; Muntane, G. & Ferrer, I. (2009). Human brain cortex: mitochondrial oxidative damage and adaptive response in Parkinson's disease and in dementia with Lewy bodies. *Free Radicals in Biology and Medicine,* Vol. 46, pp. 1574-1580, ISSN: 0891-5849

Navarro, A.; Bández, M.; Gómez, C.; Sánchez-Pino, M.; Repetto, M.G. & Boveris, A. (2010). Effects of rotenone and pyridaben on complex I electron transfer and on mitochondrial nitric oxide synthase functional activity. *Journal of Bioenergetics and Biomembranes*, Vol. 42, pp. 405-412, ISSN: 0145-479X

Ossani, G.; Dalghi, M. & Repetto, M. (2007) Oxidative damage and lipid peroxidation in the kidney of choline-defficient rats. *Frontiers in Bioscience*. Vol. 12, pp. 1174-1183, ISSN:1093-9946

Poyer, J. & McCay, P. (1971) Reduced triphosphopyridine nucleotide oxidase-catalyzed alterations of membrane phospholipids. Dependence on Fe^{3+}. *The Journal of Biological Chemistry*. Vol. 246, pp. 263-269, ISSN: 0021-9258

Rachmilewitz, D.; Stamler, J.; Karmeli, F.; Mollins, M.; Singel, D.; Loscalo, J.; Xavier, R. & Podolsky, D. (1993) Peroxynitrite induced rat colitis-a new model of colonic inflammation. *Gastroenterology*. Vol. 105. pp. 1681-1688, ISSN: 0016-5085

Repetto, M.; Reides, C.; Gomez Carretero, M.; Costa, M.; Griemberg, G., & Llesuy S. (1996) Oxidative Stress in Erythrocytes of HIV infected patients. *Clinica Chimica Acta*. Vol. 255, pp. 107-117, ISSN: 0009-8981

Repetto, M.G.; Reides, C.; Evelson, P.; Kohan, S.; Lustig, E.S. de & Llesuy, S. (1999). Peripheral markers of oxidative stress in probable Alzheimer patients. *European Journal of Clinical Investigation*, Vol. 29, pp. 643-649, ISSN: 0014-2972

Repetto, M.; María, A.; Giordano, O.; Guzmán,J.; Guerreiro, E. & Llesuy, S. (2003) Protective effect of Artemisia douglasiana Besser extracts on ethanol induced oxidative stress in gastric mucosal injury. *Journal of Pharmacy and Pharmacology*. Vol. 55, pp. 551-557, ISSN: 0022-3573

Repetto, M.G. (2008). Clinical use of chemiluminescence assays for the determination of systemic oxidative stress. In: Popov, I.; Lewin, G. (ed.), *Handbook of chemiluminescent methods in oxidative stress assessment*. Transworld Research Network: Kerala, India; pp. 163-194, ISBN: 978-81-7895-334-2

Repetto, M.G. & Ossani, G. (2008) Sequential histopathological and oxidative damage in different organs in choline deficient rats. In: Álvarez, S.; Evelson P. (ed.), *Free Radical Pathophysiology*. Transworld Research Network: Kerala, India; pp. 433-450, ISBN: 978-81-7895-311-3

Repetto, M.G.; Ferrarotti, N.F. & Boveris, A. (2010a) The involvement of transition metal ions on iron- dependent lipid peroxidation. *Archives of Toxicology*. Vol. 84, pp. 255-262, ISSN: 0340-5761

Repetto, M.; Ossani, G.; Monserrat, A. & Boveris, A. (2010b) Oxidative damage: The biochemical mechanism of cellular injury and necrosis in choline deficiency. *Experimental and Molecular Pathology*. Vol. 88, pp. 143-149. ISSN: 0014-4800.

Repetto, M. & Boveris, A. (2010) Bioactivity of sesquiterpenes: novel compounds that protect from alcohol-induced gastric mucosal lesions and oxidative damage. *Mini Reviews in Medicinal Chemistry*. Vol. 10, pp. 615-623. ISSN: 1389-5575

Repetto, M.G. & Boveris A. (2012). Transition metals: bioinorganic and redox reactions in biological systems. In: *Transition metals: uses and characteristics*. Nova Science Publishers Inc (ed.): New York, USA. pp. 349-370., ISBN: 978-1-61761-110-0

Sayre, L.M.; Perry, G. & Smith, M.A. (1999) In situ methods for detection and localization of markers of oxidative stress: application in neurodegenerative disorders. *Methods in Enzymology*. Vol. 309, pp. 133–152, ISSN: 0076-6879

Sayre, L.M.; Sha, W.; Xu, G.; Kaur, K.; Nadkarni, D.; Subbanagounder, G. & Salomon, R.G. (1996) Immunochemical evidence supporting 2-pentylpyrrole formation on proteins exposed to 4-hydroxy-2-nonenal. *Chemical Research in Toxicology*. Vol. 9, pp. 1194–1201, ISSN: 0893-228X

Schapira, A.H. (2008) Mitochondria in the aetiology and pathogenesis of Parkinson's disease. *The Lancet Neurology*. Vol. 7, pp. 97–109, ISSN:◉1474-4422

Schapira, A.H.; Cooper, J.M.; Dexter, D.; Clark, J.B.; Jenner, P. & Marsden, C.D. (1990) Mitochondrial complex I deficiency in Parkinson's disease. *Journal of Neurochemistry*. Vol. 54, pp. 823–827, ISSN: 0022-3042

Schapira, AH.; Mann, V.M.; Cooper,. JM.; Dexter, D.; Daniel, S.E.; Jenner, P.; Clark, J.B. & Marsden, C.D. (1990) Anatomic and disease specificity of NADH CoQ1 reductase (complex I) deficiency in Parkinson's disease. *Journal of Neurochemistry*. Vol. 55, pp. 2142–2145, ISSN: 0022-3042

Serra, J.A.; Domínguez, R.O.; Marschoff, E.R.; Guareschi E.M.; Famulari, A.L. & Boveris, A. (2009) Systemic oxidative stress associated with the neurological diseases of aging. *Neurochemical Research*, Vol. 34, pp. 2122–2132, ISSN/ISBN: 03643190

Shires, T. (1975) Inhibition by lipoperoxidation of amino acid incorporation by rough microsomal membranes *in vitro* and its partial reversibility. *Archives of Biochemistry and Biobiophys*. Vol. 171, pp. 695-707. ISSN: 0003-9861

Sies, H. (1991a) Oxidative stress: from basic research to clinical application. *American Journal of Medicine*. Vol. 91, pp. 31-38, ISSN: 0002-9343

Sies, H. (1991b). Role of reactive oxygen species in biological processes. *Wiener Klinische Wochenschrift*, Vol. 69, pp. 965–968, ISSN: 1613-7671

Valdez, L.B.; Zaobornij, T.; Bombicino, S.; Iglesias, D.E.; Boveris, A.; Donato, M.; D'Annunzio, V.; Buchholz, B. & Gelpi, R.A. (2011) Complex I syndrome in myocardial stunning and the effect of adenosine. *Free Radical in Biology & Medicine*. Vol. 51, pp. 1203-1212, ISSN: 0891-5849

Verstraeten, S.; Nogueira, L., Schreier, S. & Oteiza, P. (1997) Effect of trivalent metal ions on phase separation and membrane lipid packing: role in lipid peroxidation. *Archives of Biochemistry and Biobiophys*. Vol. 338, pp. 121-127, ISSN: 0003-9861

Videla, L.; Fraga, C.; Koch, O. & Boveris, A. (1983) Chemiluminescence of the in situ rat liver alter acute ethanol intoxication-effect of (+)-cyanidanol-3. *Biochemical Pharmacology*. Vol. 32, pp. 2822-2825,◉ISSN: 0006-2952

Uchida, K.; Shiraishi, M.; Naito, Y.; Tori, Y.; Nakamura, Y. & Osawa, T. (1999) Activation of stress signaling pathways by the end product of lipid peroxidation, 4-hydroxy-2-nonenal is a potential inducer of intracellular peroxide production. *Journal of Biological Chemistry*. Vol. 274, pp. 2234–2242. ISSN: 0021-9258

Uchida, K. (2003) 4-Hydroxy-2-nonenal: A product and mediator of oxidative stress. *Progress in Lipid Research*. Vol. 42, pp. 318–343. ISSN: 0163-7827

Yin, H.; Xu, L. & Porter, N.A. (2011) Free radical lipid peroxidation: mechanisms and analysis. *Chemical Reviews*. 111: 5944-5972. ISSN: 0009-2665

Lipid Peroxidation End-Products as a Key of Oxidative Stress: Effect of Antioxidant on Their Production and Transfer of Free Radicals

Hanaa Ali Hassan Mostafa Abd El-Aal

Additional information is available at the end of the chapter

1. Introduction

1.1. Oxidative stress

The term oxidative stress; is a state of unbalanced tissue oxidation refers to a condition in which cells are subjected to excessive levels of molecular oxygen or its chemical derivatives called reactive oxygen species (ROS).Under physiological conditions, the molecular oxygen undergoes a series of reactions that ultimately lead to the generation of superoxide anion (O_2^-), hydrogen peroxide (H_2O_2) and H_2O. Peroxynitrite ($OONO^-$), hypochlorus acid ($HOCl$), the hydroxyl radical ($OH.$), reactive aldehydes, lipid peroxides and nitrogen oxides are considered among the other oxidants that have relevance to vascular biology.

Oxygen is the primary oxidant in metabolic reactions designed to obtain energy from the oxidation of a variety of organic molecules. Oxidative stress results from the metabolic reactions that use oxygen, and it has been defined as a disturbance in the equilibrium status of pro-oxidant/anti-oxidant systems in intact cells. This definition of oxidative stress implies that cells have intact pro-oxidant/anti-oxidant systems that continuously generate and detoxify oxidants during normal aerobic metabolism. When additional oxidative events occur, the pro-oxidant systems outbalance the anti-oxidant, potentially producing oxidative damage to lipids, proteins, carbohydrates, and nucleic acids, ultimately leading to cell death in severe oxidative stress. Mild, chronic oxidative stress may alter the anti-oxidant systems by inducing or repressing proteins that participate in these systems, and by depleting cellular stores of anti-oxidant materials such as glutathione and vitamin E (Laval, 1996). Free radicals and other reactive species are thought to play an important role oxidative stress resulting in many human diseases. Establishing their precise role requires the ability to measure them and the oxidative damage that they cause (Halliwell and Whiteman, 2004).

Oxidative stress is involved in the process of aging (Kregel and Zhang 2007) and various chronic diseases such as atherosclerosis (Fearon and Faux 2009), diabetes (Ceriello and Motz, 2004) and eye disease (Li et al. 2009), whereas fruit and vegetable diets rich in antioxidants such as polyphenols, vitamin C, and carotenoids are correlated with a reduced risk of such chronic diseases (Dherani et al. 2008). An excessive amount of reactive oxygen/nitrogen species (ROS/RNS) leading to an imbalance between antioxidants and oxidants can cause oxidative damage in vulnerable targets such as unsaturated fatty acyl chains in membranes, thiol groups in proteins, and nucleic acid bases in DNA (Ceconi et al. 2003). Several assays to measure " total " antioxidant capacity of biological systems have been developed to investigate the involvement of oxidative stress in pathological conditions or to evaluate the functional bioavailability of dietary antioxidants. Conventional assays to determine antioxidant capacity primarily measure the antioxidant capacity in the aqueous compartment of plasma. Consequently, water - soluble antioxidants such as ascorbic acid, uric acid, and protein thiols mainly influence these assays, whereas fat - soluble antioxidants such as tocopherols and carotenoids show little inf uence over the many results. However, there are new approaches to define the total antioxidant capacity of plasma, which reflect the antioxidant network between water - and fat - soluble antioxidants. Revelation of the mechanism of action of antioxidants and their true antioxidant potential can lead to identifying proper strategies to optimize the antioxidant defense systems in the body.

1.2. Measurement of oxidative damage

A basic approach to study oxidative stress would be to measure some products such as (i) free radicals; (ii) radical-mediated damages on lipids, proteins or DNA molecules; and iii) antioxidant enzymatic activity or concentration.

1.2.1. Free radicals

Free radicals are reactive compounds that are naturally produced in the human body. They can exert positive effects (e.g. on the immune system) or negative effects (e.g. lipids, proteins or DNA oxidation). Free radicals are normally present in the body in minute concentrations. Biochemical processes naturally lead to the formation of free radicals, and under normal circumstances the body can keep them in check. If there is excessive free radical formation, however, damage to cells and tissue can occur (Wilson, 1997). Free radicals are toxic molecules, may be derived from oxygen, which are persistently produced and incessantly attack and damage molecules within cells; most frequently, this damage is measured as peroxidized lipid products, protein carbonyl, and DNA breakage or fragmentation. Collectively, the process of free radical damage to molecules is referred to as oxidative stress (Reiter et al., 1997).To limit these harmful effects, an organism requires complex protection – the antioxidant system. This system consists of antioxidant enzymes (catalase, glutathione peroxidase, superoxide dismutase) and non-enzymatic antioxidants (e.g. vitamin E [tocopherol], vitamin A [retinol], vitamin C [ascorbic acid], glutathione and uric acid). An imbalance between free radical production and antioxidant defence leads to an oxidative stress state, which may be involved in aging processes and even in some pathology (e.g. cancer and Parkinson's disease).

1.2.2. Formation of free radicals

Normally, bonds don't split in a way that leaves a molecule with an odd, unpaired electron. But when weak bonds split, free radicals are formed. Free radicals are very unstable and react quickly with other compounds, trying to capture the needed electron to gain stability. Generally, free radicals attack the nearest stable molecule, gaining its electron. When the "attacked" molecule loses its electron, it becomes a free radical itself, beginning a chain reaction. Once the process is started, it can cascade, finally resulting in the disruption of a living cell. Some free radicals arise normally during metabolism. Sometimes the body's immune system's cells purposefully create them to neutralize viruses and bacteria. However, environmental factors such as pollution, radiation, and toxins can also spawn free radicals. Normally, the body can handle free radicals, but if antioxidants are unavailable, or if the free-radical production becomes excessive, damage can occur. Of particular importance is that free radical damage accumulates with age (Packer, 1994).

1.2.3. Sources of free radicals

Free radicals have two principle sources: endogenous sources and exogenous sources. Endogenous sources of free radicals include those that are generated intracellularly, acting within the cell, and those that are formed within the cell, but are released into the surrounding area. These intracellular free radicals result from auto-oxidation and consequent inactivation of small molecules such as reduced thiols and flavins. They may also occur as a result of the activity of certain oxidases, lipoxygenases, cyclo-oxygenases, dehydrogenases and peroxidases. Electron transfer from metals such as iron to oxygen-containing molecules can also initiate free radical reactions paradoxically; antioxidants may also produce free radicals (Weir et al., 1996). A wide range of free radical molecular species are endogenous. The singlet oxygen is not a free radical but is nevertheless a reactive oxygen species and capable of causing tissue damage (Zebger et al., 2004). Exogenous sources of free radicals are environmental sources. Environmental sources of free radicals include exposure to ionizing radiation (from industry, sun exposure, cosmic rays, and medical X-rays), ozone and nitrous oxide (primarily from automobile exhaust), heavy metals (such as mercury, cadmium, and lead), cigarette smoke (both active and passive), alcohol, unsaturated fat, and other chemicals and compounds from food, water, and air. The exogenous sources of free radicals resulting from ionizing radiation play a major role in free radical production. The energy transferred into water from ionizing particles ionizes the water molecule. The water ions produced dissociate yielding free radicals (Valencia and Moran, 2004).

There are two enzymes including Aldehyd oxidase (AO) and xanthine oxidase (XO), they have a very close evolutionary relationship, based on the recent cloning of the gens and they show a high degree of amino acid sequence homology (Terao et al., 2000). They have been suggested to be relevant to the pathophysiology of a number of clinical disorders (Wright et al., 1995). **Aldehyd oxidase (AO)** commonly exists in vertebrates. Although the liver is the main site for aldehyde oxidase this enzyme has also been reported in kidney, lung, muscle, spleen, stomach, heart and brain (Beedham, 2002). The enzyme in liver of various species catalyzes the oxidation of a number of aldehydes and nitrogenous and also catalyzes the metabolism of physiological compounds such as retinaldehyde and monoamine neurotransimeters (Huang and Ichikawa, 1994). Reduction of oxygen during substrate turnover, leads to the formation of superoxide anion and hydrogen peroxide as ROS. This capacity has attracted attention to the possible role of aldehyde oxidase as a source of ROS. In vivo, it seems that aldehyde oxidase together with cytochrome P450 are quantitative, the most important cellular sources for ROS (Al-Omar et al., 2004). Additionally, the most likely sources of free radicals are **xanthine oxidase (XO)** (McCord, 1985). This enzyme is high particularly in liver and intestine. Although XO generates ROS and evidence has been presented for its role in the development of ischaemic intestinal, hepatic and renal damage (Cohen, 1992). It may also contribute to the development of lung and myocardial reperfusion injury after ischaemic episodes.

1.2.4. Production of free radicals

Free radicals are produced in a number of ways in biological systems (Halliwell and Whiteman, 2004):

a. Exposure to ionizing radiation is a major cause of free radical production. When irradiated water is ionized, and electron is removed from the molecule, leaving behind an ionized water molecule. The damaging species resulting from the radiolysis of water are the free radicals •H and •OH and hydrated electrons. They are highly reactive and have a lifetime on the order of 10 -9 to 10 -11 seconds. The hydroxyl radical is extremely reactive and is carcinogenic. Since water presents the largest number of target molecules in a cell, most of the energy transfer goes on in water when a cell is irradiated, rather then the solute consisting of protein, carbohydrate, nucleic acid, and bioinorganic molecules. Oxygen is an excellent electron acceptor and can combine with the hydrogen radical to form a peroxyl radical. Hydrogen peroxide is toxic and when present in sufficient quantities can interfere with normal cellular metabolism.

b. Enzymes and transport molecules also generate free radicals as a normal consequence of their catalytic function.

c. Auto-oxidation reactions produce free radicals from the spontaneous oxidation of biological molecules involved in non-enzymatic electron transfers.

d. Physical exercise also increases oxidative stress and causes disruptions of the homeostasis. Training can have positive or negative effects on oxidative stress depending on training load, training specificity and the basal level of training. Moreover, oxidative stress seems to be involved in muscular fatigue and may lead to overtraining.

Lipid Peroxidation End-Products as a Key of Oxidative Stress: Effect of Antioxidant on
Their Production and Transfer of Free Radicals

35

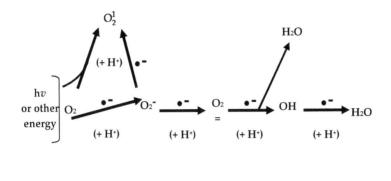

Molecular   Superoxide   Peroxide   Hydroxyl
oxygen   Radical   (H_2O_2)   radical
(Triplet)   (HO_2)

Figure 1. The active oxygen system. Molecular oxygen is reduced to water in four single-electron steps. Reduction of non-radical forms of oxygen is a "forbidden" process and thus usually involves spin-orbit coupling by a heavy metal or a halide or excitation to singlet state. An example is Fenton's reaction, the reduction of peroxide to water and hydroxyl radical by ferrous iron. Hydroxyl radical is one of the most powerful oxidizing agents known.

2. Oxidative damage to lipids (Lipid peroxidation)

The peroxidation of lipids is basically damaging because the formation of lipid peroxidation products leads to spread of free radicals reactions. The important role of lipids in cellular components emphasizes the significance of understanding the mechanisms and consequences of lipid peroxidation in biological systems. Polyunsaturated fatty acids (PUFAs) serve as excellent substrates for lipid peroxidation because of the presence of active bis-allylic methylene groups. The carbon-hydrogen bonds on these activated methylene units have lower bond dissociation energies, making these hydrogen atoms more easily abstracted in radical reactions (Davies et al., 1981). The susceptibility of a particular PUFA toward peroxidation increases with an increase in the number of unsaturated sites in the lipid chain (Nagaoka et al., 1990).

Lipid hydroperoxides are non-radical intermediates derived from unsaturated fatty acids, phospholipids, glycolipids, cholesterol esters and cholesterol itself. Their formation occur in enzymatic or non-enzymatic reactions involving activated chemical species known as "reactive oxygen species" (ROS) which are responsible for toxic effects in the body via various tissue damages. These ROS include among others hydroxyl radicals, lipid oxyl or peroxyl radicals, singlet oxygen, and peroxinitrite formed from nitrogen oxide (NO), all these groups of atoms behave as a unit and are now named "free radical". These chemical forms are defined as any species capable of independent existence that contains one or more unpaired electrons (those which occupy an atomic or molecular orbital by themselves). They are formed either by the loss of a single electron from a non-radical or by the gain of a single electron by a non-radical. They can easily be formed when a covalent bond is broken if one electron from each of the pair shared remains with each atom, this mechanism is known as homolytic fission. In water, this process generates the most reactive species, hydroxyl radical OH.. Chemists know well that combustion which is able at high temperature to rupture C-C, C-H or C-O bonds is a free-radical process. The opposite of this mechanism is the heterolytic fission in which, after a covalent break, one atom receives both electrons (this gives a negative charge) while the other remains with a positive charge.Lipid peroxidation leads to the breakdown of lipids and to the formation of a wide array of primary oxidation products such as conju- dienes or lipid hydroperoxides, and secondary products including MDA, F2-isoprostane or expired pentane, ethane or hexane. Measurement of conjugated dienes is interesting because it detects molecular reorganisation of poly- unsaturated fatty acids during the initial phase lipid peroxidation. Lipid hydroperoxide is another marker of the initial.(reaction of FR and is a specific marker of cellular damage. Other products are often used to measure oxida- stress but have the disadvantage of being secon- dary oxidation products. One of them, MDA, is during fatty acid auto-oxidation. This sub- is most commonly measured by its reaction with thiobarbituric acid, which generates thiobarbi- turic acid reactive substances (TBARS). Although of the results.MDA overestimation), this method is accepted as a the general marker of lipid peroxidation but results are subject to caution (Sies, 1997). In addition, some studcable ies tend to show that MDA is not an adapted method ucts used for such methods.

The peroxidation of lipids involves three distinct steps: initiation, propagation and termination. The initiation phase of lipid peroxidation may proceed by the reaction of an activated oxygen species such as singlet oxygen ($1O2$), $O2-$, or $HO\cdot$ with a lipid substrate or by the breakdown of preexisting lipid hydroperoxides by transition metals. In the former case, peroxidation occurs by abstraction of a hydrogen atom from a methylene carbon in the lipid substrate (LH) to generate a highly reactive carbon-centered lipid radical ($L\cdot$) (Kelly et al., 1998). In the propagation phase of lipid peroxidation, molecular oxygen adds rapidly to L at a diffusion controlled rate to produce the lipid peroxyl radical ($LOO\cdot$). The peroxyl radical can abstract a hydrogen atom from a number of in vivo sources, such as DNA and proteins, to form the primary oxidation product, a lipid hydroperoxide (LOOH). Alternatively, antioxidants such as α -tocopherol (α -TOH) can act as excellent hydrogen atom donors , generating LOOH and the relatively inert α-tocopherol phenoxyl radical (α -TO$\cdot$). In the absence of antioxidants or other inhibitors, $LOO\cdot$ can abstract a hydrogen from another lipid molecule (LH), producing another highly reactive carbon centered radical ($L\cdot$), which then propagates the radical chain as presented in Figure 2 (Waldeck and Stocke , 1996). The lipid hydroperoxide (ROOH) is unstable in the presence of iron or other metal catalysts because ROOH will participate in a Fenton reaction leading to the formation of reactive alkoxy radicals. Therefore, in the presence of irron, the chain reactions are not only propagated but amplified. Among the degradation products of ROOH are aldehydes, such as malondialdehyde, and hydrocarbons, such as ethane and ethylene, which are commonly measured end products of lipid peroxidation (Sener et al., 2004).

Lipid Peroxidation End-Products as a Key of Oxidative Stress: Effect of Antioxidant on
Their Production and Transfer of Free Radicals

37

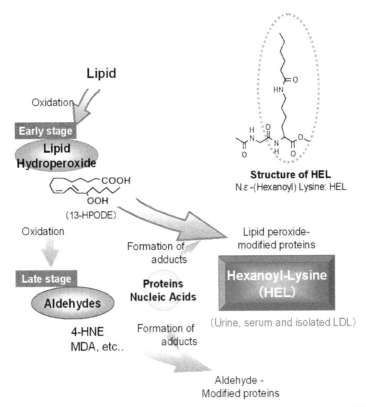

Structure of HEL
N ε -(Hexanoyl) Lysine: HEL

Abbreviations: NRP, nonradical product; LOOH, lipid hydroperoxide; α-TOH, α -tocopherol; α -TO, α -TOH radical;
LH, lipid substrate; LOO, lipid peroxyl radical. Adapted from Waldeck and Stocke(1996)

Figure 2. Overview of lipid peroxidation.

During peroxidation pathway via reactive intermediates, several end products are formed
such as aldehyde [malondialdehyde + 4-hydroxynonenal], pentane and ethane, 2,3
transconjugated diens, isoprostains and chlesteroloxides. The biological activities of MDA
and other aldehydes include cross-linking with DNA and proteins, which alters the
function/activity of these molecules. MDA + 4HNE have shown tissue toxicity. MDA can
react with amino and thiol groups, the aldehydes are more diffusible than free radicals,
which means damage is exported to distance sites. Aldehydes are quickly removed from
cells as several enzymes control their metabolism (Ustinova and Riabinin 2003).

3. Antioxidants

To minimize the negative effects of ROS generated by any pro-oxidant, endogenous
defensive mechanisms called antioxidant defense (AD) system, which utilizes enzymatic
and non-enzymatic mechanisms. Antioxidants are naturally occurring substances that

combat oxidative damage in biological entities. An antioxidant achieves this by slowing or preventing the oxidation process that can damage cells in the body. This it does by getting oxidized itself in place of the cells. Thus an antioxidant can also be termed as a reducing agent. Antioxidants are considered as important in the fight against the damage that can be done by free radicals produced due to oxidative stress. Although the human body has its own defenses against oxidative stress, these become weak with age or in the case of an illness. Although, antioxidants are sold in various forms as dietary supplements there is no clinching clinical evidence in favor of antioxidants as beneficial in maintaining health and preventing disease. However, there is a lot of anecdotal evidence that those who partake of antioxidant-rich food are better protected against problems such as heart disease, macular degeneration, diabetes, and cancer. Antioxidants are either hydrophilic or hydrophobic. Water soluble or hydrophilic antioxidants are active in the blood plasma while the water insoluble antioxidants protect the cell membranes. How do antioxidants work? Antioxidants work by bringing under control the rogue and unstable oxygen molecules that have an odd number of electrons. These oxygen molecules known as free radicals are highly reactive; they attack cells, DNA, and protein thereby accelerating the aging process. The antioxidants work in harmony and the efficacy of one antioxidant depends upon the availability and concentration of another. Essentially, antioxidants work by donating an electron to the unstable free radical. This stabilizes the free radical and converts it into a harmless compound that may safely be removed from the body. Antioxidants are segregated into two classes based on their mode of operating. They can either be chain-breaking or preventive. Chain-breaking antioxidants such as vitamins E and C halt the process of radical formation by stabilizing free radical molecules so that the chain-like process of radical formation is arrested. Preventive antioxidants such as superoxide dismutase and catalase prevent chain initiations by scavenging for initiator radicals and stabilizing them. They also stabilize transition metal radicals like iron and copper. These metals work as catalysts in the production of free radicals. Antioxidants and their various forms. Antioxidants are chiefly available to us through vitamins, enzymes, and minerals. Vitamin E is actually a group of eight tocopherols. Alpha-tocopherol is the most widely available tocopherol and also the most potent in terms of its effect on the body. Vitamin E is fat-soluble and protects cell membranes that are mainly composed of fatty acids. Vitamin C or ascorbic acid is water-soluble and it scavenges for free radicals that are present in aqueous environments within the human body. Beta carotene is also water soluble and is particularly effective in tackling free radicals in areas of low oxygen concentration. Selenium, manganese and zinc are trace elements that are important components of several antioxidant enzymes such as superoxide dismutase (SOD), catalase (CAT) and glutathione peroxidase. Enzymes work as both primary and secondary antioxidants and help repair oxidized DNA and target lipids that are oxidized. Other substances that are now being considered for their antioxidant properties include uric acid and phytochemicals found in plants.

Fruits and vegetables that have been identified as sources of powerful antioxidants help people counter the risk of heart ailments and different types of cancers. However, there is a possibility that these benefits obtained from fruits and vegetables could be a result of not just antioxidants but a mix that includes flavonoids as well. Although, clinical trials have not put forth conclusive evidence in favor of antioxidants as being helpful to our health the vast number of observational studies and anecdotal evidence offers a very strong suggestion that antioxidants are indeed of much use in keeping the body healthy. It is only a matter of time before scientists unravel the exact mechanism that governs the working of antioxidants in the body. Most nutritionists agree that the best source of antioxidants is natural food. One should try and avoid supplements if possible. It is also important to keep in mind that a high dosage of antioxidant supplements can have a detrimental effect on the body. Excessive vitamin E can lead to blood hemorrhage. Vitamin C in large amounts can cause diarrhea and also atherosclerosis. High amounts of selenium can cause hair loss and rashes on the skin. Antioxidants are thought to protect the body against the destructive effects of free radicals. Antioxidants neutralize free radicals by donating one of their own electrons, ending the electron- gain reaction. The antioxidant nutrients themselves don't become free radicals by donating an electron because they are stable in either form. They act as scavengers, helping to prevent cell and tissue damage that could lead to cellular damage and disease (Reiter, 2003; Reiter et al., 2004).

A rangeor of antioxidants are active in the body including enzymatic and non-enzymatic. All of them can to redox status. Redox status is directly linked and be intracellular or extracellular antioxidants . The body produces several enzymes, including superoxide dismutase (SOD), catalase (CAT), glutathione peroxidase (GSHPX), and glutathione reductase (GR) that neutralize many types of free radicals. Supplements of these enzymes are available for oral administration. However, their absorption is probably minimal at best. Supplementing with the "building blocks" the body requires to make SOD, catalase, and glutathione peroxidase may be more effective. These building block nutrients include the minerals manganese, zinc, and copper for SOD and selenium for GSHPX. While the nonenzymatic defense consists of substances of low molecular weight such as reduced glutathione vitamin C, vitamin E, beta-carotene, lutein, lycopene, vitamin B2, coenzyme Q10, and cysteine (an amino acid). Herbs, such as bilberry, turmeric (curcumin), grape seed or pine bark extracts, and ginkgo can also provide powerful antioxidant protection for the body. Melatonin is a hormone secreted by pineal gland and proves to be powerful antioxidant and free radical scavenger (Yang et al.,2002 and Koc et al., 2003).

3.1. Enzymatic antioxidants

Antioxidant enzymes include superoxide dismutase (SOD), catalase (CAT), glutathione peroxidase (GSHPX), and glutathione reductase (GR). Non-en-zymatic antioxidants include a variety of FR quenchers such as vitamin A (retinol), vitamin C - (ascorbic acid), vitamin E (tocopherol), flavonoids, thiols (including glutathione [GSH], ubidecarenone uric acid, bilirubin, ferritin) and micronutrients (iron, copper, zinc, selenium, mangawhich which act as enzymatic cofactors. The antioxidant system efficiency depends on nutritional ineccentric (vitamins and micronutrients) and on endoge nous antioxidant enzyme production, which can be modified by exercise, training, nutrition and agdative Moreover, the antioxidant system efficiency is important in sport physiology because exercise increases the production of FR.

3.1.1. Superoxide dismutase (SOD, EC 1.15.1.1)

Superoxide dismutase (SOD) is the major defence upon superoxide radicals and is the first defence line against oxidative stress. SOD represents a group of enzymes that catalyse the dismutation of O2– and the formation of H2O2. SOD is an enzyme (EC 1.15.1.1) discovered by McCord and Fridovich, which plays an important role in the defense mechanism of biological cells exposed to oxygen (McCord and Fridovich 1969). SOD catalyzes the dismutation of superoxide anion radical (O 2 • –) into an oxygen molecule and a hydrogen peroxide. This reaction is recognized as an antioxidant system that protects cells from superoxide toxicity. There are several types of SOD, depending on the type of metal ion. Three major isoforms of mammalian SOD have been identif ed with different tissue distributions (Zelko 2002). Cu/Zn - SOD (SOD1) exists in the cytoplasm, lysosomes, and nuclear compartments of mammalian cells (Bannister and et al., 1987; Zelko et al., 2002). In humans, the liver has a relatively high amount and activity of SOD1 (Nozik-Grayck et al., 2005). Human SOD1 is a homodimer containing one copper ion and one zinc ion in each 16 - kDa subunit which consists of 153 amino acids. The copper ion is held by interaction with imidazolate ligands of the histidine residues in SOD1 in the enzymatic active site. The zinc ion (Zn 2 +) contributes to the stabilization of the enzyme (Johnson and Giulivi , 2005).

3.1.2. Catalase (CAT, EC 1.11.1.6)

Catalse (CAT) is one of the major antioxidant enzymes (Scandalios et al., 1997). It is one of the first enzymes to be purified and crystallized and has gained a lot of attention in recent years because of its link to cancer, diabetes and aging in humans and animals (Preston et al., 2001). It is present in every cell and in particular in cell structures that use oxygen in order to detoxify toxic substances and produce H2O2. Catalase converts H2O2 into water andoxygen (Greenwald, 1990 ;Yasminch and Theologides, 1993). Catalase can also use H2O2 in order to detoxify some toxic substances via a peroxidase reaction (Mayo et al., 2003). There are many evidences that the changes of catalase activity as well as the mechanisms of its regulation are essential in the response to stress situations which catalyzes the dismutation of H2O2, forming O2 and H2O resulting good protection the cells from the toxic effects of hydrogen peroxide (Brioukhanov and Netrusor, 2004).

3.1.3. Glutathione peroxidase (GPX, EC 1.11.1.9)

GPX was discovered in 1957 ny Mills. It exists in cell cytosol and mitochondria and has the ability to transform H2O2 into waterThis reaction uses GSH and transforms it into oxidised glutathione (GSSG). GPX and CAT have the same action upon H2O2, debut GPX is more efficient with high ROS concentra-tion and CAT has an important action with lower H2O2 concentration. GPx is a glycoprotein containing a single selenocysteine residue at the active center of each subunit. To protect biological organisms from oxidative damage, GPx catalyses the reduction of hydrogen peroxide and lipid hydroperoxides to water and their corresponding alcohols, respectively, as follows (Antunes et al., 2002): ROOH + 2GSH→ROH +GSSG + H2O2 where reduced monomeric glutathione (GSH) is essential as a hydrogen donor, and GSH is oxidized to glutathione disulf de (GSSG). There are five main mammalian isozymes, which vary in the structure (amino acid sequence and subunit), tissue distribution (liver, kidney,erythrocyte, blood plasma, among others), location (cytoplasm, intestine, extracellular f uid), and substrate specif city (hydrogen peroxide and lipid hydroperoxides) (Dudek et al., 2002). GXP is a selenium dependent enzyme that is ubiquitously expressed and protects cells against oxidative damage by reducing hydrogen peroxide and a wide range of organic peroxides with reducing glutathione (Arthur, 2000). It has been suggested that GPX has anti - inflammatory activity in the cardiovascular system. An increase in cytosolic GPx is linked to a lower risk of cardiovascular disease (Blankenberg et al. 2003).

3.1.4. Glutathione reductase (GR, EC 1.6.4.2)

Glutathione Reductase (GR) is a key enzyme of glutathione metabolism and is widespread in all tissues and blood cells. It a flavin enzyme involved in the defense of the erythrocyte against hemolysis. This enzyme catalyses reduction of oxidized glutathione (GSSG) to reduced glutathione (GSH) in the presence of NADPH and maintains a high intracellular GSH/GSSG ratio of about 500 in red blood cells (Kondo et al., 1980). GR is important not only for the maintaining thr required GSH level but also for reducing protein thiols to their native state.This enzyme is conserved between all kingdoms. In bacteria, yeasts and animals, one GR gene is found, however in plant genomes two GR genes are enclosed. Under normal conditions, GSH and GR are involved in the detoxification of H2O2 generated in the light by the Mehler reaction in chloroplasts. Disturbances GH level have been correlated with oxidative stress induced by various factors including toxicity, pollutants, inflammation and different diseases particularly red blood cell defects.

3.1.5. Glutathione-S-transferase (GST, EC 2.5.1.18)

Glutathione-S-Transferase (GST) catalyzes the conjugation with glutathione of a number of electrophilic xenobiotics, including several carcinogens, mutagens and anticancer drugs (Hayes and Pulford, 1995). These electrophiles are made less reactive by conjugation with glutathione and the conjugates are thought to be less toxic to the cell. Consiquently, GSTs are believed to play an important role in the defense of cells against these zenobiotic toxins. Several antineoplastic drugs particularly the reactive electrophilic alkylating agents, can form conjugates with glutathione both spontaneously and in GST-catalyzed reactions (Awasthi et al., 1996). Morever, Some studies reported that there are a good association between cellular resistance to some anticancers drugs and expression of particular isozymes of GST (Hayes and Pulford, 1995).

3.2.1. Glutathione (GSH)

Gultathione (GSH) is a small molecule found in almost every cell (Anderson, 1997). It is the smallest intrecellular thiol (SH) molecule. Its high electron-donating capacity (high negative redox potential) combined with high intracellular concentration generate great reducing power (Kidd, 1997). Glutathione is a cysteine-containing peptide found in most forms of aerobic life. It is not required in the diet and is instead synthesized in cells from its constituent amino acids (Meister and Anderson, 1983; Meister, 1988). It can react non-enzymatically with ROS and GSH peroxidase catalyses the destruction of hydrogen peroxide and hydroperoxides resulting in its oxidation to the disulphide form (GSSG). Firstly, glutathione is the major antioxidant produced by the cell protecting it from free radicals as oxygen radicals which are highly reactive substances can damage or destroy key cell components. Its antioxidant properties result from the thiol group in its cysteine moiety is a reducing agent and can be reversibly oxidized and reduced. In cells, glutathione is maintained in the reduced form by the enzyme glutathione reductase and in turn reduces other metabolites and enzyme systems, such as ascorbate in the glutathione-ascorbate cycle, glutathione peroxidases and glutaredoxins, as well as reacting directly with oxidants. Due to its high concentration and its central role in maintaining the cell's redox state, glutathione is one of the most important cellular antioxidants. In some organisms glutathione is replaced by other thiols, such as by mycothiol in the Actinomycetes, bacillithiol in some Gram-positive bacteria, or by trypanothione in the Kinetoplastids (Meister and Larsson, 1995). Secondly, GSH is a very important detoxifing agent, enabling the body to get rid of undesirable toxins and pollutants. It forms a soluble compound with the toxin that can then be excreted through the urine or the gut. The liver and kidneys contain high levels of GSH as they have the greatest exposure to toxins. The lung are also rich in glutathione partly for the same reason. Thirdly, GSH plays a crucial role in maintaining a normal balance between oxidation and anti-oxidation. This in turn regulates many of the cell's vital functions such as the synthesis and repair of DNA, the synthesis of proteins and the activation, maintaining the essential thiol status of protein, immune function, regulate nitric oxide homeostasis, modulate the activity of neurotransmitter receptors and regulation of enzymes (Oja et al., 2000; Hogg, 2002). The lower level of GSH is related to different physiological and biochemical disturbances.

3.2.2. Ascorbic acid

Ascorbic acid or "vitamin C" is a monosaccharide oxidation-reduction (redox) catalyst found in both animals and plants (Peake , 2003). As one of the enzymes needed to make ascorbic acid has been lost by mutation during primate evolution, humans must obtain it from the diet; it is therefore a vitamin. Most other animals are able to produce this compound in their bodies and do not require it in their diets. Ascorbic acid is required for the conversion of the procollagen to collagen by oxidizing proline residues to hydroxyproline. In other cells, it is maintained in its reduced form by reaction with glutathione, which can be catalysed by protein disulfide isomerase and glutaredoxins. Ascorbic acid is redox catalyst which can reduce, and thereby neutralize, reactive oxygen species such as hydrogen peroxide. In addition to its direct antioxidant effects, ascorbic acid is also a substrate for the redox enzyme ascorbate peroxidase, a function that is particularly important in stress resistance in plants. Ascorbic acid is present at high levels in all parts of plants and can reach concentrations of 20 millimolar in chloroplasts. It act as a marked antioxidant that help in the treatement of different diseases such as cancer and cardiovascular (Coulter et al., 2006 ; Cook et al., 2007).

3.2.3. Melatonin

Melatonin (N-acetyl-5-methoxytryptamine), is synthesized from serotonin in the pineal gland which contains all the enzymes necessary for the methoxylation and acetylation reactions. Melatonin is released in mammals during the dark-phase of the circadian cycle, and declines with age (Tan et al., 2001). It is able to reduce the free radical formation which follows the interaction between transition metal ions and amyliod-beta peptide (Zatta et al., 2003). As a free radical scavenger melatonin exhibits several important properties: It has both lipophilic and hydrophilic and it passes all bio-barriers, e.g. blood brain barrier and placenta (Wakatsuki et al., 1999).

Reiter (1995) reported that melatonin seems to be more effective than other antioxidants (e.g. mannitol, glutathione and vitamin E) in protecting against oxidative damage. Thus, it may provide protection against diseases that cause degenerative or proliferative changes by shielding macromolecules, particularly DNA from such injuries. Besides its direct free radical scavenging action, melatonin functions as an indirect antioxidant by stimulating the activities of antioxidiative enzymes in addition to protecting against lipid peroxidation (Undeger et al., 2004).

Melatonin has been found to be a direct free radical scavenger and an indirect antioxidant that, may have an active role in protection against genetic damage due to endogenously produced free radicals and it may be of use in reducing damage from physical and chemical mutagens and carcinogens that generate free radicals (Bandyopadhyay et al., 2000).

3.2.4. Tocopherols and tocotrienols (vitamin E)

Vitamin E is the collective name for a set of eight related tocopherols and tocotrienols, which are fat-soluble vitamins with antioxidant properties (Roberts et al., 2007). Of these, α-tocopherol has been most studied as it has the highest bioavailability, with the body preferentially absorbing and metabolising this form. It has been claimed that the α-tocopherol form is the most important lipid-soluble antioxidant, and that it protects membranes from oxidation by reacting with lipid radicals produced in the lipid peroxidation chain reaction. This removes the free radical intermediates and prevents the propagation reaction from continuing (Cook et al., 2007). This reaction produces oxidised α-tocopheroxyl radicals that can be recycled back to the active reduced form through reduction by other antioxidants, such as ascorbate, retinol or ubiquinol.This is in line with findings showing that α-tocopherol, but not water-soluble antioxidants, efficiently protects glutathione peroxidase 4 (GPX4)-deficient cells from cell death. GPx4 is the only known enzyme that efficiently reduces lipid-hydroperoxides within biological membranes. However, the roles and importance of the various forms of vitamin E are presently unclear, and it has even been suggested that the most important function of α-tocopherol is as a signaling molecule, with this molecule having no significant role in antioxidant metabolism. The functions of the other forms of vitamin E are even less well-understood, although γ-tocopherol is a nucleophile that may react with electrophilic mutagens, and tocotrienols may be important in protecting neurons from damage. However it has a protective action against different diseases including cancer (Coulter et al., 2006).

3.3. Total antioxidant capacity

Epidemiologic studies have demonstrated an inverse association between consumption of fruits and vegetables and morbidity and mortality from degenerative diseases. The antioxidant content of fruits and vegetables may contribute to the protection they offer from disease. Because plant foods contain many different classes and types of antioxidants, knowledge of their total antioxidant capacity (TAC), which is the cumulative capacity of food components to scavenge free radicals, would be useful for epidemiologic purposes. To accomplish this, a variety of foods commonly consumed in Italy, including 34 vegetables, 30 fruits, 34 beverages and 6 vegetable oils, were analyzed using three different assays, i.e., Trolox equivalent antioxidant capacity (TEAC), total radical-trapping antioxidant parameter (TRAP) and ferric reducing-antioxidant power (FRAP). These assays, based on different chemical mechanisms, were selected to take into account the wide variety and range of action of antioxidant compounds present in actual foods. Among vegetables, spinach had the highest antioxidant capacity in the TEAC and FRAP assays followed by peppers, whereas asparagus had the greatest antioxidant capacity in the TRAP assay. Among fruits, the highest antioxidant activities were found in berries (i.e., blackberry, redcurrant and raspberry) regardless of the assay used. Among beverages, coffee had the greatest TAC, regardless of the method of preparation or analysis, followed by citrus juices, which exhibited the highest value among soft beverages. Finally, of the oils, soybean oil had the highest antioxidant capacity, followed by extra virgin olive oil, whereas peanut oil was less effective. Such data, coupled with an appropriate questionnaire to estimate antioxidant intake, will allow the investigation of the relation between dietary antioxidants and oxidative stress-induced diseases (Pellegrini et al., 2003 ; Puchau et al., 2009 ; Dilis and Trichopoulou, 2010).

4. Nutritional therapy with natural antioxidants

Antioxidants have been the focus of research on the relationship between The role of dietary factors in protecting against the change from native to oxidized LDL has received considerable attention. An overview of epidemiological research suggests that individuals with the highest intakes of antioxidant vitamins, whether through diet or supplements, tend to lower of various disease. Research examining the effects of a diet rich in fruits and vegetables on disease has been carried out using several types of study.There is strong scientific evidence to support an increase in intakes of vegetables and fruit in the prevention of disease. Further research is required to clarify which particular components of fruit and vegetables are responsible for their protective effects. Numerous epidemiological studies have indicated that diets rich in fruits and vegetables are correlated with a reduced risk of chronic diseases (German, 1999; Benzie, 2003; Hassan, 2005; Hassan and Yosef, 2009; Hassan et al., 2010). It is probable that antioxidants, present in the fruits and vegetables such as polyphenols, carotenoids, and vitamin C, prevent damage from harmful reactive oxygen species, which either are continuously produced in the body during normal cellular functioning or are derived from exogenous sources (Gate et al. 1999). The possible protective effect of antioxidants in fruits and vegetables against ROS has led people to consume antioxidant supplements against chronic diseases.

5. Nutritional factors as natural antioxidants agents in alleviating the oxidative stress induced by environmental pollutents: Some experimental studies for the author

5.1. Mitigating effects of antioxidant properties of black berry juice on sodium fluoride induced hepatotoxicity and oxidative stress in rats

Fluorosis is a serious public health problem in many parts of the world. As in the case of many chronic degenerative diseases, increased production of reactive oxygen species has been considered to play an important role, even in the pathogenesis of chronic fluoride toxicity. Black berry is closely linked to its protective properties against free radical attack. Therefore, the aim of this study was to demonstrate the role of black berry juice (BBJ) in decreasing the hepatotoxicity and oxidative stress of sodium fluoride)NaF). Results showed that NaF caused elevation in liver TBARS and nitric oxide (NO), and reduction in superoxide dismutase (SOD), catalase (CAT), total antioxidant capacity (TAC) and glutathione (GSH.(Plasma transaminases (AST and ALT), creatine kinase (CK), lactate dehydrogenase (LDH), total lipids)TL), cholesterol, triglycerides (TG), and low density lipoprotein–cholesterol (LDL–c) were increased, while high density lipoprotein–cholesterol (HDL–c) was decreased. On the other hand, BBJ reduced NaF-induced TBARS, NO, TL, cholesterol, TG, LDL-c, AST, ALT, CK and LD. Moreover, it ameliorated NaFinduced decrease in SOD, CAT, GSH, TAC and HDL-c. Therefore, the present results revealed that BBJ has a protective effect against NaF-induced hepatotoxicity by antagonizing the free radicals generation and enhancement of the antioxidant defence mechanisms. (Hassan and Yousef, 2009).

5.2. Evaluation of free radical-scavenging and antioxidant properties of black berry against fluoride toxicity in rats

Oxidative damage to cellular components such as lipids and cell membranes by free radicals and other reactive oxygen species is believed to be associated with the development of degenerative diseases. Fluoride intoxication is associated with oxidative stress and altered anti-oxidant defense mechanism. So the present study was extended to investigate black berry anti-oxidant capacity towards superoxide anion radicals, hydroxyl radicals and nitrite in different organs of fluoride-intoxicated rats. The data indicated that sodium fluoride (10.3 mg/kg bw) administration induced oxidative stress as evidenced by elevated levels of lipid peroxidation and nitric oxide in red blood cells, kidney, testis and brain tissues. Moreover, significantly decreased glutathione level, total anti-oxidant capacity and superoxide dismutase activitywere observed in the examined tissues. On the other hand, the induced oxidative stress and the alterations in anti-oxidant system were normalized by the oral administration of black berry juice (1.6 g/ kg bw). Therefore it can be concluded that black berry administration could minimize the toxic effects of fluoride indicating its free radical-scavenging and potent anti-oxidant activities. (Hassan and Fattoh, 2010)

5.3. Garlic oil as a modulating agent for oxidative stress and neurotoxicity induced by sodium nitrite in male albino rats

In the present study, we investigated the neurobiochemical alterations and oxidative stress induced by food preservative; sodium nitrite (NaNO2) as well as the role of the garlic oil in amelioration of the neurotoxicity in male albino rats. Serum and brain homogenates of the rats received NaNO2 (80 mg/kg body weight) for 3 months exhibited significant decrease in acetylcholine esterase (AChE) activity as well as the levels of phospholipids, total protein and the endogenous antioxidant system (glutathione; GSH and superoxide dismutase; SOD). In contrast, lactic dehydrogenase (LDH) activity, brain thiobarbituric acid reactive substances (TBARS) and nitric oxide (NO) levels were significantly increased. On the other hand, the oral administration of garlic oil (5 ml/kg body weight) daily for 3 months significantly improved the neurobiochemical disorders and inhibited the oxidative stress induced by NaNO2 ingestion. So, this study reveals the neural toxic effects of NaNO2 by exerting oxidative stress and retrograde the endogenous antioxidant system. However, garlic oil has a promising role in attenuating the obtained hazard effects of sodium nitrite by its high antioxidant properties which may eventually be related with the preservation of SOD activity and primary mitochondrial role against nitrite-induced neurotoxicity in rats . (Hassan et al., 2010).

5.4. *In vivo* evidence of hepato-and-reno-protective effect of garlic oil against sodium nitrite-induced oxidative stress

Sodium nitrite (NaNO2), a food color fixative and preservative, contributes to carcinogenesis. We investigated the protective role of garlic oil against NaNO2-induced abnormalities in metabolic biochemical parameters and oxidative status in male albino rats. NaNO2 treatment for a period of three months induced a significant increase in serum levels of glucose, aspartate aminotransferase (AST), alanine aminotransferase (ALT), alkaline phosphatase (ALP), bilirubin, urea and creatinine as well as hepatic AST and ALT.

However, significant decrease was recorded in liver ALP activity, glycogen content, and renal urea and creatinine levels. In parallel, a significant increase in lipid peroxidation, and a decrease in glutathione content and catalase activity were observed in the liver and the kidney. However, garlic oil supplementation showed a remarkable amelioration of these abnormalities. Our data indicate that garlic is a phytoantioxidant with powerful chemopreventive properties against chemically-induced oxidative stress (Hassan et al., 2009).

5.5. Ameliorating effect of chicory (Cichorium intybus L)-supplemented diet against nitrosamine precursors-induced liver injury and oxidative stress in male rats

The current study was carried out to elucidate the modulating effect of chicory (*Cichorium intybus L.*)-supplemente diet against nitrosamnine-induced oxidative stress and hepatotoxicity in male rats. Rats were divided into four groups and treated for 8 weeks as follow: group 1 served as control; group 2 fed on chicory-supplemented diet (10% w/w); group 3 received simultaneously nitrosamine precursors [sodium nitrite (0.05% in drinking water) plus chlorpromazine (1.7 mg/kg body weight)] and group 4 received nitrosamine precursors and fed on chicory-supplemented diet. The obtained results revealed that rats received nitrosamine precursors showed a significant increase in liver TBARS and total lipids, total cholesterol, bilirubin, and enzymes activity (AST, ALT, ALP and GGT) in both serum and liver. While a significant decrease in the levels of GSH, GSH-Rx, SOD, catalase, total protein and albumin was recorded. On the other hand, chicory-supplemented diet succeeded to modulate these observed abnormalities resulting from nitrosamine compounds as indicated by the reduction of TBARS and the pronounced improvement of the investigated biochemical and antioxidant parameters. So, it could be concluded that chicory has a promising role and it worth to be considered as a natural substance for ameliorating the oxidative stress and hepatic injury induced by nitrosamine compounds **(Hassan and Yousef, 2010).**

6. Summary

6.1. Oxidative stress

The term oxidative stress refers to a condition in which cells are subjected to excessive levels of molecular oxygen or its chemical derivatives called reactive oxygen species (ROS).Under physiological conditions, the molecular oxygen undergoes a series of reactions that ultimately lead to the generation of superoxide anion (O_2-), hydrogen peroxide (H_2O_2) and H_2O. Peroxynitrite (OONO-), hypochlorus acid (HOCl), the hydroxyl radical (OH.), reactive aldehydes, lipid peroxides and nitrogen oxides are considered among the other oxidants that have relevance to vascular biology.

Oxygen is the primary oxidant in metabolic reactions designed to obtain energy from the oxidation of a variety of organic molecules. Oxidative stress results from the metabolic reactions that use oxygen, and it has been defined as a disturbance in the equilibrium status of pro-oxidant/anti-oxidant systems in intact cells. This definition of oxidative stress implies that cells have intact pro-oxidant/anti-oxidant systems that continuously generate and detoxify oxidants during normal aerobic metabolism. When additional oxidative events occur, the pro-oxidant systems outbalance the anti-oxidant, potentially producing oxidative damage to lipids, proteins, carbohydrates, and nucleic acids, ultimately leading to cell death in severe oxidative stress. Mild, chronic oxidative stress may alter the anti-oxidant systems by inducing or repressing proteins that participate in these systems, and by depleting cellular stores of anti-oxidant materials such as glutathione and vitamin E (Laval, 1996). Free radicals and other reactive species are thought to play an important role oxidative stress resulting in many human diseases. Establishing their precise role requires the ability to measure them and the oxidative damage that they cause (Halliwell and Whiteman, 2004).

A basic approach to study oxidative stress would be to measure some products such as (i) free radicals; (ii) radical-mediated damages on lipids, proteins or DNA molecules; and iii) antioxidant enzymatic activity or concentration.

6.2. Free radicals

Free radicals are reactive compounds that are naturally produced in the human body. They can exert positive effects (e.g. on the immune system) or negative effects (e.g. lipids, proteins or DNA oxidation). Free radicals are normally present in the body in minute concentrations. Biochemical processes naturally lead to the formation of free radicals, and under normal circumstances the body can keep them in check. If there is excessive free radical formation, however, damage to cells and tissue can occur (Wilson, 1997). Free radicals are toxic molecules, may be derived from oxygen, which are persistently produced and incessantly attack and damage molecules within cells; most frequently, this damage is measured as peroxidized lipid products, protein carbonyl, and DNA breakage or fragmentation. Collectively, the process of free radical damage to molecules is referred to as oxidative stress (Reiter et al., 1997).To limit these harmful effects, an organism requires complex protection – the antioxidant system. This system consists of antioxidant enzymes (catalase, glutathione peroxidase, superoxide dismutase) and non-enzymatic antioxidants (e.g. vitamin E [tocopherol], vitamin A [retinol], vitamin C [ascorbic acid], glutathione and uric acid). An imbalance between free radical production and antioxidant defence leads to an oxidative stress state, which may be involved in aging processes and even in some pathology (e.g. cancer and Parkinson's disease).

6.3. Oxidative damage to lipids (Lipid peroxidation)

The peroxidation of lipids involves three distinct steps: initiation, propagation and termination. The initiation reaction between an unsaturated fatty acid and the hydroxyl radical involves the abstraction of a H atom from the methylvinyl group on the fatty acid. The remaining carbon-centred radical, forms a resonance structure sharing this unpaired electron among carbons 9 to 13. In the propagation reactions, this resonance structure reacts with triplet oxygen, which is a biradical having two unpaired electrons and therefore reacts readily with other radicals. This reaction forms a peroxy radical. The peroxy radical then abstracts a H atom from a second fatty acid forming a lipid hydroperoxide and leaving another carbon centered free radical that can participate in a second H abstraction. Therefore, once one hydroxyl radical initiates the peroxidation reaction by abstracting a single H atom, it creates a carbon radical product that is capable of reacting with ground state oxygen in a chain reaction. The role of the hydroxyl radical is analogous to a "spark" that starts a fire. The basis for the hydroxyl radical's extreme reactivity in lipid systems is that at very low concentrations it initiates a chain reaction involving triplet oxygen, the most abundant form of oxygen in the cell (Benderitter et al., 2003).

The lipid hydroperoxide (ROOH) is unstable in the presence of irron or other metal catalysts because ROOH will participate in a Fenton reaction leading to the formation of reactive alkoxy radicals. Therefore, in the presence of irron, the chain reactions are not only propagated but amplified. Among the degradation products of ROOH are aldehydes, such as malondialdehyde, and hydrocarbons, such as ethane and ethylene, which are commonly measured end products of lipid peroxidation (Sener et al., 2004). During peroxidation pathway via reactive intermediates, several end products are formed such as aldehyde [malondialdehyde + 4-hydroxynonenal], pentane and ethane, 2,3 transconjugated diens, isoprostains and chlesteroloxides. The biological activities of MDA and other aldehydes include cross-linking with DNA and proteins, which alters the function/activity of these molecules. MDA + 4HNE have shown tissue toxicity. MDA can react with amino and thiol groups, the aldehydes are more diffusible than free radicals, which means damage is exported to distance sites. Aldehydes are quickly removed from cells as several enzymes control their metabolism (Ustinova and Riabinin 2003).

6.4. Antioxidants

Antioxidants are thought to protect the body against the destructive effects of free radicals. Antioxidants neutralize free radicals by donating one of their own electrons, ending the electron- gain reaction. The antioxidant nutrients themselves don't become free radicals by donating an electron because they are stable in either form. They act as scavengers, helping to prevent cell and tissue damage that could lead to cellular damage and disease (Reiter, 2003).

The body produces several enzymes, including superoxide dismutase (SOD), catalase (CAT), and glutathione peroxidase (GSHPX), that neutralize many types of free radicals. Supplements of these enzymes are available for oral administration. However, their absorption is probably minimal at best. Supplementing with the "building blocks" the body requires to make SOD, catalase, and glutathione peroxidase may be more effective. These building block nutrients include the minerals manganese, zinc, and copper for SOD and selenium for GSHPX.

In addition to enzymes, many vitamins, minerals and hormones act as antioxidants in their own right, such as vitamin C, vitamin E, beta-carotene, lutein, lycopene, vitamin B2, coenzyme Q10, and cysteine (an amino acid). Herbs, such as bilberry, turmeric (curcumin), grape seed or pine bark extracts, and ginkgo can also provide powerful antioxidant protection for the body. Melatonin is a hormone secreted by pineal gland and proves to be powerful antioxidant and free radical scavenger (Yang et al.,2002 and Koc et al., 2003).

6.5. Nutritional therapy with natural antioxidants

Antioxidants have been the focus of research on the relationship between The role of dietary factors in protecting against the change from native to oxidized LDL has received considerable attention. An overview of epidemiological research suggests that individuals with the highest intakes of antioxidant vitamins, whether through diet or supplements, tend to experience 20–40% lower risk of coronary heart disease (CHD) than those with the lowest intake or blood levels (diet and disease. Research examining the effects of a diet rich in fruits and vegetables on disease has been carried out using several types of study.

There is strong scientific evidence to support an increase in intakes of vegetables and fruit in the prevention of disease. Further research is required to clarify which particular components of fruit and vegetables are responsible for their protective effects.

Numerous epidemiological studies have indicated that diets rich in fruits and vegetables are correlated with a reduced risk of chronic diseases (Banerjee and Maulik, 2002 ; Sesso et al., 2003). It is probable that antioxidants, present in the fruits and vegetables such as polyphenols, carotenoids, and vitamin C, prevent damage from harmful reactive oxygen species, which either are continuously produced in the body during normal cellular functioning or are derived from exogenous sources (Gate et al. 1999). The possible protective effect of antioxidants in fruits and vegetables against ROS has led people to consume antioxidant supplements against chronic diseases.

6.6. Some experimental studies for the author about food factors as a nutritional antioxidants agents in alleviating the oxidative stress induced by environmental pollutents

1. In vivo evidence of hepato-and-reno-protective effect of garlic oil against sodium nitrite-induced oxidative stress Int. Int. J. Biol. Sci. (5)3: 249-255. (Hassan et al., 2009)
2. Mitigating effects of antioxidant properties of black berry juice on sodium fluoride induced hepatotoxicity and oxidative stress in rats. Food Chem.Toxicol., 47 2332–2337. (Hassan and Yousef, 2010).
3. Evaluation of free radical-scavenging and antioxidant properties of black berry against fluoride toxicity in rats. Food Chem.Toxicol., 48: 1999-2004. (Hassan and Fattoh, 2010).
4. Garlic oil as a modulating agent for oxidative stress and neurotoxicity induced by sodium nitrite in male albino rats. Food Chem. Toxicol., 48: 1980-1985. (Hassan et al., 2010)
5. Ameliorating effect of chicory (Cichorium intybus L)-supplemented diet against nitrosamine precursors-induced liver injury and oxidative stress in male rats. Food Chem. Toxicol., 48: 2163-2169. (Hassan and Yousef, 2010).

Author details

Hanaa Ali Hassan Mostafa Abd El-Aal

Zoology Department, Faculty of Science, Mansoura University, Mansoura, Egypt

7. References

Antunes, F., Derick, H. and Cadenas, E. (2002). Relative contributions of heart mitochondria glutathione peroxidase and catalase to H2O2 detoxification in in vivo conditions. Free Radic. Biol. Med., 33 (9): 1260-1267

Al-Omar, M. A., Beedham, C. and al-Sarra, I. (2004). Pathological roles of reactive oxygen species and their defence mechsnisms. Saudi Pharm. J., 12:1-18.

Arthur, J. R. (2000). The glutathione peroxidase. Cell Mol. Life Sci., 57:1825 1835.

Awasthi, S., Bajpai, K. K., Piper, J. T., Singhal, S. S., Ballaatore, A., Seifert, W. E., Awasthi. Y. C. and Ansari, G. A. S. (1996): Drug metabolism. Dispos. 24: 371-374.

Bandyopadhyay, D., Biswas, K., Bandyopadhyay, U., Reiter, R. J. and Banerjee, R. K. (2000). Melatonin protects against stress-induced gastric lesions by scavenging the hydroxyl radical. J. Pineal Res., 29: 143-151.

Banerjee, S. K. and Maulik, s. K. (2002): Effect of garlic on cardiovascular disorders. J. Nutr.,1:4-36.

Bannister, J., Bannister, W. and Rotilio, G. (1987). Aspects of the structure, function, and applications of superoxide dismutase". CRC Crit. Rev. Biochem. 22 (2): 111–180.

Benderitter, M.; Vincent-Genod, L.; Pouget, J. P. and Voisin, P. (2003): The cell membrane as a biosensor of oxidative stress induced by radiation exposure: a multiparameter investigation. Radiat. Res., 159(4): 471-483.

Benzie, I. (2003). Evolution of dietary antioxidants. Comp. Biochem. Phys., 136 (1): 113–126.

Blankenberg, S., Rupprecht, H. J., Bickel, C., Torzewski, M., Hafner, G., Tiret, L., Smieja, M., Cambien, F., Meyer, J., and Lackner, K.J. (2003). Glutathione peroxidase 1 activity and cardiovascular events in patients with coronary artery disease atherogene investigators. Engl. J. Med., 349:1605-1613

Brioukhanov, A; L; and Netrusoe, a. I. (2004): Catalase and superoxide dismutase, distribution, properties and physiological role in cells of strict anaerobes. Biochem. 69:949-962.

Ceconi, C. , Boraso, A. , Cargnoni, A. and Ferrari, R., (2003). Oxidative stress in cardiovascular disease:myth or fact? Arch. Biochem. Biophys., 420: 217 – 221 .

Ceriello, A. and Motz, E. (2004). Is oxidative stress the pathogenic mechanism underlying insulin resistance, diabetes, and cardiovascular disease? The common soil hypothesis revisited .Arterioscler. Thromb. Vasc. Biol., 24 : 816 – 823 .

Cohen, P. J. (1992): Allopurinol administered prior hepatic ischaemic in the rat prevents chemiluminescence following restoration of circulation. Can. J. Anaesth., 39: 1090-1093.

Cook, NR., Albert, CM. and Gaziano, JM. (2007). A randomized factorial trial of vitamins C and E and beta carotene in the secondary prevention of cardiovascular events in women: results from the Women's Antioxidant Cardiovascular Study. Arch. Int. Med., 167 (15): 1610–1618.

Coulter, I., Hardy, M., Morton, S., Hilton, L., Tu, W., Valentine, D. and Shekelle, P. (2006). Antioxidants vitamin C and vitamin e for the prevention and treatment of cancer. J. f general int. medici.: official j. Soc. Res. Educ. in Primary Care Int. Medici., 21 (7): 735–744

Davies, AG., Griller, D., Ingold, KU., Lindsay, DA. and Walton, JC. (1981). An electron spin resonance study of pentadienyl and related radicals: homolytic fission of cyclobut-2-enylmethyl radicals. J. Chem. Soc. Perkin. Trans., II:633-641.

Dherani, M. , Murthy, GV., Gupta, SK., Young, IS. and Maraini, G, Dherani, M, Murthy, GV, Gupta, SK, Young, IS, Maraini, G, Camparini, M, Price, GM, John, N, Chakravarthy, U, Fletcher, AE. (2008). Blood levels of vitamin C,carotenoids and retinol are inversely associated with cataract in a North Indian population .Invest. Ophthalmol. Vis. Sci., 49: 3328 – 3335 .

Dilis, V. and Trichopoulou, A. (2010). Antioxidant intakes and food sources in greek adults J. Nutr., 140: 1274-1279.

Dudek, H., Farbiszewski, R., Michno, T., Łebkowski, W. J. and Kozłowski, A. (2002). Activity of glutathione peroxidase (GSH-Px), glutathione reductase (GSSG-R) and superoxide dismutase in the brain tumors 55, 5-6: 252-256.

Fearon, IM. and Faux, SP. (2009) . Oxidative stress and cardiovascular disease: novel tools give (free)radical insight . J. Mol. Cell Cardiol., 47 : 372 – 381 .

Gate L, Paul J, Ba GN, Tew KD and Tapiero H. (1999): Oxidative stress induced in pathologies: the role of antioxidants. Biomed Pharmacother,53: 169 -180.

German, JB. (1999). Food processing and lipid oxidation". Advances in experimental medicine and biology. Advanc. Experim. Medici. Biol., 459: 23–50.

Greenwald, R. A. (1990). Superoxide dismutase and catalase as therapeutic agents for human diseases. A critical review. Free Radic. Boil. Med., 8(2): 210-219.

Halliwell, B. and Whiteman, M. (2004): Measuring reactive species and oxidative damage in vivo and in cell culture: how should you do it and what do the results mean?. Br. J. Pharmacol.., 142(2):231-255.

Hassan, H. A. (2005). The ameliorating effect of black grape juice on fluoride induced testicular toxicity in adult albino rats. Egypt. J. Zool., 44:507–525.

Hassan, H. A ; El-Agmy Sh. M; Gaur, R. L; Fernando, A. Raj, M., HG and Ouhtit A. (2009): In vivo evidence of hepato-and-reno-protective effect of garlic oil against sodium nitrite-induced oxidative stress Int. Int. J. Biol. Sci. (5)3: 249-255.

Hassan, H. A and A. M. Fattoh (2010): Evaluation of free radical-scavenging and antioxidant properties of black berry against fluoride toxicity in rats. Food Chem.Toxicol.,48:1999-2004.

Hassan H. A. Hafez H. S. and Zeghebar F. E. (2010): Garlic oil as a modulating agent for oxidative stress and neurotoxicity induced by sodium nitrite in male albino rats. Food Chem. Toxicol., 48: 1980-1985.

Hassan, H. A and Yousef M. I. (2009): Mitigating effects of antioxidant properties of black berry juice on sodium fluoride induced hepatotoxicity and oxidative stress in rats. Food Chem.Toxicol., 47 2332–2337

Hassan, H. A and Yousef, M. I. (2010): Ameliorating effect of chicory (Cichorium intybus L)- supplemented diet against nitrosamine precursors-induced liver injury and oxidative stress in male rats. Food Chem. Toxicol., 48: 2163-2169.

Hayes, J. D. and Pulford, D. J. (1995): The glutathione S transferase supergene family: regulation of GST and the contribution of the isoenzymes to cancer chemoprotectiion and drug resistance. Crit. Rev. Biochem. Mol. Biol., 30: 445-600.

Hogg, N. (2002): The biochemistry and physiology of S nitrosothiols. Ann. Rev. Pharmacol. Toxicol., 42:585-600.

Huang, D. Y. and Ichikawa, Y. (1994): Two different enzymes are primary responsible for retinoic acid synthesis in rabbit liver cystosol. Biochem. Biophys. Res. Commin., 205: 1278-1283.

Johnson, F. and Giulivi, C. (2005). Superoxide dismutases and their impact upon human health. Mol. Aspects Med. 26 (4–5): 340–352.

Kelly, S. A., Havrilla, C. M., Brady, T. C., Abramo, K. H. and Levin, E. D. (1998). Review of oxidative stress in toxicology: Established mammalian and emerging piscine model systems. Environ. Health Perspect., 106:375-384.

Kidd, P. M. (1997): Glutathione systemic protectant against oxidative and free radical damage. Altern. Med. Res., 1-155-176.

Koc, M.,Taysi, S.,Buyukokuroglu, M. E. and Bakan, N. (2003). The effect of melatonin against oxidative damage during total- body irradiation in rats. Radiat. Res., 160: 251-255.

Kondo, T., Dale, O.L. and Beutler, E. (1980): Glutathione transport by inside-out vesicles from human erythrocytes. Proc. Nat. Acad. Sci. Biochem., 77: 6359-6362.

Kregel, KC. and Zhang, HJ . (2007) . An integrated view of oxidative stress in aging: basic mechanisms,functional effects, and pathological considerations . Am. J. Physiol. Regul. Integr. Comp. Physiol .,292: R18 – 36 .

Laval, J. (1996): Role of DNA repair enzymes in the cellular resistance to oxidative stress. Pathol. Biol., 44: 14-24

Mayo, J. C., Tan, D. X., Sainz, R. M., Lopez-Burillo, S. and Reiter, R, J. (2003). Oxidative damage to Catalase induced by peroxyl radicals: Functional protection by melatonin and other antioxidants. Free Radic. Res., 37: 543-553.

McCord, J. M. (1985): Oxygen-derived free radicals in postischemic tissue injury. N. Eng. J. Med., 312: 159-163.

Meister, A. (1988). Glutathione metabolism and its selective modification. J. Biol. Chem., 263(33):17205-17208.

Meister, A. and Anderson, M. N. (1983). Glutathione. Ann. Rev. Biochem., 52: 611-660.

Meister, A. and Larsson, A. (1995). Glutathione synthetase deficiency and other disorders of the gamma-glutamyl cycle. In: Scriver, C. R.; Kinzler, K. W. ; Valle, D., et al., eds. The

Metabolic and molecular bases of inherited diseases. New York: Mc Graw- Hill, 1461-1477.

Nagaoka, S., Okauchi, Y., Urano, S., Nagashima, U. and Mukai, K. (1990). Kinetic and ab initio study of the prooxidant effect of vitamin E: hydrogen abstraction from fatty acid esters and egg yolk lecithin. J. Am. Chem. Soc., 112:8921-8924.

Nozik-Grayck, E., Suliman, H. and Piantadosi, C. (2005). Extracellular superoxide dismutase. Int. J. Biochem. Cell Biol., 37 (12): 2466–2471.

Oja, S. S., Janaky, R., Varga, V. and Saranasaari, P. (2000): Modulation of glutamate receptor functions by glutathione. Neurochem. Int., 37: 299-306.

Packer, L. (1994). Methods in Enzymology; Oxygen Radicals in Biological Systems, Part C, Academic Press, 1994.

Peake, J. (2003). Vitamin C: Effects of exercise and requirements with training. Int. J. Sport Nutr. Exerc. Metab., 13 (2): 125–151.

Pellegrini, N. Serafini, M., Colombi, B., Del Rio, D., Salvatore, S., Bianchi, M. and Brighenti, F. (2003): Total antioxidant capacity of plant foods, beverages and oils consumed in italy assessed by three different in vitro assays. J. Nutr. 133:2812-2819.

Preston, T.J., Muller. W.J. and Singh, G. (2001): Scavenging of extracellular H2O2 by catalase inhibits the proliferation of HER-2/Neu-transformed Rat-1 fibroblasts through the induction of a stress response. J. Biological Chem., 276: 9558-9564.

Puchau, B., Ángeles, M., Zulet, de Echávarri, A. G., Hermsdorff, H. H. M., and Martínez, J. A., (2009). Dietary total antioxidant capacity: A novel indicator of diet quality in healthy young adults J. Am. Coll. Nutr., 28: 648-656.

Reiter, R. J. (1995). Oxygen radical detoxification processes during aging: the functional importance of melatonin. Aging (Milano), 7: 340-351.

Reiter, R. J. (2003). Melatonin: clinical relevance. Best Pract. Res. Clinic. Endocr. Metabol., 17 (2):273–285.

Reiter, R. J.; Carneiro, R. C. and Oh, C. S. (1997): Melatonin in relation to cellular antioxidative defense mechanisms. Horm. Metab. Res., 29: 363-372.

Reiter, R. J.,Tan, D. X., Gitto, E., Sainz, R. M., Mayo, J.C., Leon, J., Manchester, L. C.,Vijayalaxm, Kilic, E. and Kilic, U. (2004). Pharmacological utility of melatonin in reducing oxidative cellular and molecular damage. Pol. J. Pharmacol.,56(2):159-170.

Roberts, LJ., Oates, JA. and Linton, MF. (2007). The relationship between dose of vitamin E and suppression of oxidative stress in humans. Free Radic. Biol. Med., 43 (10): 1388–1393.

Scandalios, J. G., Guan, l. and Polidoros, A. (1997): Catalase in plants.: Gene structure, properties, regulation and expression In J. G. Scandalios Ed., Oxidative stress and the molecular biology of antioxidant defenses Pp 343-406.

Sener, G.; Paskaloglu, K.; Toklu, H.; Kapucu, C.; Ayanoglu-Dulger, G.; Kacmaz, A. and Sakarcan, A. (2004): Melatonin ameliorates chronic renal failure-induced oxidative organ damage in rats. J Pineal Res., 36(4):232-241.

Sesso, H.D.; Liu, S.; Gaziano, J.M. and Buring, J. E. (2003): Dietary Lycopene, Tomato-Based Food Products and Cardiovascular Disease in women. J. Nutr., 133: 2336-2341.

Sies, H. (1997). Oxidative stress: Oxidants and antioxidants. Exp. Physiol., 82(2): 291-295.

Tan, D. X., Manchester, L. C.,Burkardt, S., Sainz, R. M., Mayo, J. C., Kohen, R., Shohami, E., Huo, Y. S., Hardeland, R. and Reiter, R. J. (2001). N-acetyl-N-formyl-5-methoxyknuramine, a biogenic amine and melatonin metabolite, functions as a potent antioxidant. FASEB. J., 15: 2294-2296.

Terao, M., Kurosaki, M., Saltini, G., Demontis, S., Marini, M., Salmona, M. and Garattini, E. (2000): Cloning of the cDNAs coding for two novel molybdoflavoproteins showing high similarity with aldehyde oxidase and xanthine oxidoreductase. J. Biol. Chem., 275: 30690-30700.

Undeger, U., Giray, B., Zorlu, A. F., Oge, K. and Bacaran, N. (2004). Protective effects of melatonin on the ionizing radiation induced DNA damage in the rat brain. Exp. Toxicol. Pathol., 55: 379-384.

Ustinova, A. A. and Riabinin, V. E. (2003): Effect of chronic gamma-irradiation on lipid peroxidation in CBA mouse blood serum. Radiat. Biol. Radioecol., 43(4):459-463.

Wakatsuki, A.,Okatani, Y., Izumiya, C. and Lkenoue, N. (1999). Melatonin protect against ischemia and reperfusion- induced oxidative lipid and DNA damage in fetal rat brain. J. Pineal Res., 26: 147-152.

Waldeck, A. R. and Stocker, R. (1996). Radical-initiated lipid peroxidation in low density lipoprotein. Insights obtained from kinetic modeling. Chem. Res. Toxicol., 9: 954-964.

Weir, E.K., Archer, S.L. and Reeves, J.T. (1996). Nitric Oxide and Radicals in the Pulmonary Vasculature, Futura Publishing, Armonk, N.Y.

Wilson, J. (1997): Antioxidant defense of the brain: a role for astrocytes (review). Can. J. phys. Pharmacol., 75: 1149 – 1163.

Wright, R. M., Vaitaitis, G. M., Weigel, L.K., Repine, T. B., McManaman, J. L. and Repine, J.E. (1995): Identification of the candidate ALS2 gene at chromosome 2q33 as a human aldehyde oxidase gene. Redox Report 1:313-321.

Yang, J.; Lam, E. W.; Hammad, H. M.; Oberley, T. D. and Oberley, L. W. (2002): Antioxidant enzyme levels in oral squamous cell carcinoma and normal human oral epithelium. J. Oral Pathol. Med., 31: 71-77.

Yasminch, W. and Theologides, A. (1993). Catalase as a removing scavenger of hydrogen peroxide: a hypothesis. J. Lab. Clin. Med., 122(1): 110-114.

Zatta, P.,Tognon, G. and Carampin, P. (2003). Melatonin prevent free radical formation due to the interaction between beta-amyloid peptides and metal ions [Al(III), Zn(II), Cu(II), Mn(II), Fe(II)]. J. Pineal. Res., 35: 98-103.

Zebger, I.,Snyder, J. W., Andersen, L. K., Poulsen, L., Gao, Z., Lambert, J. D., Kristiansen, U. and Ogilby, P. R. (2004). Direct optical detection of singlet oxygen from a single cell. Photochem. Photobiol., 79(4): 319-322.

Chapter 3

Lipid Oxidation in Homogeneous and Micro-Heterogeneous Media in Presence of Prooxidants, Antioxidants and Surfactants

Vessela D. Kancheva and Olga T. Kasaikina

Additional information is available at the end of the chapter

1. Introduction

The human body is constantly subjected to a significant oxidative stress as a result of the misbalance between antioxidative protective systems and the formation of strong oxidizing substances, including free radicals. The stress can damage DNA, proteins, lipids and carbohydrates and could cause negative effect to intracellular signal transmission. Antioxidants could be promising agents for management of oxidative stress-related diseases. Oxygen is essential for all living organisms, but at the same time it is a source of constant aggression for them. In its ground triplet state ($^3O_2$) oxygen has weak reactivity, but it can produce strongly aggressive and reactive particles such as singlet state oxygen ($^1O_2$), hydroperoxides (H_2O_2), superoxide anion ($O_2^{\cdot-}$), hydroxylic radical ($OH^{\cdot}$) and various peroxide ($LO_2^{\cdot}$) and alkoxy radicals ($LO^{\cdot}$). It is well known that the latter lead to an oxidative degradation of biological macromolecules, changing their properties and thus the cell structure and functionality. The free radicals formation in the hydrophobic parts of the biological membranes initiates radical disintegration of the hydrocarbon "tails" of the lipids. This process is known as lipid peroxidation (Figure 1) [1-3].

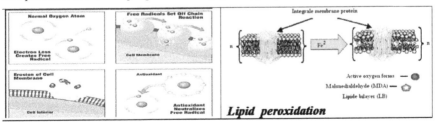

Figure 1. Erosion of cell membrane, antioxidant neutralizes free radicals and lipid peroxidation

In recent decades, many communications have been devoted to the significant role of physical factors, which control the structure of nutrition systems, in the chemistry of lipid oxidation in oil/water (O/W) emulsions [4,5]. In this case, the rate of lipid oxidation strongly depends on the physical properties of interfaces, because they affect the character of the interaction between water-soluble compounds of transition metals and hydroperoxides located inside and on the surface of emulsion droplets. For example, positively charged and high viscosity interfaces that hinder the contact between iron ions and hydroperoxides inhibit oxidation of fat emulsions [6,7]. Another example of the influence of physical factors on the oxidation is the antioxidant polar paradox [6-9], which is based on the fact that nonpolar antioxidants are efficient in O/W emulsions, because they are located in emulsion droplets together with oxidizable lipids. Polar antioxidants are more efficient in W/O emulsions because they are concentrated at the interfaces [10-12]. In the frame of this chapter, the main features of lipid oxidation in homogeneous and micro - heterogeneous oil media, formed by surfactants (W/O microemulsions) will be discussed.

2. Kinetic model of homogeneous lipid oxidation

The kinetic model is based (Table 1) on the reactions and corresponding rate constants known for the oxidation of methyl linoleate (MeLi), because linoleic esters are easily oxidizable components of many natural lipid systems and thus determines the oxidizability of the lipid substrates.

The kinetic scheme of liquid-phase (homogeneous) oxidation of lipids (LH) includes reactions 1-12.

In the presence of an initiator (I), i.e. initiated oxidation, the formation of radicals occurs with a constant initiation rate $R_{IN}=2k_1[I]$. Under autoxidation conditions ([I]=0), the rates of radical formation in the reactions of LH with O_2 (reaction 6) and decomposition of lipid hydroperoxides (LOOH) (reactions 7 and 8) increase as LOOH is accumulated. The chain termination occurs due to recombination or disproportionation of radicals (reactions 9, 11 and 12). The scheme of inhibited oxidation includes reactions 13-19, known for the phenolic antioxidants, AH.

The calculations was performed for three groups of AH, which differ in their activity in reaction 13 with peroxyl radicals LOO: group I (of the type of alpha-tocopherol) with $k_{13}=1.5$ $10^6 M^{-1}s^{-1}$; group II (AH of type of unhindered phenols, e.g. hydroquinone), with $k_{13}=1.5$ $10^5 M^{-1}s^{-1}$; and group III (AH of the type of sterically hindered phenols, e.g. butylated hydroxyl toluene, BHT with $k_{13}=1.5$ $10^4 M^{-1}s^{-1}$. The effect of AH regeneration in reaction 16 was considered for rapidly ($k_{16}=4$ $10^7 M^{-1}s^{-1}$) and slowly ($k_{16}=3$ $10^3 M^{-1}s^{-1}$) reacting phenoxyl radicals A·. Reaction 19 is the chain transfer by the inhibitor radical. It can be a hydrogen abstraction from the substrate molecule with regeneration of the inhibitor. Chain transfer can occur as the addition of A· to unsaturated bonds of polyene compounds, in this case, the inhibitor is not regenerated [13]. We examined both cases, rate constants k_{19} were varied within the 0-100 $M^{-1}s^{-1}$ range (taking into account the published data). In this series of calculations, we accepted that reaction 16 occurs as disproportionation.

No	Reaction	$k/M^{-1}s^{-1}$, MeLi	Refs.	$k/M^{-1}s^{-1}$, Limonene
1	$I \rightarrow I\cdot + I\cdot$	$5\ 10^{-6}$	13	$5\ 10^{-6}$
2	$I\cdot + O_2 \rightarrow IO_2\cdot$	$5\ 10^{6}$	13	$5\ 10^{6}$
3	$IO_2\cdot + LH \rightarrow LO_2\cdot + IOOH$	$1\ 10^{3}$	13	$1\ 10^{3}$
4	$L\cdot + O_2 \rightarrow LO_2\cdot$	$5\ 10^{6}$ $1\ 10^{8}$	13 15	$1.5\ 10^{7}$
5	$LO_2\cdot + LH \rightarrow LO_2\cdot + LOOH$	90 100	13 15,16	14
6	$LH + O_2 \rightarrow L\cdot + HO_2\cdot$	$5.8\ 10^{-11}$	13	$1.2\ 10^{-14}$
7	$LOOH + LH \rightarrow L\cdot + LO_2 + H_2O$	$2.3\ 10^{-7}$	13	$4\ 10^{-8}$
8	$2LOOH \rightarrow LO\cdot + LO_2 + H_2O$	$2.4\ 10^{-6}$	13	$1\ 10^{-6}$
9	$2\ LO_2\cdot \rightarrow Alc + Ket$	$1\ 10^{5}$ $4.4\ 10^{6}$ $1\ 10^{7}$	17 13,18 15	$3.5\ 10^{6}$
10	$LO\cdot + LH \rightarrow Alc + Ket$	$1\ 10^{7}$ $1\ 10^{5}$	19 13	$1.7\ 10^{7}$
11	$LO\cdot + LO_2\cdot \rightarrow Ket + LOOH$	$5\ 10^{6}$	13	
12	$LO_2\cdot + IO_2\cdot \rightarrow Alc + Ket$	$5\ 10^{6}$	13	$5\ 10^{6}$
13	$AH + LO_2\cdot \rightarrow A\cdot + LOOH$	$1.5\ 10^{4} - 1.5\ 10^{6}$ $2\ 10^{6}$	13,16 15	
14	$AH + LO\cdot \rightarrow A\cdot + Alc$	$1\ 10^{7}$	13	
15	$AH + IO_2\cdot \rightarrow A\cdot + IOOH$	$1.5\ 10^{5}$	13	
16	$2A\cdot \rightarrow P1\ (+AH)$	$3\ 10^{3} - 4\ 10^{7}$	13,16,18	
17	$A\cdot + LO_2\cdot \rightarrow P2$	$2.5\ 10^{6}$ $3\ 10^{8}$	20 13,16	
18	$A\cdot + LO\cdot \rightarrow P3 +AH + Ket$	$3\ 10^{8}$	13	
19	$A\cdot + LH \rightarrow L\cdot + (AH/P4)$	$0-100$ 0.07	13 21	

Note: The rate constants (k_0 correspond to the oxidation of MeLi at 60°C; in reaction 1, 2 and 4, k are presented in s^{-1}. Initial concentrations: $[LH]=2.9M$, $[LOOH]_0=10^{-5}M$, $[I]=4\ 10^{-3}M$, $[AH]_0=10^{-4}M$, $[O_2]=10^{-3}M$=const; oxidation usually occurs at a constant oxygen pressure, therefore $[O_2]$ is included in the corresponding rate constants: $k_2=k_4=k_6=k_i[O_2]$.

Table 1. The Approximate Rate Constants of the Different Reactions Involved in the Autoxidation of Methyl Linoleate [13] and Limonene [14] in initiated oxidation, autoxidation and inhibited oxidation (at 60°C).

The main kinetic parameters	Initiated oxidation	Lipid autoxidation
Rate of initiation (R_{IN})	Constant and well-controlled $R_{IN}=2k_1[I]$	$R_{IN}=2k_6[LH][O_2] + 2k_7[LH][LOOH]+2k_8[LOOH]^2$
Rate of oxidation (R_0) and (R_A)	$R_0=k_p [LH](R_{IN}/2k_t)^{0.5}$	
Rate of non-inhibited oxidation (R_0)	$R_A=k_p[LH]R_{IN}/nk_A[AH]_0$	$R_0=k_p [LH](R_{IN}/k_t)^{0.5}$
	$k_p = k_5 ; k_t = k_9$	$R_A=k_p[LH]R_{IN}/nk_A[AH]_0$
Rate of inhibited oxidation (R_A)		
Oxidizability parameter	$a = k_p/(2k_t)^{0.5}$	$a = k_p/(2k_t)^{0.5}$
Inhibition degree (ID)	$ID = v_0/v_A$	$ID = R_0/R_A$
Induction period (IP)	$IP=n[AH]_0/R_{IN}$	$IP=n[AH]_0/R_{IN}$
Antioxidant efficiency	nk_A	$PF=IP_A/IP_0$ and $RAE=(IP_A-IP_0)/IP_0$

Table 2. The main kinetic parameters of initiated oxidation and lipid autoxidation

Under other equivalent conditions, the bimolecular decay of 2A· by disproportionation in which AH is regenerated gives a considerable advantage in retardation effects as compared with the situation where no regeneration occurs (recombination of 2A·). The presence of the second hydroxyl group in the aromatic ring results in higher k_{16}. In this case, an increase in the induction period related to AH regeneration is most pronounced.

Lipid oxidation is one of the important reactions in biology. Chemical reaction kinetics considers two aspects: the rate of reaction and effective factors – temperature concentration of reactants and products. This knowledge is an essential prerequisite for modeling the lipid oxidation, the shelf life of stored foods, durability of low density proteins, and so on.

3. Effect of pro-oxidants (ROH) leading to acceleration of lipid hydroperoxides (LOOH) decomposition

3.1. Kinetic modeling of lipid oxidation for different mechanism of LOOH decomposition

A kinetic analysis of non-inhibited lipid (LH) autoxidation for different mechanisms of hydroperoxides (LOOH) decay is proposed [22]. It is based on using of mathematical simulation methods of LH autoxidation kinetics. Kinetic schemes of LH autoxidation for some different ways of hydroperoxides decay - mono-molecular, pseudo-mono-molecular and/or bimolecular mechanism are presented. This analysis permits establishing the influence degree of different hydroperoxides decay mechanisms on the kinetic parameters, characterizing the substrate oxidizability. The proposed kinetic analysis has been applied to the methyl linoleate, MeLi) autoxidation at 60°C.

The kinetic model that describes the lipid hydroperoxides decomposition taking into account the possibility of monomolecular (LOOH), pseudo-monomolecular (LOOH + LH) and bimolecular (2 LOOH) mechanisms in both cases: in presence of an oxygen (O_2) and in its absence, i.e. in an inert atmosphere (N_2) is illustrated by Scheme 1. In these equations:

$$[LOOH] = \frac{1 + K_1[LH]}{2K_2}\left[\sqrt{1 + \frac{4K_2[T]}{(1 + K_1[LH])}} - 1\right]$$

$$[T] = \left(\sqrt{[T_0]} + \frac{1}{2}\frac{kp}{\sqrt{kt}}[LH]\sqrt{k_i}\ t\right)^2$$

$$C = \frac{4K_2}{(1 + K_1[LH])^2}$$

$$d = \frac{e_0 k_{30} + e_1 k_{31}K_1[LH]}{1 + K_1[LH]} = k_i$$

$$[O_2] = \frac{k_p}{\sqrt{k_t}}[LH]\sqrt{k_i[T_0]}\ t + \frac{1}{4}(\frac{k_p}{\sqrt{k_t}}[LH])^2 k_i t$$

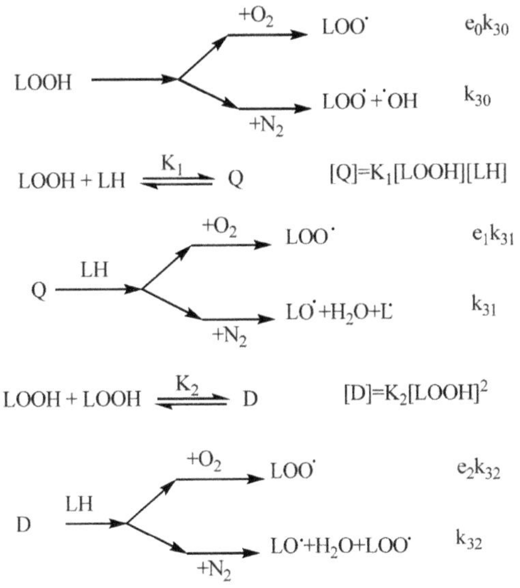

$$[T] = [LOOH] + [Q] + [D] = [LOOH](1 + K_1[LH] + K_2[LOOH])$$

LH: is linoleic acid with its allylic hydrogen
LOO•: peroxide radical
LOOH: lipid hydroperoxides
K_1 and K_2: are the equilibrium constants for complexes Q and D, respectively
[T]: summary concentration of LOOH
k_{30}, k_{31} and k_{32}: are the corresponding rate constants
e_0, e_1, e_2: are the corresponding radicals yield

Scheme 1. Kinetic scheme of lipid hydroperoxide decomposition reactions

The kinetic scheme 2 is significantly simplified and readily solved assuming a quasi-steady-state for LOO•, rapid achievement of equilibrium and neglected of the loss of Q and LOOH since their decomposition rate constants are low. There are marked: **C** - is the ratio between the equilibrium constants of bi- and pseudo-mono-molecular mechanisms of LOOH decomposition, needed to be marked for the solution of the equation.

$$\text{Chain Generation} \xrightarrow{+O_2, +LH} LOO^\bullet + HO_2^\bullet \qquad R_{IN}$$

$$LOO^\bullet + LH \xrightarrow{+O_2} LOO^\bullet + LOOH \qquad k_p$$

$$LOOH \xrightarrow{+O_2, +LH} LOO^\bullet + \text{products} \qquad e_0 k_{30}$$

$$LOOH + LH \rightleftharpoons Q \xrightarrow{+O_2, +LH} LOO^\bullet + \text{products} \qquad e_1 k_{31}$$

$$LOOH + LOOH \rightleftharpoons D \xrightarrow{+O_2, +LH} LOO^\bullet + \text{products} \qquad e_2 k_{32}$$

$$2 LOO^\bullet \longrightarrow \text{products} \qquad k_t$$

R_{IN}: the rate of chain generation, k_p: rate constants of chain propagation, k_t: rate constants of chain

Scheme 2. Kinetic scheme of lipid autoxidation by Kancheva and Belyakov [22]

Figures 2-5 presents kinetics of different mechanisms of lipid hydroperixides decomposition. In Figure 6 it is shown, that k_i doesn't change with growing of MeLi concentration from 0.3 to 1.7 M, when the concentration of MeLi hydroperoxydes is smaller than $5 \cdot 10^{-3}$ M. It is established, that MeLi hydroperoxides decay is in agreement with a first order reaction and pseudo-mono-molecular mechanism (a reaction between hydroperoxides and non-oxidized lipid substrate; LOOH + LH).

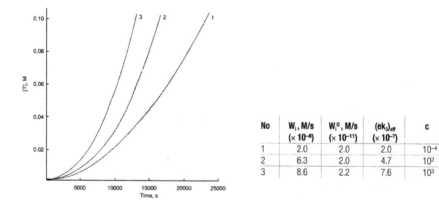

No	W_i, M/s ($\times 10^{-8}$)	W_i^0, M/s ($\times 10^{-11}$)	$(ek_3)_{eff}$ ($\times 10^{-7}$)	c
1	2.0	2.0	2.0	10^{-4}
2	6.3	2.0	4.7	10^2
3	8.6	2.2	7.6	10^3

Figure 2. Influence of dimer formation equilibrium constant (K_2) on the kinetics of MeLi autoxidation at 60°C, when e_2k_{32} has a great value ($e_2k_{32}=2 \cdot 10^{-6}$), $[T_0]=10^{-4}M$, $d=2 \cdot 10^{-7}$ and $k_p/(k_t)^{1/2} = 6 \cdot 10^{-2}$

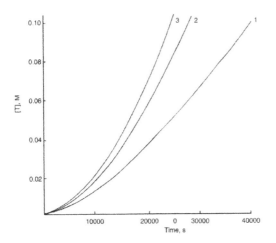

No	c	$(ek_3)_{eff}$ $(\times 10^{-7})$
1	10^3	5.54
2	10^2	1.25
3	20	1.71

Figure 3. Influence of dimer formation equilibrium constant (K_2) on the kinetics of MeLi autoxidation at 60°C, when e_2k_{32} has a small value ($e_2k_{32}= 10^{-10}$), $[T_0]=10^{-4}M$, $d=2 \ 10^{-7}$ and $k_p/(k_t)^{1/2} = 6 \ 10^{-2}$(i.e. very small value of e_2)

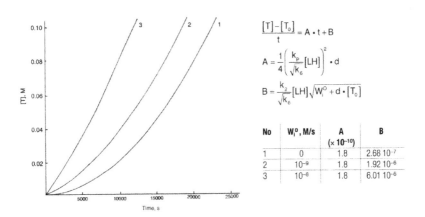

$$\frac{[T]-[T_0]}{t} = A \cdot t + B$$

$$A = \frac{1}{4}\left(\frac{k_p}{\sqrt{k_6}}[LH]\right)^2 \cdot d$$

$$B = \frac{k_2}{\sqrt{k_6}}[LH]\sqrt{W_i^o + d \cdot [T_0]}$$

No	W_i^o, M/s	A $(\times 10^{-10})$	B
1	0	1.8	$2.68 \ 10^{-7}$
2	10^{-9}	1.8	$1.92 \ 10^{-6}$
3	10^{-8}	1.8	$6.01 \ 10^{-6}$

Figure 4. Influence of the substrate (MeLi) diluting with an inert solvent (concentrations of 25, 50 and 100%) at 60°C ($e_2k_{32}=2 \ 10^{-6}$)

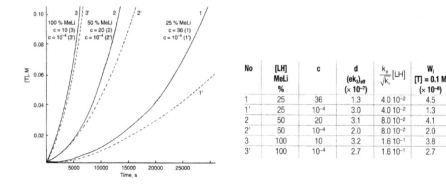

Figure 5. Kinetic curves of inhibited oxidation and autoxidation of MeLi at 60°C, when there is no dimerization of lipid hydroperoxides ($K_2=0$)

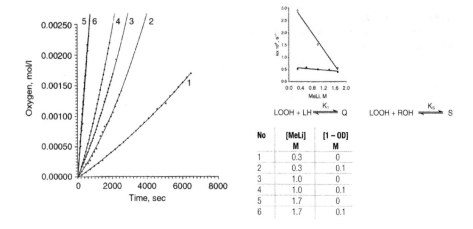

Figure 6. Effect of ROH (0.1M, 1-Octadecanol, 1-OD) on the kinetics of hydroperoxide accumulation of MeLi at 60°C, at different MeLi concentrations (0.3, 1.0 and 1.7M) and dependence of the effective initiation rate constants k_i on [LH] in the absence (1) and in presence of 0.1M 1-Octadecanol (2).

It is shown, that in presence of a lipid hydroxyl compound k_i^{ROH} is strongly growing with the decrease of MeLi (LH) concentration (Figure 6). This is explained with the competition of reactions (LOOH + LH and LOOH + ROH). Some different mechanisms, which are possible for reaction between LOOH and ROH, were discussed (Scheme 3).

$$LOOH + ROH \underset{}{\overset{K_S}{\rightleftharpoons}} S$$

$$[S] = K_s[LOOH][ROH]$$

$$S \xrightarrow{+O_2} LOO^{\cdot} + products \qquad e_s k_s$$

$$[T]=[LOOH]+[Q]+[S] =[LOOH](1+K_1[[LH]+K_s[ROH])$$

$$[LOOH] = [T]/(1+K_1[LH] +K_s[ROH])$$

$$[Q] = K_1[LH] [T]/(1+K_1[LH]+K_s[ROH]$$

$$[S] =K_s[ROH] [T] /(1+K_1[LH]+K_s[ROH])$$

Ks: is the equilibrium constant for complex S, initial rate of decomposition of complex S ($e_s k_s$)
Lipid hydroxy compounds (LOH) from the oxidized lipid substrate (LH) is formed during the whole oxidation process:
LOH and H_2O with the rate e_0 k_{30} [LOOH] from the hydroperoxides of substrate LH
LOH and H_2O with the rate e_1 k_{31} [Q] from the Q‖
LOH and H_2O with the rate e_3 k_{33} [S] from the S

Scheme 3. Kinetic scheme of lipid hydroperoxides (LOOH) decay in presence of ROH

$$A = \frac{1}{4}\left(\frac{k_p}{\sqrt{k_t}}[LH]\right)^2 k_i$$

$$k_i = \frac{4A}{\left(\frac{k_p}{\sqrt{k_t}}[LH]\right)^2}$$

$$B = \frac{k_p}{\sqrt{k_t}}\sqrt{k_i[T_o]}$$

$$[T_o] = \left(\frac{B}{\frac{k_p}{\sqrt{k_t}}[LH]\sqrt{k_i}}\right)^2$$

There are presented some different mechanisms of the interaction between lipid hydroperoxides (LOOH) and hydroxy compounds (ROH):

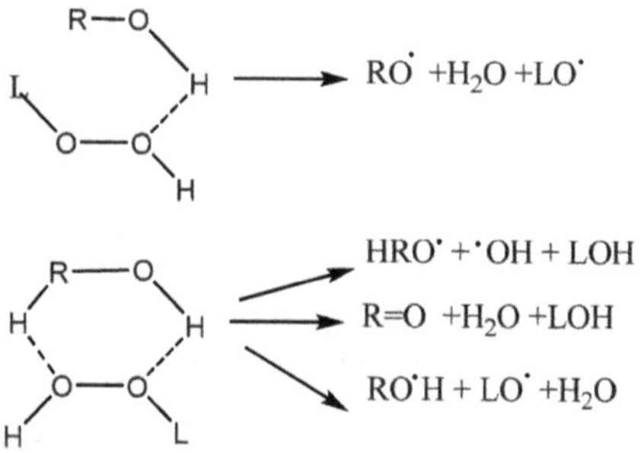

Scheme 4. Additional lipid hydroperoxides decomposition in presence of an antioxidant (AH)

$$LOOH + AH \underset{K_P}{\rightleftharpoons} P \xrightarrow[LH, O_2]{k_{iP}} LO_2^{\bullet}$$

$$K_P = \frac{[P]}{[LOOH][AH]}$$

K_P: the equilibrium constant for complex P,

k_{iP}-initiation rate constant of P decomposition

Total hydroperoxides concentration [T] in presence of ROH and AH

$[T] = [LOOH]+[Q]+[S]+[P]= [LOOH](1+K1[LH]+KS[ROH]+KP[AH])$

Scheme 5. Equilibrium constant of complex formation between an antioxidant (AH) and lipid hydroperoxides (ROH)

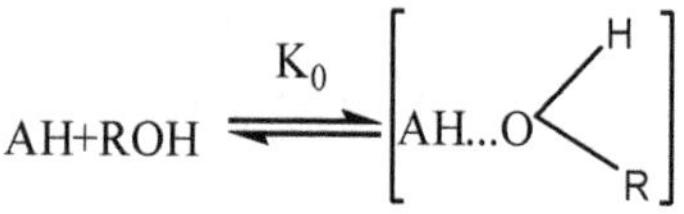

$$AH+ROH \underset{K_0}{\rightleftharpoons} \left[AH...O \begin{matrix} H \\ \diagdown \\ R \end{matrix} \right]$$

$$K_0 = \frac{[AH]_0 - [AH]}{[AH][ROH]}$$

$$[AH] = \frac{[AH]_0}{1+K_0[ROH]}$$

Scheme 6.

It has been proven [23,24] that fatty alcohols with different chain length, mono- and diacylglycerols increase the rate of LOOH decomposition into free radicals and thus accelerated lipid oxidation in absence of an antioxidant. In presence of phenolic antioxidants ROH make complexes basing on H bond formation and thus decrease the antioxidant efficiency of them [25,26]. DL-alpha –tocopherol and butylated hydroxyl toluene demonstrate the best antioxidant efficiency in presence of ROH [26]. Taking into account that ROH are formed during the proceeding, transportation and storage of lipids and lipid containing products as a result of hydrolysis, it is of importance to know how to improve their oxidative stability.

4. Antioxidants – Inhibitors of lipid oxidation

The introduction of antioxidants in the affected body normalizes not only the peroxide oxidation, but also the lipid content. Antioxidants used in oncology are effective in the first stages as mono-therapy with antioxidants at high concentrations and at the last stages mainly as additives in the complex tumor therapy - the antioxidant is in low concentrations. In this respect the medical treatment of most of diseases includes formulations based on a combination of traditional drugs with targeted functionality and different antioxidants [3,27].

The activity of antioxidants depends on complex factors including the nature of the antioxidants, the condition of oxidation, the properties of substrate, being oxidized and the stage of oxidation [2,3,27-33].

Capacity of antioxidants has at least two sides: the antioxidant potential, determined by its composition and properties of constituents and is the subject of food chemistry, and the biological effects, depending, among other things, on bioavailability of antioxidants, and is a medico-biological problem.

4.1. Classification of antioxidants [2,16,30-36]

Depending on their mechanism of action:

Antioxidants, inhibiting lipid oxidation by trapping lipid peroxide radicals– they are aromatic compounds with a weak O-H, N-H bonds (phenols, amines, aminophenols, diamines etc.

Antioxidants, inhibiting the oxidation process by trapping alkyl radicals – they are quinones, methylene quinones, which are effective in low oxygen concentration.

Hydroperoxide decomposers – these compounds react with hydroperoxides without formation of free radicals.

Metal chelators – oxidation process can be inhibited by addition of compounds, forming complexes with metal ions and thus made them inactive towards hydroperoxides. In this groups are hydroxyl acids, flavonoids etc.

Antioxidants with multistage action – systems containing such kind of compounds (alcohols and amines) inhibitors can be regenerated during the oxidation process.

Inhibitors with combined action – inhibitor molecule has two or more functional groups, each of them react in different reactions.

Depending on their nature:

Natural antioxidants – usually with low toxicity (with some exception), wide spectrum of biological and antioxidant activities.

Synthetic - they are with a high antioxidant activity. However, antioxidants for application in foods and additives or supplements they must pass additional criteria (no toxicity, safety, healthy, low cost, etc.)

Depending of their biological activity:

Bio-antioxidants – compounds with both biological and antioxidant activities. Last decades there is a growing interest to the nature-like bio-antioxidants. The most important known bio-antioxidants are flavonoids and phenolic acids.

Antioxidants without a biological activity - some even natural antioxidants can show a toxic activity and for that reason they must be tested.

Depending on the number of phenolic groups:

Monophenols – the known compounds are butylated hydroxyl toluene (BHT), Tocopherol (TOH), p-coumaric acid (p-CumA), ferulic acid (FA), sinapic acid (SA) etc.

Biphenols – the known compounds are caffeic acid (CA), hydroquinone (HQ), tertbutylated hydroquinone (TBHQ) etc.

Polyphenols –flavonoids: quercetin (Qu), rutin (Ru), luteolin (Lu), kampferol (Kf), isorhamnetin (Isorh) etc.

The inherent compositional and structural complexity of real foods and in vivo studied means that systematic studies of lipid oxidation must first be carried out in model systems. The following models were applied to explain the structure-activity relationship of different phenolic antioxidants: model 1, a DPPH assay used for the determination of the radical scavenging capacity (AH+DPPH•→A·+DPPH-H); model 2, chemiluminescence (CL) of a model substrate RH (cumene and diphenylmethane) used for determination of the rate constant of a reaction with model peroxyl radicals (AH+RO₂·→A·+ROOH); model 3, lipid autoxidation (LAO) used for the determination of the chain-breaking antioxidant efficiency and reactivity (AH+LOO·→A·+LOOH; A·+LH(+O₂)→AH+LOO·); and model 4, theoretical methods used for predicting the activity (predictable activity by statistical and/or quantum-chemical calculations).

4.2. Structure of the antioxidants

By combination of different experimental methods: DPPH test, lipid autoxidation kinetics, chemiluminescence kinetics and quantum chemical calculations it has been proven that the prooxidant activity of chalcones is due to the possible reaction of phenoxyl radicals formed with oxygen and formation of dioxiethanes, [37]Vasil'ev *et al.*, 2009:

New bis-coumarins are found to have anti-HIV activity [38]. Together with their antioxidant capacity (Fig. 8C) they are one of the most important bio-antioxidants nowadays.

The studied simple dihydroxy-coumarins are natural (Cum0) and nature-like synthetic compounds with a wide range of biological activities against cancer, inflammatory, cardio-vascular diseases, diabetes etc. Together with the strong antioxidant activity and synergistic effect with Tocopherol, they are very important for the practical application, [40]Kancheva *et al.*, 2010a.

A. Benzoic acids	Abbr.	R3		R5	Activity	Methods
	HBA	H		H	Weak	LAO,CL,Theor
	VanA	H		OCH₃	Weak	LAO,CL,Theor
	SyrA	OCH₃		OCH₃	Moderate	LAO,CL,Theor
	DHBA	OH		H	Strong	LAO,CL,Theor
	GA	OH		OH	Strong	LAO, DPPH,Theor
B.Cinnamic acids	p-CA	H		H	Weak	LAO, CL,DPPH, Theor
	FA	OCH₃		H	Moderate	LAO, CL,DPPH, Theor
	SA	OCH₃		OCH3	Moderate	LAO, CL,DPPH, Theor
	CA	OH		OH	Strong	LAO, CL, DPPH, Theor.
	PHC	Prenyl		H	Strong	LAO
	DPHC	Prenyl		Prenyl	Moderate	LAO
C. N-cinnamic acids amides	Abbr.	R		R5	Activity	Method
	N1	CH(CH₂C₆H₅)-COOC(CH₃)₃		H	Moderate	LAO
	N2	CH(CH₂C₆H₄-F-m)-COOCH₃		H	Moderate	LAO
	N3	CH(CH₃)-COOC(CH₃)₃		H	Moderate	LAO
	N4	CH(CH₂C₆H₄-OH-p)-COOCH₃		H	Moderate	LAO
	N5	CH(CH₂C₆H₄-OH-p)-COOCH₃		OCH₃	Strong	LAO
	N6	CH(CH₂C₆H₄-F-m)-COOCH₃		OCH₃	Strong	LAO
	N7	CH(CH₂C₆H₅)-COOC(CH₃)₃		OCH₃	Strong	LAO
D.Hydroxy-chalcones	Abbr.	R2	R3 R4		Activity	Methods
	Ch1	H	OH H		Weak	LAO, CL, DPPH, Theor
	Ch2	H	H OH		Moderate	LAO, CL, DPPH,Theor
	Ch3	H	OH OH		Strong	LAO, CL, DPPH Theor
	Ch4	H	OH OCH₃		Weak	LAO, CL, DPPH,Theor
	Ch5	OH	H H		Weak	LAO, CL, DPPH,Theor
	Ch6	OH	OCH₃ H		Weak	LAO, CL, DPPH,Theor

Table 3. The main structures and activities with methods for benzoic acids, cinnamic acids, N-cinnamic acids amides and chalcones

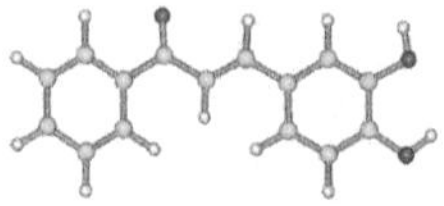

Chalcone Ch3(ΔH_f^0 = -58.3 kcal/mol) 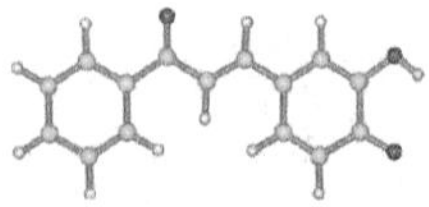　　　*Radical formed from Ch 3 (ΔH_f^0 = -29.0 kcal/mol)*

(a)

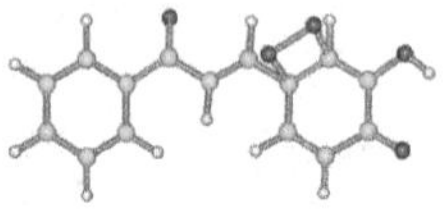

D1 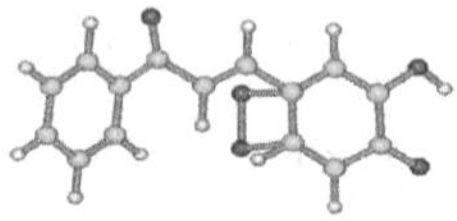　　　　　　　　　　　　　　*D2*

Dioxiethanes D1 and D2
ΔH_f^0(D1$^\bullet$) = -19. 9 kcal/mol, ΔH_f^0(D2$^\bullet$) = -27.1 kcal/mol

(b)

Figure 7. a) Optimized structures of Chalcone Ch3 (7a) and its aryl radical (7b); b) Optimized structures of Dioxiethanes D1 and D2 formed from Ch3 radical and oxygen.

E.Simple Coumarins	Abbr.	R3	R4	R5	R6	R8	Activity	Method
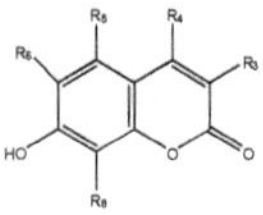	Cum0	H	H	H	OH	H	Strong	LAO
	Cum1	H	CH₃	H	OH	H	Strong	LAO
	Cum2	H	CH₃	H	H	OH	Strong	LAO
	Cum3	EtCOOMe	CH₃	H	H	OH	Strong	LAO
	Cum4	MeCOOEt	CH₃	H	H	OH	Strong	LAO
	Cum5	H	CH₃	OH	H	H	Weak	LAO
	Cum6	H	CH₃	H	H	H	Weak	LAO
	Cum7	H	OH	H	H	H	Weak	LAO

F.Bis-Coumarins	Abbr.	R3	R4	R5	Activity	Method	Abbr.	R3
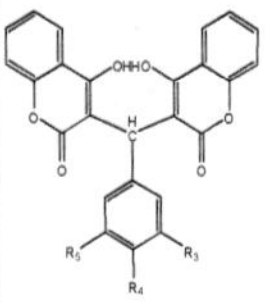	Bis-Cum1	OH	OH	H	Strong	LAO	Bis-Cum1	OH
	Bis-Cum2	OCH₃	OH	OCH₃	Moderate	LAO	Bis-Cum2	OCH₃
	Bis-Cum3	OCH₃	OH	NO₂	Weak	LAO	Bis-Cum3	OCH₃
	Bis-Cum4	OCH₃	OCH₃	H	Weak	LAO	Bis-Cum4	OCH₃
	Bis-Cum5	OCH₃	OCH₃	OCH₃	Weak	LAO	Bis-Cum5	OCH₃

Table 4. Simple and Bis-Coumarins

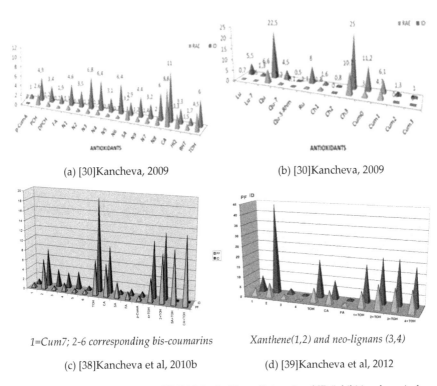

(a) [30]Kancheva, 2009

(b) [30]Kancheva, 2009

1=Cum7; 2-6 corresponding bis-coumarins

Xanthene(1,2) and neo-lignans (3,4)

(c) [38]Kancheva et al, 2010b

(d) [39]Kancheva et al, 2012

Figure 8. The main kinetic parameters PF, RAE (antioxidant efficiency) and ID (inhibition degree) of lipid autoxidation in presence of different antioxidants (for abbreviation see corresponding tables)

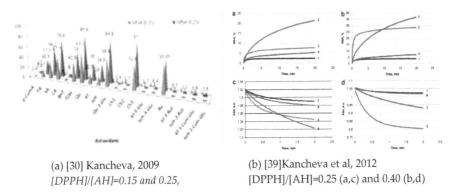

(a) [30] Kancheva, 2009
[DPPH]/[AH]=0.15 and 0.25,

(b) [39]Kancheva et al, 2012
[DPPH]/[AH]=0.25 (a,c) and 0.40 (b,d)

Figure 9. Radical scavenging activity (%) of studied compounds at different [DPPH]/[AH] ratio

Nature-like neo- and xhanthene-lignans recently synthesized showed activity agains cardio-vascular, inflammatory and cancer diseases. Together with their excellent capacity to scavenge free radicals and to inhibit lipid autoxidation these bio-antioxidants are of great importance for the practie, as individuals and in binary mixtures with TOH [39].

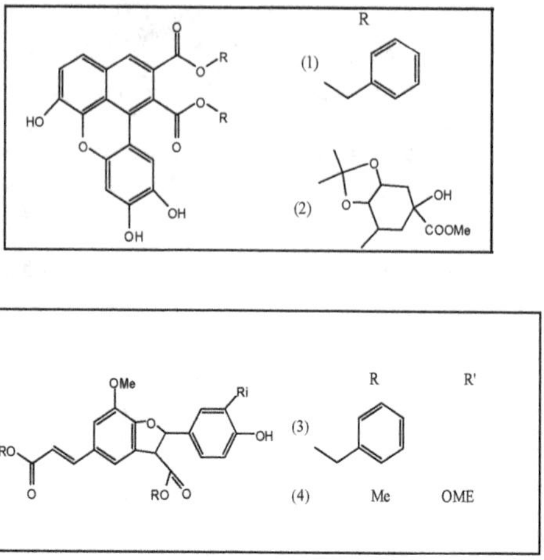

Figure 10. Structures of Xanthene (MF1, MF2) and neo-lignans (MF3, MF4)

The highest values of radical scavenging activity (%RSAmax) and largest rate constants for reaction with DPPH radical were obtained for xanthenes and neo-lignans (compounds 2 and 3, Fig. 7B). Comparison of %RSAmax with that of standard antioxidants DL-a-tocopherol (TOH), caffeic acid (CA) and butylated hydroxyl toluene (BHT) give the following new order of %RSA max: TOH(61.1%) > CA(58.6%) > 3(36.3%) > 2(28.1%) > 4(6.7%) > 1(3.6%) = BHT(3.6%). On the basis of a comparable kinetic analysis with standard antioxidants a new order of the antioxidant efficiency were obtained: **PF:** 2(7.2) ≥ TOH(7.0) > CA(6.7) > 1(3.1) > 3(2.2) > FA(1.5) > 4(0.6); and of the antioxidant reactivity: **ID:** 2(44.0) >> TOH(18.7) >> CA(9.3) >> 1(8.4) > 3(2.8) >FA(1.0) > 4(0.9) [36].

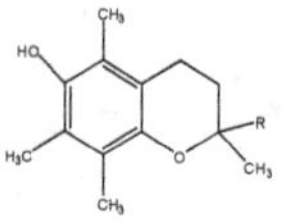 R=CH₃ 2,2,5,7,8-pentamethyl-chroman-3-ol (Chroman C1) R=Phytyl; alpha-tocoperol (TOH)	R₁=H; R₂=OH –tert-butylated-hydroquinone (TBHQ) R₁=t-But; R₂=CH₃ –tert-butyl-hydroxytoluene (BHT) 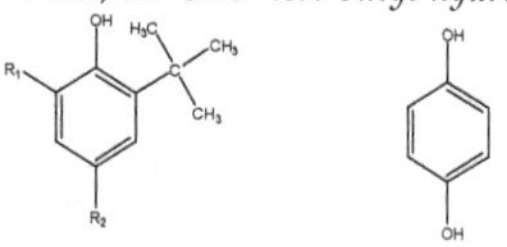 Hydroquinone (HQ)
ChrC1 and TOH -Strong activity, LAO, CL, DPPH,Theor	TBHQ, HQ -strong activity, BHT –weak/moderate activity; LAO, CL, DPPH,Theor

Table 5. Standard Antioxidants

![flavonoid structure with R3', R4', R7, R5, R6, R3 positions, rings A B C]	R3'	R4'	R3	R5	R7	%RSA exper	%QSAR theor	LAO Activity
Quercetin (Qu)	OH	OH	OH	OH	OH	62.2	88.40	Strong
Qu-3-O-Glu	OH	OH	O-Glu	OH	OH	63.9	88.40	Strong
Qu-3-O-Rhm	OH	OH	O-Rhm	OH	OH	59.0	88.40	Strong
Qu-3-O-Rut	OH	OH	O-Rut	OH	OH	62.2	88.40	Strong
Qu-7-O-Glu	OH	OH	OH	OH	O-Glu	nd	nd	Strong
Luteolin (Lu)	OH	OH	H	OH	OH	nd	nd	Strong
Lu-7-O-Glu	OH	OH	H	OH	O-Glu	nd	nd	Strong
Kampferol (Kf)	H	OH	OH	OH	OH	54.4	88.40	Strong
Kf-3-O-Glu	H	OH	O-Glu	OH	OH	1.7	12.75	Weak
Kf-3-O-Rut	H	OH	O-Rut	OH	OH	0.7	12.75	Weak
Kf-3-O-Cum-Glu	H	OH	O-Cum-Glu	OH	OH	0.8	12.75	Weak
Isorhamnetin (Isorh)	OCH3	OH	OH	OH	OH	19.2	88.40	Strong
Isorh-3-O-Glu	OCH3	OH	O-Glu	OH	OH	4.4	12.75	Weak
Isorh-3-O-Rut	OCH3	OH	O-Rut	OH	OH	2.1	12.75	Weak
Isorh-3-O-Cum-Glu	OCH3	OH	O-Cum-Glu	OH	OH	2.8	12.75	Weak

Glu: D-glucoside; Rut: rutinoside; Glu-Com: p-coumaroyl-glucosides;

%QSARtheor=3.954+75.950.I$_{3',4'-di-OH \, or \, 3-OH}$ + 8.499.I$_{5-OH}$ – by statistical analysis (QSAR) of Amic *et al* [43]

(I=1 for 3',4'-di-OH and/or3-OH) and I=1 for 5-OH); %QSAR=3.95+8.5+75.95 (for Qu all derivatives, Kf, Isrh) – 88.40;

%QSAR=3.95+8.5 (for Kf 3Oderivatives and Isrh 3Oderivatives) – 12.75

Table 6. Substitution pattern of the series of flavonoids examined for their radical scavenging activity [41-43]

4.3. Synergism, additivism and/or antagonism of binary mixtures of phenolic antioxidants [30,31,35-37,41-45]

It is known that in the literature usually are published data about mixtures without or with synergism between the components. Separation of different effects of binary mixtures (synergism, additivism and/or antagonism) of different antioxidants was made for the first time by Denisov [32]. The latest gives possibility to make differences about different effects of binary mixtures, not only to be separated as mixtures without or with a synergism.

Synergism – is observed when the inhibiting effect of the binary mixtures (IP_{1+2}) is higher than the sum of the induction periods of the individual phenolic antioxidants ($IP_1 + IP_2$) i.e. $IP_{1+2} > IP_1 + IP_2$. The percent of the synergism is presented by the following formulae % *Synergism = $100[IP_{1+2} – (IP_1 + IP_2)]/ (IP_1 + IP_2)$.*

Additivism - is observed when the inhibiting effect of the binary mixtures (IP_{1+2}) is equal to the sum of the induction periods of the phenolic antioxidants alone ($IP_1 + IP_2$) i.e. $IP_{1+2} = IP_1 + IP_2$.

Antagonism - is observed when the inhibiting effect of the binary mixtures (IP_{1+2}) is lower than the sum of the induction periods of the individual phenolic antioxidants ($IP_1 + IP_2$) i.e. $IP_{1+2} < IP_1 + IP_2$.

Binary mixtures (1:1)	[AH] mM	IP_{1+2} h	IP_1 h	IP_2 h	Effects, %	Ref
Qu (1) + Lu(2)	0.1	7.5±0.8	9.9±0.9	2.2±0.2	Antagonism	41
	0.5	12.3±0.9	24.5±0.6	6.3±0.4	Antagonism,	41
Qu (1) + Ru(2)	0.1	8.3±0.8	9.9±0.9	2.7±0.2	Antagonism	41
	0.5	21.5±0.6	24.5±0.6	2.8±0.2	Antagonism	41
Qu-7(1) + Lu-7(2)	0.5	2.0±0.2	1.5±0.2	1.8±0.2	Antagonism	41
	1.0	2.1±0.2	1.0±0.2	1.4±0.2	Additivism	41
Qu (1)+ α-TOH(2)	0.1	29.7±1.5	9.9±0.9	10.5±0.9	Synergism,46%	41
Ru (1) + α-TOH(2)	0.1	24.9±1.5	2.7±0.2	10.5±0.9	Synergism,87%	41
#Myr (1) +α-TOH(2)	0.1	10.5±0.9	4.7±0.3	3.2±0.2	Synergism,33%	3
	0.3	20.5±1.5	8.9±0.9	5.5±0.5	Synergism,42%	3
	0.6	31.1±1.5	16.3±0.9	7.4±0.5	Synergism,14%	3
CA (1) + α-TOH(2)	0.1	20.4±1.5	9.8±0.9	10.5±0.9	Additivism	3
SA(1) +α- TOH(2)	0.1	16.1±0.9	5.3±0.5	10.5±0.9	Additivism	3
BHT(1) +α- TOH(2)	0.1	21.5±1.5	7.5±0.5	10.5±0.9	Synergism,19%	3
TBHQ(1) + αTOH(2)	0.1	26.1±1.5	7.9±0.5	10.5±0.9	Synergism,42%	3
Cum₁(1) + αTOH(2)	0.1	11.8±0.9	1.5±0.2	10.5±0.9	Additivism	40
Cum₆ (1) + αTOH(2)	0.1	14.2±0.9	2.0±0.2	10.5±0.9	Synergism,14%	40
Cum₄(1) + αTOH(2)	0.1	12.7±0.9	7.1±0.5	10.5±0.9	Antagonism	40
BisCum1(1) + αTOH(2)	0.1	12.6±0.9	7.9±0.5	10.5±0.9	Antagonism	38
BisCum3(1) + αTOH(2)	0.1	6.1±0.5	2.2±0.2	10.5±0.9	Antagonism	38
MF1(1)+α-TOH(2)	0.1	10.0±0.9	4.0±0.3	10.5±0.9	Antagonism	39
MF2(1)+α-TOH(2)	0.1	15.0±0.9	9.2±0.9	10.5±0.9	Antagonism	39
MF3(1)+α-TOH(2)	0.1	14.0±0.9	2.8±0.2	10.5±0.9	Synergism,5.3%	39
MF4(1)+α-TOH(2)	0.1	13.8±0.9	0.75±0.05	10.5±0.9	Synergism,22%	39
#SA(1) + α-TOH(2)	0.1	45.0±1.0	8.5±0.5	21.0±1.5	Synergism,52%	44
Lipid substrate oxidized TGSO, 80°C , only # TGL,100°C						

Table 7. Effects of equimolar (1:1) binary mixtures of studied antioxidants without and with alpha-tocopherol (α-TOH)

Synergism obtained for different binary mixtures are explained taking into account that the during the oxidation process the antioxidant molecules of both strong antioxidants may be regenerated, which leads to higher antioxidant efficiency of the mixture, than of the individual compounds. The regeneration of both antioxidant molecules of compounds with catecholic moiety, QH_2 (4-hydroxy-bis-coumarin, caffeic acid (CA) and MF1-MF3) and of tocopherol (TOH) is possible as a result of the following possible reactions:

a.   reaction of an antioxidant radical (QH• or TO•) with other antioxidant molecule:
-    semiquinone radical (QH•) and tocopherol (TOH)
QH• + TOH →QH$_2$ + TO• (regeneration of QH$_2$ by H transfer)
-    between tocopheryl radical (TO•) and QH$_2$
TO• + QH$_2$ →TOH + QH• (regeneration of TOH by H transfer)
b.   homo-disproportionation reaction of two equal radicals:
2QH• → QH$_2$ + Q regeneration of QH$_2$ (Q is quinone)
2TO• →TOH +T=O regeneration of TOH (T=O is tocopheryl quinone)
c.   cross-disproportionation reaction of different radicals:
QH• + TO • → QH$_2$ + T=O regeneration of QH$_2$
TO• + QH• → TOH + Q regeneration of TOH.

As a result the oxidation stability of lipid sample increases, because the both antioxidants with strong efficiency are regenerated during the oxidation process.

In case of binary mixture of TOH with monophenolic antioxidants (AH), predominantly TOH molecule will be regenerated during the following reactions:

-    tocopheryl radical (TO•) with monophenolic antioxidant (AH)
TO• + AH →TOH + A•
-    homo-disproportionation of tocopheryl radicals (TO•)
2TO• →TOH +T=O
-    cross-disproportionation of phenoxyl radical (A•) and tocopheryl radical (TO•)
A• + TO• →TOH+ A$_{-H}$ or A• + TO• →T=O+ AH (depending on AH structure)

These reactions demonstrate that during oxidation process initial molecules of individual antioxidants are regenerated by different mechanisms in the binary mixtures. Nevertheless both binary mixtures may be used as effective antioxidant compositions. It is proven that the positions of phenolic hydroxyl groups in 4-hydroxy-bis-coumarins are of significance for their antioxidant activity and mechanism of action. Comparable kinetic analysis showed that the antioxidant efficiency (PF) and reactivity (ID) depend significantly from the substitution of the phenolic ring.

5. Surface-active compounds – surfactants (S)

Surfactants (S) are amphiphilic substances which adsorb at interface and decrease an excess of free energy (surface tension, γ) of interface. Surfactants may act as detergents, wetting agents, emulsifiers, foaming agents, and dispersants. A surfactant molecule contains both a water insoluble hydrophobic component and a water soluble hydrophilic component (polar head).

Surfactant solutions are one of the simplest examples of self-assembling soft nano-systems, whose micro-aggregates (micelles) are of 1–500 nm in size [10,11]. Micelles are prevalent in naturally occurring and biological catalytic reactions Micelles are formed by those surfactants which possess rather long bulky hydrophobic part along with strong hydrophilic head. Such surfactants form direct micelles in water and other polar solvent, and reverse micelles in organic solution.

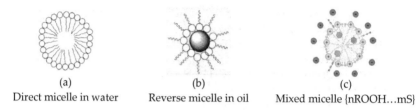

(a)   (b)   (c)
Direct micelle in water   Reverse micelle in oil   Mixed micelle {nROOH...mS}

Figure 11. Micelles formed by surfactants in polar and nonpolar media

The phenomenon of micellar catalysis has been known for a rather long time and applied in many processes, but the significant influence of surfactants on the lipid and hydrocarbon oxidation has been found and studied only in recent decades [6-8, 46-48]. Specific features of micellar catalysis for oxidation processes have two causes: (1) Hydroperoxides (LOOH), which are formed as the primary oxidation products, are amphiphilic and surface active, in contrast to initial oils; (2) There is spontaneous allocation of amphiphilic compounds in every heterogeneous and colloid system resulting in the reduction in the total free energy of a system, including the interface boundaries (the rule of polarity equalization). In the presence of surfactants in oxidized oil hydroperoxides and surfactants form mixed micelles {nLOOH...mS}(Fig. 11c). Using the measurement of the interphase tension [49], nuclear magnetic resonance (NMR), and dynamic light scattering [50], it was shown the association of LOOH and a surfactant in combined micelles in which LOOH plays the role of a co-surfactant. The average self-diffusion coefficient for hydroperoxide decreases with growth of the surfactant (CTAC) concentration up to the equalization with the surfactant diffusion coefficient, when all LOOH is bound in mixed micelles {nLOOH⋯mS}. The mixed micelle effective size calculated by the Stocks–Einstein equation was ~2 nm [50]. The size determined by DLS for mixture cumene hydroperoxide and CTAB are about 20 nm. Hydroperoxide facilitates the colloid dilution of CTAB in organic medium.

5.1. Effect of surfactants on lipid oxidation.

The comparison of the effects of different surfactants on lipid and hydrocarbon oxidation reveals that cationic surfactants (CS) promote hydroperoxide destruction resulting in the formation of free radicals [46,47] and the oxidation as a whole (Fig.12a,b).

By means of NMR and GC–MS, it is shown that in the presence of CS (CTAC and CTAB)) cumene hydroperoxide decomposes into dimethyl phenylcarbinol, acetophenon, and dicumylperoxide which are known as resulting from the radical decomposition of hydroperoxide. In the presence of anionic SDS, cumene hydroperoxide decomposes without radical formation into phenol and acetone [48].

The kinetics of oxidation of sunflower (TGSO) and olive (TGOO) oil triacylglycerols (Fig.12a) and natural olefin (limonene) (Fig.12b) in the presence of surfactants show that cationic surfactants (CS) promote the oxidation, whereas the anionic sodium dodecylsulfate (SDS) has no influence in the case of limonene and SDS demonstrates a weak retardation in TGOO oxidation (Fig.12a.) The chain breaking inhibitor α-tocopherol completely suppress the limonene oxidation accelerated by CTAC and CTAB (Fig.12b). Before and after the induction periods, the oxidation rate is described by the well known equation for the liquid-phase chain oxidation of hydrocarbons and lipids (see above):

$$R_{02} = a\left[LH\right]\cdot R_{IN}^{0.5} \tag{1}$$

where R_{IN} is the radical initiation rate. R_i can be calculated from the duration of the induction period (τ), caused with α-tocopherol (Fig.12) as follows: $R_{IN} = 2[InH]/\tau$.

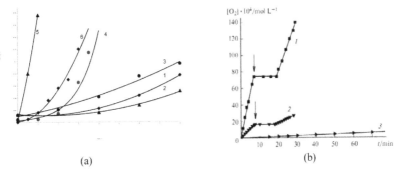

(a)　　　　　　　　　(b)

Figure 12. (a). Effect of surfactant on the LOOH formation during autoxidation of TGOO at 100oC and TGSO at 80oC: 1, 4- without additives; 3, 5 - 0.1M CTAB; 2,4- 0.1 M SDS; 6 - 0.04 M 1-OD [48]; (b) Effect of surfactant on the oxygen absorption during 1M limonene oxidation at 60oC; 1 – 1mM CTAC; 2 – 0,8 mM CTAB; 3 – 1mM SDS; 3- without additives; [LOOH]=22mM; Arrows show the moments of introducing α-tocopherol: 1-0,8 mM; 2 – 0,2 mM [51].

The rate constants for the propagation (k_P) and termination (k_t) of the oxidation chain for limonene and its oxidizability parameter $a = k_P/(2k_t)^{0.5}$ are known at the temperatures 30-80°C [52]. So, the radical initiation rate can be calculated on the base of the measured value of oxygen uptake rate: $R_i = (R_{02}/a[LH])^2$. In the case of limonene both formulas give very close R_i values. These data show that the mechanism of the catalytic action of CS on the oxidation processes consists in the increase of the chain initiation rate, caused by acceleration of hydroperoxides decomposition into free radicals.

To estimate the mixed micelles {nLOOH… mS} as a free radical initiator quantitatively, cumene hydroperoxide and hydrogen peroxide, which are produced in industrial scale, and limonene hydroperoxide were taken, and natural polyphenol quercetin was used as a radical acceptor. The interaction of LOOH with quercetin (Qu) in the presence of surfactants (S) is described by reactions:

nLOOH + mS ⊚ {nLOOH … mS} → LO₂˙
Qu + LO₂˙ → products,

The initiation rate is equal to: $R_i = -2d[Qu]/dt$. Quercetin is characterized by an intense absorption band in a visible region of the electronic spectrum, it is soluble in water and organic solvents and readily reacts with free radicals; i.e. it is a convenient kinetic probe to study the processes of radical generation in various media. The generated peroxyl radicals come into the volume and can initiate chain oxidation, polymerization, or other radical processes. In the presence of oxygen the concentrations of hydroperoxide and other polar products increase during the accelerated oxidation, and this, in its turn, influences the structure and properties of micelles. The instability of micelles and also the varied composition of their polar cores do not allow applying the known and frequently used pseudophase approach to the analysis of the lipid oxidation kinetics in the presence of surfactants. The comparison of the activities of LOOH and cationic surfactants in the generation of radicals can be conducted on the basis of the specific rates of radical initiation $\varpi_i = R_i / ([LOOH] \cdot [S])$. Data in the Table 6 show that cationic surfactants catalyze free radical initiation from hydroperoxide both in water and in organic solutions that is both in direct and reverse mixed micelles.

It means that the CS–hydroperoxide system can be used both as a lipophilic and a hydrophilic initiator. Small amounts of LOOH and CS provide significant radical generation rates (10^{-8}–10^{-7} Ms^{-1}), which are inaccessible at low temperatures for the known azoinitiators. By their activity in the generation of radicals in organic media the surfactants can be arranged in the following order, which indicates the essential role of counter ions in the catalytic action of CS:

CTAC ≈ TDTAC > CTAB ≈ CPB > DCDMAB > CTAHS

Surfactant	Cumene hydroperoxide, 37°C		Hydrogen peroxide, 37°C		Limonene hydroperoxide, 60°C
	ϖ_i, (M·s)$^{-1}$ Organics	ϖ_i, (M·s)$^{-1}$ Water	ϖ_i, (M·s)$^{-1}$ Organics	ϖ_i, (M·s)$^{-1}$ Water	ϖ_i, (M·s)$^{-1}$ Organics
CTAC	$2{,}1 \cdot 10^{-3}$	$3{,}7 \cdot 10^{-4}$	$0{,}67 \cdot 10^{-3}$	$0{,}14 \cdot 10^{-3}$	$27 \cdot 10^{-3}$
CTAB	$1{,}9 \cdot 10^{-3}$	$1{,}9 \cdot 10^{-4}$	$0{,}2 \cdot 10^{-3}$	$0{,}14 \cdot 10^{-3}$	$3{,}6 \cdot 10^{-3}$
CTAHS	$0{,}17 \cdot 10^{-3}$	$1{,}1 \cdot 10^{-4}$	≈0	≈0	$2{,}5 \cdot 10^{-3}$
DCDMAB	$1{,}5 \cdot 10^{-3}$	$2{,}8 \cdot 10^{-4}$	$2{,}1 \cdot 10^{-3}$	$0{,}37 \cdot 10^{-3}$	$3{,}6 \cdot 10^{-3}$
CPB	$1{,}9 \cdot 10^{-3}$	$3{,}3 \cdot 10^{-4}$	$0{,}2 \cdot 10^{-3}$	$0{,}14 \cdot 10^{-3}$	$3{,}6 \cdot 10^{-3}$
TDTAC	$2{,}1 \cdot 10^{-3}$	$3{,}7 \cdot 10^{-4}$	$0{,}47 \cdot 10^{-3}$	$0{,}13 \cdot 10^{-3}$	-
SDS	≈0	≈0	≈0	≈0	≈0
Lecithin	≈0	≈0	≈0	≈0	≈0

Table 8. Specific rates of radical initiation in the system: cationic surfactant + hydroperoxide in chlorbenzene and in water solution [47]

Along with hydroperoxide, water, and other polar oxidation products, catalytic and inhibiting components can be concentrated in mixed micelles {nLOOH·mS}. The combination of cationic surfactants with transition metal compounds known as homogeneous catalysts of the hydrocarbon oxidation was found to demonstrate synergism, i.e., for the mixture of components the oxidation rate (R_Σ) exceeds the sum of the rates in the experiments with separately used components (R_{Me} and R_S): $\beta = R_\Sigma/(R_{Me} + R_S) > 1$. The ethylbenzene is oxidized selectively into acetophenon and water catalyzed with the combination of CTAB and cobalt acetylacetonate [53]. Under similar conditions limonene is oxidized with the primary formation of a carbonyl compound (carvon) [51].

Let us look on the mixed micelles of hydroperoxide and cationic surfactant once more. In mixed micelle, peroxide bond is localized in the interphase which has very strong intensity of electric field, about $5 \cdot 10^5 V/m$. It affects peroxide bond, weakens it and facilitates decomposition into free radicals. Apparent activation energies of hydroperoxide decay decrease to 50-60 kJ/mol in mixed micelles from ~ 100 kJ/mol for thermal decay [51].

In the case of anionic surfactant the direction of electric field is different and decomposition into radicals is not facilitated. On the contrary, alkali metal alkyl sulfates [47,48] and alkyl phosphates [54,55] act as antioxidants to retard or completely suppress the oxidation process.

Nonionic surfactants form neutral micelles which have no electric field. May be, by that reason nonionic surfactants do not affect free radical formation in hydroperoxide decay, although they form mixed micelles {LOOH…S} with nonionic surfactant as well.

5.2. Phospholipid oxidation.

Phospholipids (PL) are natural surfactants, which are widely used in the production of food, drug, and cosmetics. PL are the basic lipid components of plasmatic cell membranes and membranes of subcellular organelles of animals, plants, and microorganisms. Phosphatidylcholines (1,2-diacyl-sn-glycero-3-phosphocholines, lecithins) are the most widely used; they are present in large amounts in myocardium, liver, kidneys, and egg yolk [56]. In lecithin molecules, anionic phosphate and cationic choline (tetraalkylammonium) groups are connected via a zwitterionic bond to form a neutral polar head. Hydrocarbon moiety represents residues of fatty acids, whose composition depends on the type of PLs (egg, soybean, fish, etc.). Lecithins have a zwitterionic structure in a wide pH range.

Unsaturated fatty acid residues of PL are readily oxidized with atmospheric oxygen as well as nonpolar unsaturated lipids. The primary products of PL oxidation are mainly isomeric hydroperoxides [9,12,57-59]. Lecithins are easily dissolved in organic solvents to yield compact reverse micelles [60,61]. In aqueous solutions, lecithin forms multilamellar liposomes or vesicles under the action of ultrasound dispersion [12,56,62]. Using the DLS method, it was found that, at egg lecithin concentrations 10–90 mg/mL, the size of micro-aggregates observed are equal to 5-6 nm in organic solvents and in water, liposomes are formed with a wide size distribution of 60–1000 nm.

PC oxidation in the presence of azoinitiators or transition metals occurs via free radical chain mechanism. The formation of micro-aggregates both in organic and water media results in a nonlinear dependence of the rate of oxygen absorption on substrate concentration (at constant initiation rate). The deviations from the linearity were observed at concentrations of egg lecithin above 5 mg/mL, corresponding to the formation of micro-aggregates. The rate increment caused by a further increase in the concentration markedly decreases [60,61]. It is possible that a partial shielding of active C–H bonds, which interact with peroxyl radicals from an initiator in solvent bulk results in a relative decrease in the oxidation rate.

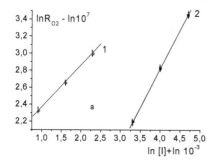 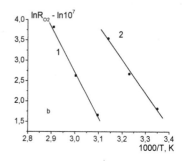

Figure 13. a) Dependences of lecithin (45mg/mL) oxidation rates on the initiator concentration in logarithmic coordinates: 1 – in n-decane solution, 60°C, I – azobisisobutyronitrile, AIBN; 2 – in water, 37°C, I – azodiiso-butyramidine-dihydrochloride (AAPH); b) Temperature dependence of the rate of lecithin (45mg/mL) oxidation plotted in Arrhenius coordinates: 1 – in chlorbenzene, [AIBN]= 5mM; 2 – in water, [AAPH]=55mM.

It was shown in [60,61] that the dependences of PC oxidation rates on initiation rates differ in organic and water solutions. In organic solvents, R_{O2} is proportional square root of R_i, whereas in water, $R_{O2} \sim R_i$. It can be seen from Fig.13a, where the dependences of egg lecithin oxidation rates on the corresponding initiator concentration (it means $R_{O2} \sim R_{IN}^n$, because $R_{IN} = k_i[\text{initiator}]$) are presented: in organics (chlorobenzene) n = 0,5 and n ≈ 1 in water media.

So, the rate of PL oxidation in an aqueous medium cannot be described by Eq. (1), common for lipid and hydrocarbon oxidation. Therefore, even in a narrow concentration range, it is unreasonable to compare the oxidizability of PL in water and an organic medium using parameter $a = k_p/(2k_t)^{0.5}$. Nevertheless, in many studies devoted to the oxidation of phospholipids in various media, Eq. (1) was applied to describe the rate of oxidation (absorption of oxygen [63-66] or accumulation of hydroperoxides [9,58,67]) and to determine the oxidizability of PLs or individual phosphatidylcholines [68]. The majority of these works was carried out at the physiological temperature (37°C). The measurements were performed in different ranges of the overall concentrations of PL and with different initiators and inhibitors used to determine the initiation rates; therefore, the conclusions were very different right up to the opposite ones.

According to [58,68] the oxidizability of PLs in aqueous dispersions is lower than that in organic solvents by an order of magnitude; it is higher in reverse micelles than in molecular alcohol solutions [58]. In [65,68], it was assumed that the micro-heterogeneity of PL solutions and the dispersity of colloidal solutions do not influence the oxidizability of unsaturated lipids in both aqueous and organic media.

A comparison of the experimentally measured rate of O_2 absorption during PL oxidation in aqueous solutions with the corresponding values obtained in an organic solvent at the same temperature, mass concentration of PL, and the initiation rate, which is governed by the contents of water- (AAPH) and oi-soluble (AIBN) azo-initiators, respectively, in the volumes of the solvents demonstrates the following [61]. At a temperature of 45°C, radical initiation rate of $22.5 \cdot 10^{-8}$ M/s, and PL concentration of 45 mg/mL, the rates of oxygen absorption in water and chlorobenzene are $3.5 \cdot 10^{-6}$ and $2.1 \cdot 10^{-6}$ M/s, respectively. A comparison suggests that, in the presence of a source of radicals, PL organized into multilamellar liposomes is ~1.5_fold faster oxidized in water than in the organic solution of reverse micelles. In order to explain this result, for micro-heterogeneous systems, one must introduce the concept of the effective (apparent) concentration of an oxidized substrate. In a system of multilamellar liposomes, the effective concentration of the oxidized substrate is higher than that in a system of reversed micelles occurring in an organic medium; therefore, a higher oxidation rate is observed at the same temperature and the rate of radical initiation. The rates of PL oxidation both in organics and in water, initiated by corresponding initiator, increase with temperature according to the Arrhenius equation (Fig.10b). The effective activation energy of AAPH-initiated PL oxidation in an aqueous solution (74 kJ/mol) is lower than the activation energy of AAPH decomposition (112 kJ/mol). Hence, a radical chain mechanism of PL oxidation in water is more complex than described above mechanism of model oil oxidation (Table 1). In a micro-heterogeneous medium, in addition to individual radicals and molecules, reagents that are included into microaggregates (liposomes) and characterized by reactivity different from that of molecular_dispersed particles in corresponding reactions are involved in the stages of chain initiation, propagation, and termination. Cross-disproportionation reactions of different radicals occur to result in the imitation of a linear chain termination.

α-Tocopherol is well known the most effective lipid antioxidant [1-3]. Lecithin liposome oxidation in the presence of α-tocopherol demonstrates that marked induction periods may be observed when α-tocopherol is added inside liposome during preparation. The incomplete suppression of O_2 absorption by α-tocopherol may be indicative of the PC oxidation inside of liposome without migration of peroxyl radicals into the bulk solvent.

It turns out that catecholamines dopamine, adrenaline and noradrenaline are much more strong and effective inhibitors for PC oxidation than α-tocopherol (Compare Fig.14 and Fig.15). Evidently, the positive charge of catecholamines at neutral pH facilitates their adsorption and protective action on the surface of negative charged liposomes [61].

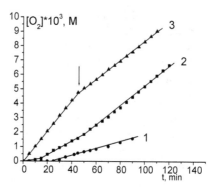

Figure 14. Kinetic curves for O_2 absorption during AAPH- initiated (55 mM) oxidation of PL (45 mg/ml) in water at 37°C in the presence of (1) 0.33 and (2) 0.05 mM α-tocopherol incorporated upon preparation of liposomes and (3)0.05 mM α-tocopherol introduced directly into solution

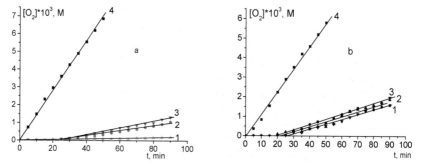

Figure 15. Effect of 0.1 mM (1) adrenalin, (2) dopamine, (3) noradrenalin on AAPH-initiated (55 mM) oxidation of PL (45 mg/ml) at 37°C in (a) water and (b) in phosphate buffer with pH 7.4; (4) no additives.

It must be noted that in the phosphate buffer, induction periods τ are nearly equal for all of the catecholamines (Fig.15b), while, in an aqueous solution, adrenalin provides a longer inhibition of the oxidation than dopamine and noradrenalin do (Fig.15a). The analysis of the ratios between the rates of radical initiation and the durations of the induction periods testified that, for all catecholamines in the buffer solution, the stoichiometry of inhibition, which is numerically equal to the number of radicals corresponding to one acceptor molecule, is n =$(R_{IN}\cdot\tau)/[CA]_0$ = 2, which is characteristic of catechols. In an aqueous and a physiological solution (0.9% NaCl), dopamine and noradrenalin exhibit n= 2, while for adrenaline, n= 4. Moreover, the adrenalin-containing mixture acquires a pink color in water.

Figure 16a illustrates variations in the optical absorption spectra of adrenalin solutions in water and a physiological solution (0.9% NaCl) during its free-radical oxidation initiated by AAPH. It can be seen that the oxidation results in the formation of a colored product with an absorption maximum at 480 nm. In the phosphate buffer of pH 7.4, adrenalin also undergoes transformations (Fig. 16b); however, they yield no colored product. The spectral characteristics of the pink product ($\varepsilon = 4.02 \times 10^3$ M^{-1} cm^{-1} at 480 nm) correspond to adrenochrome (3-hydroxy-1-methyl-2,3-dihydro-1H-indole-5,6-dion), which is formed via the abstraction of four hydrogen atoms from adrenaline [68].

Figure 16. Variations in UV spectrum of 0.1 mM adrenalin solution in the process of its oxidation at 37°C (a) in water and 0.9% NaCl solution and (b) in phosphate buffer with pH 7.2.

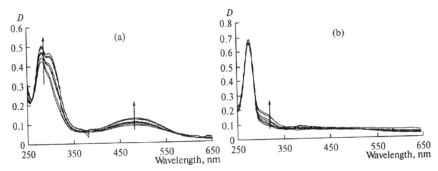

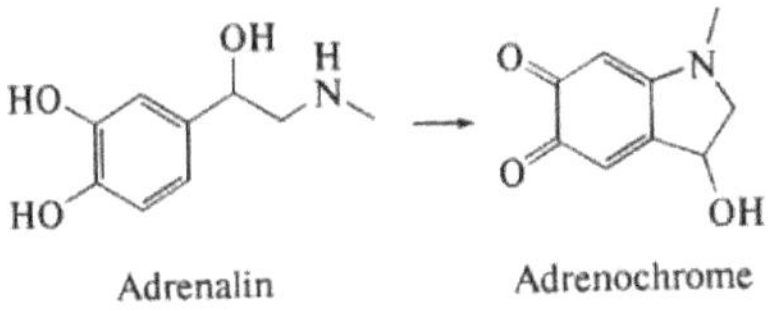

Adrenalin   Adrenochrome

Figure 17. Structures of adrenalin and adrenochrome

It is interesting that, in an aqueous solution, under the conditions of AAPH-initiated free-radical oxidation, adrenalin is quantitatively transformed into adrenochrome. However, in the phosphate buffer solution adrenochrome is not formed.

6. Concluding remarks

Radical scavenging activity towards DPPH radical gives information only about the H-donating capacity of the studied compounds and some preliminary information for their possibility to be used as antioxidants. Antioxidant activity is capacity of the compound to shorten the oxidation chain length as a result of its reaction with peroxyl radicals. For that reason we mean as antioxidant activity the chain-breaking activity of the compounds. This comparable study showed a good correlation between experimental antioxidant activity of compounds under study and their predictable activity by using TLC DPPH radical test.

It has been demonstrated that phenolic compounds with catecholic moiety are the most powerful scavengers of free radicals and they may be used as effective chain-breaking antioxidants. The highest antiradical and antioxidant activity of phenolic antioxidants with catecholic moiety is explained by possible mechanism of homo-disproportionation of their semiquinone radicals formed.

Thus regeneration of the antioxidant molecule during the oxidation process is possible. It has been found for the first time that only substitution in the aromatic nucleus of the studied bis-coumarins and xanthenes-lignans is responsible for their antioxidant activity.

It must be noted that antioxidants' activity depends significantly not only on their structural characteristics, but also on the properties of the substrate being oxidized and the experimental conditions applied. Structural characteristics of the complex system: oxidizing substrate - antioxidant must be considered. On the basis of this comparable analysis, the most effective individual antioxidants and binary mixtures were proposed for highest and optimal lipid oxidation stability.

Author details

Vessela D. Kancheva
Lipid Chemistry Department, Institute of Organic Chemistry with
Centre of Phytochemistry, Bulgarian Academy of Sciences, Sofia, Bulgaria

Olga T. Kasaikina
N.N. Semenov Institute of Chemical Physics, Russian Academy of Sciences, Moscow, Russia

Acknowledgement

The financial support of Project No BG051PO001/3.3-05-0001 "Science and Business" - Ministry of Education and Sciences in Bulgaria for publishing this chapter in the Lipid Peroxidation monograph open access is gratefully acknowledged.

7. References

[1] Kancheva V (2010a) Antioxidants. Structure - activity relationship. In: "Antioxidants - Prevention and Healthy Aging", Ed. by F. Ribarova, SIMELPRESS Publ., Sofia, Bulgaria, Chapter 1, pp 56-72.

[2] Kancheva V (2010b) Oxidative stress and lipid oxidation. In: "Antioxidants - Prevention and Healthy Aging", Ed. by F.Ribarova, SIMELPRESS Publ., Sofia, Bulgaria, Chapter 3 pp 233-238.

[3] Kancheva VD (2012) Phenolic antioxidants of natural origin – structure activity relationship and their beneficial effect on human health. In: "Phytochemicals and Human Health: Pharmacological and Molecular Aspects", Nova Science Publishers Inc., USA, Ed. A.A.Farooqui, Chapter I, pp 1-45.

[4] Huang SW, Frankel EN, Schwarz K, Aeschbach R, German JB (1996) Antioxidant activity of alpha-tocopherol and Trolox in different lipid substrates: Bulk oils and oil-in-water emulsions, *J.Agric. Food Chem.* 44: 444–452.

[5]   Huang S W and Frankel EN (1997) Antioxidant activity of tea catechins in different lipid systems, *J. Agric. Food Chem.* 45:3033–30383.

[6]   Mc'Clements DJ Food Emulsions: Principles, Practice and Techniques, Boca Raton, FL: Taylor and Francis/CRC Press, 2005.

[7]   Chaiyasit W, Elias R J, Mc'Clements DJ, Decker EA (2007) Role of Physical Structures in Bulk Oils on Lipid Oxidation, *Critical Reviews in Food Science and Nutrition.* 47: 299-317.

[8]   Chen B, McClements DJ, Decker EA (2011) Minor Components in Food Oils: A Critical Review of their Roles on Lipid Oxidation Chemistry in Bulk Oils and Emulsions , *Critical Reviews in Food Science and Nutrition.* 51: 901-916. http://dx.doi.org/10.1080/10408398.2011.606379

[9]   Porter NA, Weber BA, Weenen H, Khan JA. (1980) Autoxidation of Polyunsaturated Lipids. Factors Controlling the Stereochemistry of Product Hydroperoxides, *J. Am. Chem. Soc.*102: 5597-5601.

[10] Dynamics of Surfactant Self-Assemblies. Micelles, Microemulsions, Vesicles and Lyotropic Phases, Zana, R., Ed., Boca Raton, FL: Taylor and Francis/CRC Press, 2005.

[11] Hamley I.W. (2000) Introduction to Soft Matter: Polymers, Colloids, Amphiphils and Liquid Crystals, Chichester: Wiley.

[12] Roschek BJ, Tallman KA, Rector CL, Gillmore JG, Pratt DA, Punta C, Porter NA. (2006) Peroxyl Radical Clocks, J. Org. Chem. 71: 3527–3532.

[13] Kasaikina OT, Kortenska VD, Yanishlieva NV (1999) Effect of chain transfer and recombination/disproportionation of inhibitor radicals on inhibited oxidation of lipids, *Russ. Chem. Bull.* 48: 1891-1996.

[14] Pisarenko LM, Krugovov DA, Kasaikina OT (2008) A kinetic model for limonene oxidation. *Russ. Chem. Bull.* 57:80-86.

[15] Naumov VV, Vasil'ev RF (2003) Antioxidant and prooxidant effect of tocopherol. *Kinet. Catal.* 44: 101-105.

[16] Roginsky VA (1988) Phenolic antioxidant-efficiency and reactivity, Naukam Moscow, Russia (in Russian).

[17] Barkley L R C, Baskin K A, Locke S J, Vinquist M R (1989) Absolute rate constants for lipid peroxidation and inhibition in model biomembranes, *Can. J.Chem.* 68: 2258-2269.

[18] Burton GW, Doba T, Gabe EJ, Hughes L, Lee FL, Prasad L, Ingold KU (1985) Antioxidants in biological molecules. 4. Maximizing the antioxidant activity of phenols. *J. Am. Oil Chem. Soc.* 107: 7053-7065.

[19] Small RD, Scaiano JC, Patterson LK.(1979) radical process in lipids:\a laser photolysis study of t-butoxy radical reactivity toward fatty acids. *Photochem. Photobiol.* 29:49-51.

[20] Kaouadji MN, Jore D, Ferradini C, Patterson LK (1987) Radiolytic scanning of vitamin E/vitamin C oxidation reduction mechanisms. *Bioelectrochem. Bioenerg.* 18: 59-70.

[21] Roginsky VA (1990) Kinetics of polyunsaturated fatty acid esters oxidation inhibited by substituted phenols. Kinet. Catal. 31: 546-552.

[22] Kortenska-Kancheva VD, Belyakov VA (2005) Simulation of lipid oxidation kinetics in various mechanism of hydroperoxides decomposition, *Riv. Ital. delle Sost. Grasse* 82: 177-185.

[23] Belyakov VA, Kortenska VD, Rafikova VS, Yanishlieva NV (1992a) Kinetics of the inhibited oxidation of model lipid systems. Role of fatty alcohols. Kinetics and Catalysis, 33: 617-622;

[24] Belyakov VA, Kortenska VD, Rafikova VS, Yanishlieva NV (1992b) Mechanism of lipid autoxidation in presence of fatty alcohols and partial glycerols. Kinetics and Catalysis 33: 756-771.

[25] Kortenska VD, Yanishlieva NV, Roginskii VA (1991) Kinetics of inhibited oxidation of lipids in the presence of 1-Octadecanol and 1-Palmitoylglycerol, *J. Am. Oil Chem. Soc.* 68: 888-890.

[26] Kortenska VD, Yanishlieva NV (1995) Effect of the phenol antioxidant type on the kinetics and mechanism of inhibited lipid oxidation in the presence of fatty alcohols, *J. Sci. Food Agric.* 68:117-126.

[27] Kortenska VD, Yanishlieva NV, Kasaikina OT, Totzeva IR, Boneva MI, Rusina IF (2002) Phenol antioxidant efficiency in presence of lipid hydroxy compounds in various lipid systems, *European J. Lipid Science and Technology* 104: 513-519.

[28] Burlakova E B (2007) Bioantioxidants. Molecular Cell Biophysics. Russ.Chem. J. 51:3-12 (in Russian).

[29] Pobedimskiy DG, Burlakova EB (1993) Mechanism of Antioxidant Action in Living Organisms in Atmospheric Oxidation and Antioxidants. Ed. J. Scott. 2: 223-235.

[30] Kancheva VD (2009) Phenolic antioxidants – radical scavenging and chain breaking activities. Comparative study. Eur. J. Lipid Sci. Technol. 111: 1072-1089.

[31] Kancheva VD (2010) New bioantioxidants strategy. T3D-2010 International Symposium on Trends in Drug DSiscivery and Development 5th-8th January 2010 Delhi (India), Book of AbstractsIL-50.

[32] Denisov ET, Denisova TG: Handbook of antioxidants. Bond dissociation energies, rate constasnts, activation energies and enthalpies of reactions. Second Ed. CRS Press, New York, Washington, D.C. 2001.

[33] Shahidi F: Natural antioxidants, a review in: Natural antioxidants. Chemistry, Health effects and applications, Shahidi, F. (ed) AOCS, Press, Champaign, Illinois, USA, 1999, 1-11.

[34] Vermeris W, Nikolson R. Phenolic compound biochemistry. Springer 2006, The Nederlands, pp 267.

[35] Frankel E N (1998) Lipid oxidation, The Oily Press, Dundee (Scotland) 1998, pp. 303.

[36] Kamal-Eldin A, Makinen M, Lampi A-M (2003) The Challenging Contribution of Hydroperoxides to the Lipid Oxidation Mechanism, in Lipid Oxidation Pathways (Kamal-Eldin, A., ed.), Chapter 1, AOCS Press, Champaign, II pp 1-36.

[37] Vasilev RF, Kancheva VD, Fedorova GF, Batovska DI, Trofimov AV (2010) Antioxidant activity of chalcones. The chemiluminescence determination of the reactivity and quantum – chemical calculation of the energies and structures of reagents and intermediates. Kinetics and Catalysis 51: 507-515.

[38] Kancheva VD, Boranova PV, Nechev JT, Manolov II (2010) Structure-activity relationship of new 4-hydroxy-bis-coumarins as radical scavengers and chain-breaking antioxidants. Biochimie 92:1138-1146.

[39] Kancheva VD, Saso L, Angelova SA, Foti MC, Slavova-Kazakova A, Draquino C, Enchev E, Firuzi O, Nechev J (2012) Antiradical and antioxidant activities of new bioantioxidants. Biochemie 94: 403-415.

[40] Kancheva VD, Saso L, Boranov PV, Khan A, Saron M ...., Parmar V S (2010a) Structure-activity relationship of dihydroxy-4methyl-coumarins as powerful antioxidants. Correlation between experimental & theoretical data and synergistic effect. Biochemie 92: 1089-1100.

[41] Kancheva VD, Taskova R, Totseva I, Handjieva N (2007) Antioxidant activity of extracts, fractions and flavonoid constituents from Carthamus lanatus L., Riv. Ital. delle Sost. Grasse, 84: 77-86.

[42] Kancheva VD, Dinchev D, Tsimidou M, Kostova I, Nenadis N (2007) Antioxidant properties of Tribulus Terrestris from Bulgaria and radical scavenging activity of its flavonoid components, Riv. Ital. delle Sost. Grasse 54: 10-19.

[43] Amic D., Davidovic-Amic D, Beslo D, Trrinajstic N (2003) Structure-radical scavenging activity relationships of flavonoids, Croatia Chemica Acta, CCACAA, 76: 55-61.

[44] Kortenska VD, Yanishlieva NV, Kasaikina OT, Totzeva IR, Boneva MI, Rusina IF (2002) Phenol antioxidant efficiency in presence of lipid hydroxy compounds in various lipid systems, European J. Lipid Science and Technology 104: 513-519.

[45] Kasaikina O T, Golyavin A A, Krugovov D A, Kartasheva Z S, Pisarenko L M (2010) Micellar catalysis in the lipid oxidation Mosc. Univ. Chem. Bull. (Engl. Transl.) 65: 206-209.

[46] Kasaikina OT, Kortenska VD, Kartasheva ZS (1999) Hydrocarbon and lipid oxidation in microheterogeneous systems formed by surfactants or nanodispersed Al2O3, SiO2, and TiO2, Coll. Surf. A: Physicochem. Eng. Asp.149: 29–38.

[47] Kasaikina OT, Kartasheva ZS, Pisarenko LM (2008) Effect of surfactants on the liquid phase oxidation of hydrocarbon and lipids. Rus. J. Gen. Chem. (Engl. Transl.)78:1533–1544.

[48] Kasaikina O T, Kancheva V D, Maximova T V, Kartasheva S Z, Vedutenko VV, Yanishlieva NV, Kondratovich VG, Totseva IR (2006) Catalytic Effect of Amphiphilic Components on the Lipid Oxidation and Lipid Hydroperoxide Decomposition. Oxid. Commun. 29:574-584.

[49] Trunova NA, Kartasheva ZS, Maksimova TV, Bogdanova YuG (2007) Cumene hydroperoxide decomposition in the system of direct and reverse micelles formed by cationic surfactants. Colloid. J. (Engl. Transl.) 69(5): 655–659.

[50] Krugovov DA, Pisarenko LM, Kondratovich BG, Shchegolikhin AN, Kasaikina OT (2009) Catalysis of limonene oxidation by cationic surfactants in combination with transition metal acetylacetonates, Petr. Chem. (Engl. Transl.)49:120–126.

[51] Pisarenko L M, Krugovov D A, Kasaikina O T (2004) . Influence of Cationic Surfactants on the Limonene Oxidation, Russ. Chem. Bull., Int. Ed. 53:2210-13.

[52] Maksimova TV, Sirota TV, Koverzanova EV, Kashkai AM (2001) Effect of Surfactants on the Ethylbenzene Oxidation. 4. Catalysis of Oxidation by CTAB in Combination with Co(acac)2, Petr. Chem. (Engl. Transl.) 41(5): 389-395

[53] Bakunin VN, Popova ZV, Oganesova EYu , Parenago (2001) The change of hydrocarbon medium structure in the course of liquid phase oxidation, *Petr. Chem. (Engl. Transl.)* 41: 41-46.

[54] Yoshimoto M, Miyazaki Y, Umemoto A, Walde P, Kuboi R, Nakao K (2007) Phosphatidylcholine vesicle-mediated decomposition of hydrogen peroxide, *Langmuir* 23: 9416-9422.

[55] Gregor Cevc. *Phospholipids Handbook*, Marvel Dekker Inc., New York, 1993.

[56] Xu L, Davis TA, Porter NA (2009) Rate Constants for Peroxidation of Polyunsaturated Fatty Acids and Sterols in Solution and in Liposomes, *J. Am. Chem. Soc.* 131: 13037–13044.

[57] Wang X, Ushio H, Ohshima T (2003). Distributions of Hydroperoxide Positional Isomers Generated by Oxidation of 1-Palmitoyl-2-arachidonoyl-sn-glycero-3-phosphocholine in Liposome and in Methanol Solution, *J. Lipids* 38: 65-72.

[58] Gupta R, Muralidhara HS, Davis HT (2001) Structure and phase behavior of phospholipids-based micelles in nonaqueous media, *Langmuir* 17: 5176-5183.

[59] Mengele EA, Kartasheva ZS, Plashchina IG, Kasaikina O.T. (2008) Specific features of lecithin oxidation in organic solvents, *Colloid J.* 70: 753-758.

[60] Mengele E.A., Plashchina I.G., Kasaikina O.T. (2011) Kinetics of lecithin oxidation in liposome water solutions, *Colloid J.* 73: 701-707.

[61] Barclay LRC, MacNeil J M, VanKessel JA, Forrest BJ, Porter NA, Lehman LS, Smith IKJ, Ellington JC (1984) Autoxidation and aggregation of phospholipids in organic solvents *J. Am. Chem. Soc.* 106: 6740-6747.

[62] Burton GW, Doba T, Gabe EJ, Hughes L, Lee FL, Prasad L, Ingold KU (1985) Autoxidation of biological molecules. 4. Maximizing the antioxidant activity of phenols, *J. Am. Chem. Soc.* 107:7053-7065

[63] Roginsky V (2003) Chain breaking antioxidant activity of natural polyphenols as determined during the chain oxidation of methyl linoleate in Triton X-100 micelles, *Arch. Biochem. Biophysics* 414: 261-267.

[64] Roginsky V, Tikhonov I. (2010) Natural polyphenols as chain-breaking antioxidants during lipid peroxidation, *Chem. Phys. Lipids* 163: 127133.

[65] Nakamura T., Maeda H.A. (1991) Simple assay for lipid hydroperoxides based on triphenylphosphine oxidation and high performance liquid chromatography, *Lipids* 26: 765-768.

[66] Barclay LRC, Ingold KU (1981) Autoxidation of Biological Molecule. 2. The Autoxidation of a Model Membrane. A Comparison of the Autoxidation of Egg Lecithin Phospha-tidylcholine in Water and in Chlorobenzene *J. Am. Chem. Soc.* 103: 6478-6485.

[67] Roginsky V, Lissi EA (2005) Review of methods to determine chain-breaking antioxidant activity in food *Food Chem* 92: 235-254.

[68] Green, S. (1956) Mechanism of the catalytic oxidation of adrenaline by ferritin. *J. Biol. Chem.* 220: 237−255.

Chapter 4

Iron Overload and Lipid Peroxidation in Biological Systems

Paula M. González, Natacha E. Piloni and Susana Puntarulo

Additional information is available at the end of the chapter

1. Introduction

Fe is an essential element for the growth and well-being of almost all living organisms, except for some strains of lactobacillus, where the role of Fe may be assumed by another metal [1]. It is involved in many biological functions since by varying the ligands to which it is coordinated, Fe has access to a wide range of redox potentials and can participate in many electron transfer reactions, spanning the standard redox potential range. It is also involved in O_2 transport, activation, and detoxification, in N_2 fixation and in several of the reactions of photosynthesis [2]. However, there are problems in the physiological management of Fe, since in spite of its overall abundance, usable Fe is in short supply because at physiological pH under oxidizing conditions, Fe is extremely insoluble. Anytime Fe exceeds the metabolic needs of the cell it may form a low molecular weight pool, referred to as the labile iron pool (LIP), which catalyzed the conversion of normal by-products of cell respiration, like superoxide anion (O_2^-) and hydrogen peroxide (H_2O_2), into highly damaging hydroxyl radical ($\bullet OH$) through the Fenton reaction (reaction 1) or by the Fe^{2+} catalyzed Haber-Weiss reaction (reaction 2), or into equally aggressive ferryl ions or oxygen-bridged Fe^{2+}/Fe^{3+} complexes. Fe^{3+} can be reduced either by O_2^- (reaction 3) or by ascorbate leading to further radical production.

$$Fe^{2+} + H_2O_2 \xrightarrow[(Fe)]{} Fe^{3+} + HO^- + \bullet OH \tag{1}$$

$$O_2^- + H_2O_2 \Rightarrow O_2 + HO^- + \bullet OH \tag{2}$$

$$Fe^{3+} + O_2^- \Rightarrow Fe^{2+} + O_2 \tag{3}$$

Defense against the toxic effect of Fe and O_2 mixtures is provided by two specialized Fe-binding proteins: the extracellular transferrin (Tf) and the intracellular ferritin (Ft). Both retain Fe in the form of Fe^{3+} which unless mobilized will not be able to efficiently catalyze the production of free radicals. Fe is stored mainly intracellularly, where its potentially damaging effects are greatest.

The marine ecosystem can be seen as an integrative system with many factors that interact with the biota. Natural variables such as temperature, winds, precipitations, tide flows, currents, human activities, affect metal deposition into the sea. Once metals become bioavailable, they can enter the food web starting with the primary producers, and also in heterotrophic organisms at the bottom of the marine food chain, such as benthic filter feeders. Metals follow a bioaccumulation process inside the animals, depending on the animal's detoxification capacities and on exogenous Fe availability.

In plants, Fe concentrations increased during seed maturation, and by immunodetection experiments it was indicated that Ft concentration of seeds also increased with maturity, containing up to 1800 atoms of Fe per molecule [3]. This seed Fe could be stored for future use during seedling growth, as has been proposed by Hyde et al. [4], avoiding toxicity. Over growth, the oxidative stress depends upon a wide array of factors related to an enhanced radical production due to several metabolic pathways activated during the initial water uptake, including mitochondrial O_2 consumption. On the other hand, excess Fe effects seem to be limited mostly to the hydrophobic domain of the cell following different profile than during physiological development.

In the last decade or so, important advances have been made in the knowledge of conditions that involve Fe-overload in humans. Those conditions would include short term processes, as organ or tissue ischemia-reperfusion and local inflammation, as well as progressive pathologies essentially affecting the central nervous system. In the first case, the de-compartamentalization of Fe would lead to the expansion of the LIP and the increase of the oxidative damage. In the second case, it has been described an increase in Fe levels in the substantia nigra of Parkinsonian brains [5], Hallervorden-Spatz syndrome [6] and in mitochondria of Friedrich's ataxia cerebella [7]. Hereditary hemochromatosis is a very common genetic defect in the Caucasian population, with an autosomal recessive inheritance. It is characterized by inappropriately increased Fe absorption from the duodena and upper intestine, with consequent deposition in various parenchymal organs, notably the liver, pancreas, heart, pituitary gland and skin [8]. Fe overload is characterized by the presence of several clinical manifestations such as: increased susceptibility to infections, hepatic dysfunction, tumors, joint diseases, myocardiopathy, and endocrine alterations. Fe overload has been also observed (a) if dietary Fe is excessive, such as in the severe Bantu siderosis, reported in the Bantu tribe of Africa who drink acidic beer out of Fe pots, (b) in other inherited diseases, such as congenital atransferrinemia (lacking circulating Tf), and (c) during the medical treatment of thalassemia. Moreover, clinical and epidemiologic observations indicated that increased Fe storage status is a risk factor in several diseases such as porphyria cutanea tarda and sudden infant death syndrome, among others.

Oxidative damage to lipids had been studied over several decades, and it had been characterized in terms of the nature of the oxidant, the type of lipid, and the severity of the oxidation. Many stable products are formed during the process and accordingly, the assays developed to assess these products to evaluate lipid peroxidation include many techniques. The most currently used assay is the determination of malondialdehyde (MDA) formation with the thiobarbituric acid reactive substances test (TBARS). However, electron paramagnetic resonance (EPR) spectroscopy has shown the capacity of detecting, in the presence of exogenous traps, the presence of the lipid radical formed during peroxidation, by yielding unique and stable products. EPR, also known as Electron Spin Resonance (ESR) is at present the only analytical approach that permits the direct detection of free radicals. This technique reports on the magnetic properties of unpaired electrons and their molecular environment [9].

This chapter will be dedicated to overview the Fe-related alterations in oxidative metabolism in photosynthetic and non-photosynthetic organisms after experimental exposure to excess Fe employing different protocols of administration. Data assessing lipid peroxidation post-treatment both, as TBARS generation and/or EPR detection of lipid radicals, are reviewed in a wide range of biological systems.

2. Fe overload in aquatic organisms

Fe content in the upper earth's crust is around 6% [10]. The Fe concentration in sediments influences the Fe concentration in the associated surrounding seawater. However, the concentration of dissolved Fe (defined as Fe that can diffuse through a membrane of less than 0.45 μm) in open-oceanic waters is extremely low (< 56 ng/l) [11]. Natural parameters that augment the Fe levels in coastal and central oceanic areas are: aeolian deposition of dust, river discharge, washout of dust particles in the atmosphere by rainfall, ground water discharge, glacial melting, volcanic sediments, coastal erosion and up-welling of Fe-rich deep waters over hydrothermal vents [12]. Human activities also have a great impact on Fe levels, especially around coastal areas. Chemical and mining industries, disposal of waste metal, ports, aeolian deposition of atmospheric dust from polluted areas, are some of the human activities bringing Fe and other metals to the marine ecosystem. Therefore waters from different regions may have different Fe concentrations. Fe was recognized as a bioactive element [13] and a deficiency in Fe had been suggested to limit primary productivity in some ocean regions [14,15]. Fe uptake is strictly required for phytoplankton development since the photosynthetic apparatus contains numerous loci for Fe. Moreover, it was pointed out that it is critical to avoid Fe overload in water with low organic matter content under aquarium conditions to prevent Fe-dependent toxicity [16].

Over a decade ago, Estévez et al. [17] studied the effect of *in vivo* Fe supplementation to the green algae *Chlorella vulgaris* in terms of the establishment of oxidative stress conditions. Growth under laboratory conditions increased with Fe availability up to 90 μM with increases in biomass, suggesting that Fe supply at concentrations lower than 90 μM could be considered limiting for algal growth. However, Kolber et al. [18] pointed out that in their field experiments in the equatorial Pacific, 2 days following Fe enrichment, photosynthetic energy conversion efficiency began to decline. It was also indicated that some algal cultures showed deleterious effects if exceeding an Fe threshold (14-28 μM) in unpolluted freshwater [16]. Between 90 and 200 μM Fe in *C. vulgaris* cultures, there was no effect on growth with increased Fe additions and further increases on Fe availability led to a drastic decrease in the growth of the cultures (Table 1). The increase of Fe at the intracellular level showed a linear dependence with the concentration of added Fe below 200 μM Fe, however concentrations between 200 and 500 μM Fe added to the medium led to a less active increase in intracellular Fe (Table 1), suggesting an intracellular control for Fe uptake. Thus, the data presented by Estévez et al. [17] under laboratory conditions suggested the possibility that excess Fe could be responsible for the decrease in *C. vulgaris* growth by inducing oxidative stress. Accordingly, when *C. vulgaris* cells were incubated with an EPR-spin trapping for lipid radicals (α-(4-pyridyl 1-oxide)-N-t-butyl nitrone, POBN), a POBN-spin adduct was observed. The spin adduct EPR spectra exhibit hyperfine splitting that were characteristic for POBN/lipid radicals, a_N = 15.56 G and a_H = 2.79 G, possibly generated from membrane lipids as a result of β-scission of lipid-alkoxyl radicals [19,20]. Quantification of lipid radical EPR signals in algal cells indicated that Fe supplementation significantly increased radical content in the membranes supplemented with the higher Fe dose, as compared to cells supplemented with 90 μM Fe (Table 1). These results indicate that lipid peroxidation was increased by Fe availability. In this context, even though an increased content of antioxidants has been detected in *C. vulgaris* cells exposed to increased Fe, the damaging potential of Fe excess in the cell did not seem to be efficiently controlled by the activity of the antioxidants [17].

Fe added (μM)	Chlorophyll content (μM)	Intracellular Fe (nmol $(10^7$ cell$)^{-1}$)	Lipid radicals (pmol $(10^7$ cell$)^{-1}$)
0	1.2	4 ± 1	nd
50	4.5*	18 ± 2*	nd
90	7.0*	28 ± 3*	6 ± 2
200	6.8*	62 ± 3*	nd
300	6.0*	65 ± 7*	nd
500	1.2	85 ± 10*	36 ± 9**

[1]Taken from [17]. nd stands for not determinated.
C. vulgaris cultures were supplemented with up to 500 μM Fe (EDTA:Fe, 2:1). Development was followed measuring chlorophyll content and each experimental value represents chlorophyll content of the cultures after 12 days of development. Intracellular Fe content as a function of the Fe addition to the incubation medium was spectrophotometrically measured. Data are expressed as means ± SE of 4-6 independent experiments, with two replicates in each experiment. Lipid radicals were detected and quantify by EPR.
*significantly different from value without Fe added, p $\leq$ 0.05. ANOVA.
**significantly different from value in the presence of 90 μM Fe added, p $\leq$ 0.05. ANOVA.

Table 1. Fe supplementation effect on *C. vulgaris* culture after 12 days of development[1]

It has been postulated that if as a result of ozone loss, UV-B flux at the surface of the earth increases, negative impacts on biological organisms will be inevitable since UV-B radiation causes a multitude of physiological and biochemical changes in photosynthetic organisms, probably related to oxidative stress [21,22]. Estévez el at. [17] exposed to 30 kJ/m^2 UV-B *C.*

vulgaris cells grown at up to 500 μM Fe. They observed that either 50 or 90 μM Fe did not alter significantly cell morphology. However, 30 kJ/m² UV-B exposure of algal cultures grown at 500 μM Fe affected cellular internal structure and there were no signs of cellular division. Exposure of *C. vulgaris* cells to 30 kJ/m² UV-B during lag phase did not significantly affect the content of lipid radicals in log phase of development under conditions of standard supplementation of Fe (90 μM) (Figure 1). This parameter was significantly increased by the addition of 500 μM Fe during development of the cultures in the absence of UV-B irradiation. Exposure of the cultures grown at 500 μM Fe to 30 kJ/m² UV-B during log phase led to a further increase in the content of lipid radicals in the membranes. In conclusion, even though exposure of *C. vulgaris* cells to UV-B under Fe standard concentration did not lead to cellular oxidative alterations, increase in Fe availability (500 μM Fe) was responsible for a substantial increase in lipid deterioration in the membranes by oxidative stress. These data strongly suggest that oxidative stress triggered by an excess content of Fe could affect cellular growth and have a negative biogeoimpact to phytoplankton when exposed to other environmental conditions.

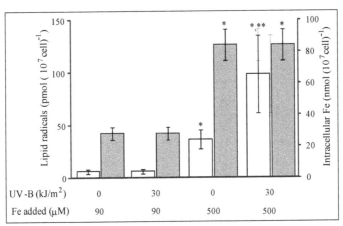

Algal cultures (2×10⁵ cells ml⁻¹) were added with Fe starting day 0 of growth, and were assayed on day 12 (log phase). Data are expressed as means ± SE of 4-6 independent experiments, with 2 replicates in each experiment.
*significantly different from values for cells grown in the presence of 90 μM Fe without exposure to UV-B, p ≤ 0.05. ANOVA.
"significantly different from values for cells added with 500 μM Fe without exposure to UV-B, p ≤ 0.05. ANOVA. (Statview SE+, v 1.03, Abacus Concepts Inc, Berkeley, CA).
Lipid radicals were detected and quantified by EPR and intracellular Fe by the use of acid solutions to digest the cells which were measured spectrophotometrically after reduction with thioglycolic acid followed by the addition of bathophenanthroline.

Figure 1. Effect of Fe addition on UV-B-dependent lipid radical (□) and intracellular Fe content (■ in algae cells. Taken from [17].

Marine animals incorporate Fe bound to inorganic particles or to organic matter during food ingestion. Further, dissolved Fe is absorbed over the respiratory surfaces and mantle tissue in filter-feeding molluscs. The extrapallial water around these tissues is constantly exchanged with the surrounding seawater. Marine invertebrates are less tolerant of metal accumulation than vertebrates and can be affected at lower metal concentrations. Bivalves are widely used as sentinel organisms in marine pollution monitoring programs, due to their sessile and filtering habits, and their ability to bioaccumulate organic pollutants and metals in their tissues [23]. The exposure of marine molluscs to metals has been shown to induce oxidative stress through the formation of reactive O_2 species (ROS) and reactive nitrogen species (RNS), leading to lipid peroxidation. Bivalves have also been used as models for the study of the effect of Fe supplementation. Viarengo et al. [24] treated the mussel *Mytilus galloprovincialis* with 300-600 µg Fe/l (as $FeCl_3$) and observed a significantly Fe accumulation in the digestive gland (DG) (190 ± 25, 394 ± 131 and 412 ± 146 µg Fe/l in 0, 300 and 600 µg Fe/l supplemented animals, respectively). The TBARS content was measured in animals treated with 600 µg Fe/l, and a significant increase was observed among control and treated mussels. Lately, Alves de Almeida et al. [25] exposed mussels from *Perna perna* species to 500 µg/l Fe (as $FeSO_4$) and it was reported that mussels exposed to Fe for 12, 24 and 72 h presented increased phospholipid hydroperoxide glutathione peroxidase (PHGPx) activity, and no differences in MDA levels. However, at 120 h of Fe exposure both, MDA and PHGPx were significantly higher than control. Such increased MDA levels agree with previous findings by Viarengo et al. [24]. The negative correlation observed between PHGPx activity and MDA levels after Fe exposure, supports an interpretation that PHGPx protects tissues from lipid peroxidation. Thus, the exposure of mussels to Fe along with a concomitant increase in •OH formation would be involved in the modulation of PHGPx activity, however the precise mechanism remains unclear. Also, exposure of mussels to 500 µg/l of Fe caused no changes in other antioxidant enzymes such as glutathione S-transferase and glutathione peroxidase. These data suggest that PHGPx have a role in the susceptibility of DG of mussels against lipid peroxidation, and that exposure to transition metals such as Fe could lead mussels to stimulate PHGPx in order to prevent lipid peroxidation. Thus, the authors postulated that the evaluation of MDA levels in parallel with antioxidant defenses, such as PHGPx, could be considered as a potential new biomarker of toxicity associated with contaminant exposure in marine organisms.

Recently, González et al. [26] investigated the oxidative effects produced by the *in vivo* Fe exposure of the bivalve *Mya arenaria*. The soft shell clams were collected on an intertidal sand flat near Bremerhaven, Germany, and the bivalves were placed in small aquaria containing 500 µM Fe (EDTA:Fe, 2:1). Exposure to 500 µM Fe in natural seawater resulted in a significant increase in DG total Fe content (Table 2). After 2 days of exposure to Fe, TBARS content showed a significant increase by approximately 3.8-fold as compared to control values. This increase was followed by a decrease to control values at treatment day 7 and afterwards TBARS concentration increased constantly until day 17 (Table 2). The LIP in DG tissue increased on day 7 of exposure to high dissolved Fe concentration. By day 9, the LIP increase was accompanied by a significant induction of the oxidative stress signals, ROS and ascorbyl radical content and correlated with the final increase of TBARS content in tissues. Once the LIP has increased, the catalytically active Fe is able to efficiently catalyze Fenton [27,28] and Haber-Weiss reactions [29,30] and consistently and drastically accelerated accumulation of TBARS. Contrary, oxidative stress effects measured on day 2 of treatment cannot be attributed to a significant increase of the LIP, since neither total Fe content nor the LIP were enhanced over the initial values in the 0 day exposure group. However, the H_2O_2 scavenging antioxidant, catalase (CAT), increased after 2 days of treatment compared to controls (day 0) but the activity went back to control level on day 7 of exposure. Catalase activity was, however, increased again on day 9 of exposure compared to controls [26]. It was postulated that the initial phase of elevated oxidative stress, occurring before significant Fe accumulation could be attributed to indirect effects under the experimental exposure conditions. Metabolic rates were not measured, but it is possible that Fe exposure triggers an initial stress response including accelerated respiration as the animals are pumping to rid themselves of the inflowing Fe enriched seawater. H_2O_2 is a good candidate for triggering cellular responses since it is a stable species [27]. H_2O_2 diffuses freely into the tissue and leads the oxidative stress, and further increases causes oxidative damage, assessed as TBARS content. H_2O_2 induced oxidative stress may have triggered the endogenous antioxidant system in such a manner that by day 7 of exposure to Fe excess the TBARS content was reduced to the starting values. Even though the superoxide dismutase (SOD) activity was not changed, induction of other protective mechanisms, such as metallothioneins, might act as effective transient control of heavy metal effects during the initial phase of exposure [24,25].

Time (days)	Total Fe content (ng Fe/mg FW)	TBARS (pmol/mg FW)	LIP (ng LIP/mg FW)
0	39 ± 4	57 ± 8	3.8 ± 0.4
2	48 ± 8	218 ± 14***	5.3 ± 1.3
7	42 ± 6	75 ± 13	7.2 ± 0.3*
9	66 ± 4**	157 ± 14***	14.2 ± 1.1*
17	106 ± 3**	226 ± 20***	10.4 ± 0.7*

²Taken from [26].
*significantly different from the value at day 0 with p < 0.05,
**p < 0.01 and
***p < 0.001. ANOVA.
Experimental bivalves were placed in small aquaria containing 13 l (1 l/animal) of natural seawater of 23-26‰ at 10°C, and 500 μM Fe (Fe:EDTA, 1:2).

Table 2. Fe supplementation effect on lipid peroxidation in *Mya arenaria*²

Other studies evaluate the impact of nutritional Fe on Fe level and concentrations of MDA in tissues. Baker et al. [31] analyzed the Fe in the diet of the African catfish, *Clurims gariepinus*. This fish model is of particular relevance when considering that *C. gariqinus* is typically cultured in earth-ponds, and these may be high in dissolved Fe content. Additionally, catfish may consume mud-burrowing organisms to supplement their diet, with incidental associated silt consumption, and therefore further metal loading. After 5 weeks of feeding the animals with a diet supplemented with Fe (6354.4 mg Fe/kg), the total

Fe content was measured in muscle, liver and blood-plasma and no significant differences with control animals were found, suggesting the possibility of efficient regulation of Fe status by the fish. MDA determination in tissues revealed that there was significantly more MDA in livers and hearts of fish fed high Fe diets than in controls, and no significant difference was found in skeletal-muscle. Values of MDA concentration were higher in Fe-stressed liver tissue comparative to other tissues, possible because hepatic tissue is lipid-rich making the liver a target organ for lipid peroxidation. The relative lack of response in skeletal muscle may have resulted from decreased abundance of polyunsaturated fatty acids within this tissue, and these findings are consistent with those of Desjardins et al. [32].

All together these data show that Fe in aquatic ecosystems could be a major stressor having a main role in lipid peroxidation not only in unicellular species, such as algae, but also in higher organisms, such as invertebrates and vertebrates. These kind on analyses should be performed before consider ecological strategies which may involved Fe fertilization in seawater [33-35], to increase primary production in the oceans as an answer to global temperature increments. These actions may drastically modify marine communities in ocean layers triggering oxidative reactions, which should be properly considered due to the fact that Fe may be profitable or unfavorable, depending of its usefulness as a micronutrient or as a catalyzer of free radical reactions.

3. Fe overload in soybean seeds

Plants have developed several mechanisms to maintain fairly constant internal concentrations of mineral nutrients over a wide range of external concentrations. To avoid Fe-dependent oxidative cellular damage, Fe^{2+} is either incorporated into the mineral core of Ft [36] which is located exclusively in the plastids [37] or reoxidized by O_2 and chelated by organic acids [38]. Bienfait et al. [39] reported that plants grown on Fe-EDTA formed a substantial pool of free space Fe in the roots and that Fe could be mobilized upon Fe-free growth in order to be transferred to the leaves. During growth in water culture at pH 5 to 6, a free space pool of 500 to 1000 nmol/g FW was formed in roots of bean grown in the presence of Fe-EDTA 20 μM and a pool of 20 to 50 nmol/g FW in roots without Fe supplementation. Like Ft in the cell, the free space Fe^{3+} precipitate is not only an immobile result of a defensive action against an excessive Fe supply; the plant may also use it as storage form of Fe that can be mobilized [39]. Even more, Caro and Puntarulo [40] indicated that O_2 radical generation depends on total Fe content, however it could mostly reflect Fe content in the free space. In soybean, Fe^{3+} reduction is an obligatory step in Fe uptake, and this is probably true for all strategy I plants [41]. Both total Fe content and the *in vitro* rate of Fe reduction were higher in roots grown in the presence of exogenously added Fe (up to 500 μM) than in roots grown in absence of supplemented Fe (Table 3). However, no visual differences (e.g. evidence of damage) between any of the roots or growth (assessed as the fresh weight of the roots, 0.21 ± 0.01 g/root) have been observed at the studied range of Fe supplementation. Total Fe content in soybean roots exposed to 50 and 500 μM Fe-EDTA, was higher than in roots grown in absence of supplemented Fe

(Table 3) and lipid oxidation, assessed as the content of TBARS, were not significantly affected by Fe supplementation up to 500 μM, to the incubation medium (Table 3). However, Fe supplementation to the roots did affect α-tocopherol content that was significantly decreased in the homogenates and the microsomes isolated from roots supplemented with Fe, as compared with values in roots developed in absence of Fe [40]. These data suggest that *in vivo* Fe supplementation could increase O_2 radical generation in soybean roots that was adequately control.

	No added Fe	500 μM added Fe	Ref
Soybean roots			
Total Fe content (μg/g FW)	0.07 ± 0.01	0.15 ± 0.02*	[40]
Fe-EDTA reduction rate (nmol/min/mg prot)	1.4 ± 0.1	3.1 ± 0.6*	[40]
TBARS (nmol MDA eq/mg)	5.7 ± 0.7	5.7 ± 0.7	[40]
Soybean embryonic axes			
Total Fe content (nmol/mg DW)	1.3 ± 0.2	3.9 ± 0.8*	[42]
Fe-EDTA reduction rate (nmol/min/mg DW)	15 ± 1	22 ± 2	[42]
Ft (μg Ft/ g DW)	34 ± 11	27 ± 10	[42]
Ft Fe content (Fe atoms/molec Ft)	1054 ± 111	494 ± 103*	[42]
LIP (pmol/mg DW)	50 ± 10	310 ± 50*	[42]
TBARS (nmol MDA eq/mg)	0.4 ± 0.1	0.3 ± 0.1	[75]

*significantly different from values without Fe addition, p ≤ 0.05. ANOVA.

Table 3. Fe supplementation effects in soybean after 24 h of incubation

Robello et al. [42] reported that total Fe content in soybean embryonic axes exposed to 500 uM Fe-EDTA was higher than in axes grown in absence of supplemented Fe after 24 h of incubation. However, neither Fe reduction rate nor growth assessed, either as the fresh weight or the dry weight of the embryonic axes, were significantly affected by Fe supplementation to the incubation medium. Membrane integrity was no affected by the supplementation with 50 and 500 μM Fe:EDTA (1:2) since electrolyte leakage at 24 h and 48 h of imbibition was not significantly different from electrolyte leakage found in non-supplemented Fe axes (15.3 ± 0.7 and 8.0 ± 0.3%, after 24 h and 48 h of incubation with 500 μM Fe, as compared to 12.4 ± 0.4 and 8.6 ± 0.6%, after 24 h and 48 h of incubation in the absence of added Fe, respectively). Moreover, as it was previously reported in soybean roots [43], Fe accumulation was not followed by Ft accumulation in soybean embryonic axes upon growth. Without any significant change in the content of Ft in the embryonic axes incubated for 24 h upon Fe supplementation, a 53% decrease in the Fe content per molecule of Ft was observed in the presence of 500 μM Fe (Table 3). These data differed from previous observations showing Fe induction of Ft synthesis and accumulation in soybean [44], however, the nature of the model employed by Lescure et al. [44], cells in suspension grown heterotropically, could alter the kinetic of the response. In this regard, it should not be discarded that a transient increase in Ft content could occur under these experimental conditions before 24 h of imbibition. The observed rapid decrease in Fe content per molecule of Ft, as compared to non-added Fe conditions, could reflect an early loosing of Ft molecules altered by free radicals, or a reduction of its capacity of binding Fe, or both. The increase in the protein sensitivity to proteases would lead to an early degradation, as compared to axes grown in a non-added Fe medium. The increased rate of ROS generation could be due to the significant increase in the LIP under conditions of Fe supplementation. However, it is important to point out that the substantial increase in the total Fe content in axes grown in the presence of 500 μM Fe for 24 h, as compared to seeds grown in non-added Fe medium, could not be allocated as the measured increase in the LIP that would represent only the 10% of the increase in the total Fe content. Besides the LIP critical importance as initiator of free radical reactions and the decisive requirement of keeping Fe concentration as low as possible to minimize cellular deterioration, the role of other soluble and insoluble Fe-storage proteins, the formation and contribution of Fe-nitrosyl complexes, glutathione, nitric oxide, etc. should be considered among other non-protein agents, as possible candidates to handle Fe transport and storage under stress conditions since TBARS content was not significantly affected in Fe overloaded soybean embryonic axes (Table 3). Beside the apoplastic space [45], Lanquar et al. [46] identified the vacuole as a major compartment for Fe storage in plant seeds and showed that retrieval of the Fe stored in vacuoles is an essential step for successful germination in a wide range of environments.

On the other hand, recently Simontacchi et al. [47] summarized assays performed to characterize lipid radical-dependent oxidation in photosynthetic organisms where EPR was successfully employed to evaluate not only lipidperoxidation but also to analyze the relative scavenging capacity of plant extracts, the effects of both, natural environmental challenges and oxidative stress situations, in several model and biological systems. Further studies should be oriented in this direction to explore the critical effect of Fe overload on radical-dependent pathways that play a major role in plant metabolism.

4. Fe overload in mammals

Fe overload in mammals has been often associated with injury, fibrosis, and cirrhosis in the liver followed by cardiac disease, endocrine abnormalities, arthropathy, osteoporosis and skin pigmentation [48]. Several mechanisms has been proposed whereby excess hepatic Fe causes cellular injury, but Fe-induced peroxidative injury to phospholipids of organelle membranes is a potential unifying mechanisms underlying the major theories of cellular injury in Fe overload [49]. With progressively increasing Fe deposition, the capacity to maintain Fe in storage forms is exceeded resulting in a transient increase in the hepatic LIP [50]. Moreover, Fe-catalyzed generation of ROS has been implicated in the pathogenesis of many disorders including atherosclerosis [51,52], cancer [53], ischaemia reperfusion injury [54,55] besides in Fe overload [56], such as haemochromatosis [57].

Several experimental models of Fe overload have been developed. In the dietary model used by Dabbagh et al. [58] rats were fed for 10 weeks a chow diet enriched with 3% (w/w) reduced pentacarbonyl Fe (a 99%, w/w, pure form of elemental Fe). Dietary Fe overload resulted in significant increases in hepatic Fe levels; with no difference in Fe content in serum (Table 4). Lipid peroxidation was assessed by measuring TBARS and F2-isoprostanes. The latter are a series of prostaglandin-F2-like compounds derived from the free-radical-catalyzed, non-enzymatic peroxidation of arachidonic acid [59] and the *in vivo* levels of F2-isoprostanes have been shown to increase dramatically in acute hepatotoxicity [60]. Direct evidence for moderately increased lipid peroxidation products in liver was reported after dietary Fe overload. In addition to hepatic oxidative damage, Fe overload also caused changes in the plasma lipid profile. These data suggest that in this rat model of Fe overload, oxidative stress is associated with depletion of endogenous antioxidants in plasma and liver, and although no conclusive evidence for lipid peroxidation in plasma was found, hepatic F2-isoprostane levels were significantly increased in treated rats.

Experimental Fe overload in rats using dietary supplementation with carbonyl Fe is a well established model, where Fe deposition results mainly in the hepatocytes in a periportal distribution, as observed in idiopathic hemochromatosis [48]. Galleano and Puntarulo [61] used the dietary carbonyl-Fe model carried out on male Wistar rats that were fed during 6 weeks with either a) control chow diet, or b) control chow diet supplemented with 2.5% (w/w) carbonyl-Fe. Both, Fe and TBARS content, were increased in liver (Table 4). However, mild dietary Fe overload increased Fe content in plasma but did not lead to a significant increase in TBARS probably because Fe content after dietary Fe supplementation was increased less dramatically in plasma than in liver (88% and 15-fold, respectively), suggesting that plasma mechanisms for sequestering catalytically active Fe were fully operative (Table 4). Under these conditions, TBARS content in plasma does not seem to be a good indicator of oxidative stress conditions in the liver, and more sensitive techniques should be used in plasma to assess Fe-dependent oxidative stress.

Cockell et al. [62] used sucrose-based modified AIN-93G diets formulated to differ in Fe (35 mg/kg and 1500 mg/kg for control and Fe overloaded diets). Weanling male Long-Evans rats were fed these diets for 4 weeks and killed. Fe content was measured in plasma and liver. No differences in plasma between control and treated groups were found, meanwhile a significantly increase in liver between control and treated groups was observed. Since TBARS content in livers was significantly increased in Fe overloaded animals, hepatic Fe concentrations in this study were correlated positively with increases in TBARS. However, Fischer et al. [63] showed that Fe overloaded diets did not significantly alter other oxidative stress indices, such as DNA double-strand breaks or NF-κB activation despite observed increases in hepatic lipid peroxidation.

	Fe content		TBARS	
	control	Fe-overload	control	Fe-overload
Pentacarbonyl Fe, diet 3% (w/w)				
Liver [58]	104 ± 15(a)	1391 ± 242*(a)	-	-
Plasma [58]	134 ± 55(c)	124 ± 46(c)	nd	nd
Carbonyl Fe, diet 2.5% (w/w)				
Liver [61]	69 ± 16(a)	1091 ± 178*(a)	0.45 ± 0.05(b)	0.58 ± 0.01*(b)
Plasma [61]	179 ± 43(c)	336 ± 57*(c)	0.6 ± 0.1(d)	0.6 ± 0.2(d)
Sucrose-basemodified AIN-93G, diet 1500 mg/kg				
Liver[62]	218 ± 46(e)	895 ± 376**(e)	0.54 ± 0.07(b)	0.78 ± 0.19**(b)
Plasma [62]	2.72 ± 1.74(i)	3.82 ± 1.21(i)	-	-
Fe-dextran, ip 500 mg/kg				
Liver [68]	257 ± 11(e)	1837 ± 205*(e)	40 ± 1(f)	110 ± 30*(f)
Plasma [70]	126 ± 20(g)	1538 ± 158*(g)	0.7 ± 0.1(h)	2.7 ± 0.1*(h)
Kidney [49]	14 ± 3(e)	113 ± 15*(e)	29 ± 2(f)	37 ± 3*(f)

Letters indicate the units for each parameter as follows: (a) μg Fe/g FW; (b) nmol/mg prot; (c) μg/dl; (d) nmol/ml; (e) μg Fe/g DW; (f) pmol/min/mg prot; (g) μg Fe/dl; (h) nmol/l; (i) mg/l.
*significantly different from control values p < 0.01,
**p < 0.001, ANOVA.
nd stands for not-detectable.

Table 4. Fe effects in different organs and plasma employing several models of Fe overload

Fe-dextran treatment seems as a good model for the study of Fe toxicity resembling the pathological and clinical consequences of acute Fe overload in humans [48]. Fe supplied as Fe-dextran, is initially taken up by Kupffer cells, and when their storage capacity is exceeded the metal is accumulated by parenchymal cells producing a mild Fe overload. The increased Fe content alters the Kupffer cell functional status by inducing a progressive increase in macrophage-dependent respiration at earlier times after treatment. The effect is sensitive to macrophage inactivation by GdCl₃ pretreatment, decreases the respiratory response of the Kupffer cell to particle stimulation, plays a role in the development of liver injury, and seems to condition the impairment of hepatic respiration observed at later times after Fe overload [64]. Other pathological situations that increase oxidative conditions in the cell, could enhance Fe-dependent damage. As an example, hyperthyrodism increases the susceptibility of the liver to the toxic effects of Fe, which seems to be related to the development of a severe oxidative stress status in the tissue, thus contributing to the concomitant liver injury and impairment of Kupffer cell phagocytosis and particle-induced respiratory burst activity [65]. It was also shown that acute Fe overload was responsible for oxidative stress in rat testes with a concurrent decrease of antioxidant content [66,67]. The oxidative stress has been developed using Fe-dextran intra peritoneal (ip) administration as 500 mg/kg body weight and killed after 20 h.

Spontaneous organ chemiluminescence (CL) reflects the rate of lipid peroxidation reactions through the detection of the steady-state level of excited species and is considered to be an useful technique to evaluate oxidative stress *in vivo*. Galleano and Puntarulo [68] reported an association between Fe content and light emission in rats exposed to Fe-dextran after 2-6 h. Presumably, with progressively increasing Fe deposition, the capacity of maintaining Fe in storage forms is exceeded resulting in a transient increase in the hepatic LIP. However, at longer times (20 h) the significant increase in cytosolic Fe is limited, and CL goes back to control values. Moreover, cytochrome P_{450} inactivation is an early event and precedes other enzyme inactivation [68]. Data included in Table 4 show that liver Fe content was increased by 7-fold after 8 h of Fe-dextran administration, and TBARS generation rate was enhanced by 3-fold (6 h after ip) suggesting that liver is deeply affected by acute Fe-overload.

Mammalian red blood cells are particularly susceptible to oxidative damage because (i) being an O_2 carrier, they are exposed uninterruptedly to high O_2 tension, (ii) they have no capacity to repair their damaged components, and (iii) the haemoglobin is susceptible to autoxidation and their membrane components to lipid peroxidation. Red blood cells, however, are protected by a variety of antioxidant systems which are capable of preventing most of the adverse effects of oxidative stress, under normal conditions [69]. Galleano and Puntarulo [70] reported, employing the ip Fe-dextran model of Fe overload, that 20 h after Fe-dextran injection Fe concentration in plasma of treated rats showed approximately 12-fold increase, and TBARS content in plasma showed a 285% increase as compared to control values (Table 4). On the other hand, *in vitro* studies showed that Fe can stimulate the peroxidation of erythrocytes membrane lipids. Since red blood cells from Fe overloaded rats are continuously being exposed to an increase Fe content, no differences in TBARS content were detected in red blood cells from control rats as compared to erythrocytes from Fe overloaded rats, suggesting high resistance to oxidative stress of these cells.

Galleano et al. [71] also employed this model to comparatively studying Fe overload in kidney. Fe content in whole kidney was 8-fold increased (Table 4), and 5-fold increased in kidney mitochondria (16 ± 5 to 78 ± 1 nmol/mg prot for control and treated animals, respectively). Even thought TBARS content showed no significant differences after Fe administration, in Fe-treated rats TBARS production rate by kidney homogenates was higher in treated animals than in kidneys from control rats (Table 4). The authors suggested that Fe-dextran treatment does not affect kidney integrity, even though increases in lipid peroxidation rate occurs. α-tocopherol, one of the most efficient antioxidant in the hydrophobic phase, appeared to be effective in controlling Fe-dextran dependent damage in kidney.

Brain tissue is thought to be very sensitive to oxidative stress. Neurons are enriched in mitochondria and possess a very high aerobic metabolism, which makes these tissues susceptible to ROS-dependent damage than other organs. Moreover, low levels of some antioxidant enzymes, high contents of polyunsaturated fatty acids in brain membranes, and high Fe content may combine their effects to make the brain a preferential target for oxidative stress-related degeneration [72]. Maaroufi et al. [73] developed a chronic Fe overload model consisting in a daily 3 mg Fe/kg administrated in adult rats during 5 days. These treatments resulted, 16 days after treatment, in a significant Fe accumulation in the hippocampus, cerebellum, and basal ganglia. Lately, Maaroufi et al. [74] studied rats which received daily one ip injection of 3 mg FeSO$_4$/kg dissolved in sodium chloride 0.9% (or vehicle) during 21 consecutive days, and this accumulation was correlated to behavioral deficits. No increase levels of the TBARS content in different brain structures were observed in any brain region investigated. This observation suggested that chronic Fe administration had induced adaptive responses involving stimulation of the antioxidant defenses since, both SOD and CAT activities, were increased after treatment.

Thus, different forms and quantities of Fe administrated to rats, supplemented either as diets or ip, lead to an increase in Fe content in several tissues and plasma. This Fe increase seems to be associated with an increase in lipid peroxidation. The underlying mechanisms of tissue damage are unclear, but they probably depend on the Fe administration protocol. Even though lipid damage was observed in many cases after Fe overload, antioxidant capacity seems to play a crucial role in controlling the impairment mechanisms.

5. Concluding remarks

Fe metabolism is very complex since Fe is both, an essential element and a toxic compound that has to be carefully kept under a regulated concentration in a living cell. Toxic Fe activity is due to its ability of catalyzing free radical reactions. The most efficient Fe fraction to act as a free radical promoter is that forming the LIP. LIP content is the resultant of multiple dynamic equilibrium between the Fe incorporated to the cell, utilized and intracellularly stored. We have briefly reviewed the role of Fe on the oxidative damage to lipid membranes employing both *in vitro* and *in vivo* models of Fe overload in several biological systems. Much progress has still to be made in order to understand the nature and function of the LIP, the mechanisms of the Fe-catalyzed reactions *in vivo*, the contribution of Fe to oxidative stress and disease, and the development of appropriate chemotherapeutic strategies. Thus, alterations in Fe metabolism should be carefully analyzed before evaluating cellular responses to either damaging agents or xenobiotics of biomedical or ecological impact since Fe is a double-faced element that can be either good or bad to the cell, depending on whether it serves as a micronutrient or as a catalyst of free radical reactions.

Since a tight metabolic organization is required to successfully face oxidative external conditions in invertebrates, anthropogenic contamination with Fe could be toxic for animals that are adapted to their natural environment. As it could be understood from the data presented here, it is strongly suggested that natural habitats should be strictly preserved even though absolute Fe content did not seem to reach critical values to avoid cellular deterioration.

Mobilization of Fe stored in plant seeds is an essential step for germination in a wide range of environments. The analysis of these aspects would provide information that could be the key to understand Fe nutrition in plants, and will allow the designing and engineering of crop plants requiring minimal fertilizer input, contributing to a more ecological agricultural practice under optimal and sub-optimal environmental conditions avoiding reaching Fe overload conditions that would jeopardize successful plant development.

Moreover, therapeutic strategies should be designed to chelate either Fe from the LIP or Fe loosely bound to Ft to avoid Fe-related oxidative damage. Focus in chemical-related aspects of the Fe-chelator complexes should help to fulfill the new drugs designing expectances to control Fe toxicity in humans that through promoting lipid peroxidation could severely affect human health.

Author details

Paula M. González, Natacha E. Piloni and Susana Puntarulo[*]
Physical Chemistry-PRALIB, School of Pharmacy and Biochemistry, University of Buenos Aires, Junín, Buenos Aires, Argentina

Acknowledgement

This study was supported by grants from the University of Buenos Aires and CONICET. S.P. is career investigator from CONICET, and P.M.G. and N.P. are fellows from CONICET.

6. References

[1] Harrison PM, Arosio P (1996) The ferritins: molecular properties, iron storage function and cellular regulation. Biochim. Biophys. Acta 1275: 161-203.

[2] Crichton RR, Ward RJ (1992) Iron metabolism. New perspectives in view. Biochemistry 31: 11255-11264.

[3] Laulhère JP, Lescure AM, Briat JF (1988) Purification and characterization of ferritins from maize, pea, and soyabean seeds. J. Biol. Chem. 263: 10289-10294.

[4] Hyde BB, Hodge AJ, Kahn A, Birnstiel ML (1963) Studies on phytoferritin: I. identification and localization. J. Ultrastruct. Res. 9: 248-258.

[5] Dexter DT, Wells FR, Agid F, Agid Y, Lees AJ, Jenner P, Marsden D (1987) Increased nigral iron content in postmortem parkinsonian brain. Lancet 8569: 1219-1220.

[6] Ponka P (2004) Hereditary causes of disturbed iron homeostasis in the central nervous system. Ann. N.Y. Acad. Sci. 1012: 267-281.

[7] Chaston TB, Richardson DR (2003) Iron chelators for the treatment of iron overload disease: relationship between structure, redox activity, and toxicity. Am. J. Hematol. 73: 200-10.

[8] Limdi JK, Crampton JR (2004) Hereditary haemochromatosis. QJM 97: 315-324.

[9] Tarpey MM, Wink DA, Grisham MB (2004) Methods for detection of reactive metabolites of oxygen and nitrogen: in vitro and in vivo considerations. Am. J. Physiol. Regul. Integr. Comp. Physiol. 286R: 431-444.

[10] Wedepohl KH (1995) The composition of the continental crust. Geochim. Cosmochim. Acta 59 (7): 1217-1232.

* Corresponding Author

[11] Rue EL, Bruland KW (1995) Complexation of iron(III) by natural organic ligands in the Central North Pacific as determined by a new competitive ligand equilibration/adsorptive cathodic stripping voltammetric method. Mar Cehm. 50: 117-138.

[12] Watson AJ (2001) Iron limitation in the oceans. In: Turner DR, Hunter KA, editors. The biogeochemistry of iron in seawater. New York: Wiley and Sons. pp 85-121.

[13] Bruland, KW, Donat JR, Hutchins DA (1991) Interactive influences of bioactive metals on biological production in oceanic waters. Limnol. Oceanogr. 36: 1555-1577.

[14] Martin JH, Fitzwater SE, Gordon RM (1990) Iron deficiency limits phytoplankton growth in Antarctic waters. Global Biogeochem. Cycles 4: 5-12.

[15] Martin JH, Fitzwater SE, Gordon RM, Hunter CN, Tanner SJ (1993) Iron, primary productivity and carbon nitrogen flux studies during the JGOFS North Atlantic Bloom Experiment. Deep-Sea Res. 40: 115-134.

[16] Brand LE, Sunda WG, Guillard RRL (1983) Limitation of phytoplankton reproductive rates by zinc, manganese and iron. Limnol. Oceanogr. 28: 1182-1198.

[17] Estévez MS, Malanga G, Puntarulo S (2001) Iron-dependent oxidative stress in *Chlorella vulgaris*. Plant Sci. 161: 9-17.

[18] Kolber ZS, Barber R, Coale KH, Fitzwater SE, Greene RM, Johnson KS, Lindley S, Falkowski PG (1994) Iron limitation of phytoplankton photosynthesis in the equatorial Pacific Ocean. Nature 371: 145-149.

[19] North JA, Specto AA, Buettner GR (1992) Detection of lipid radicals by electron paramagnetic resonance spin trapping using intact cells enriched with polyunsaturared fatty acids. J. Biol. Chem. 267: 5743-5746.

[20] Jurkiewicz BA, Buettner GR (1994) Ultraviolet light-induced free radical formation in skin: an electron paramagnetic resonance study. Photochem. Photobiol. 59: 1-4.

[21] Malanga G, Puntarulo S (1995) Oxidative stress and antioxidant content in *Chlorella vulgaris* after exposure to ultraviolet-B radiation. Plant Physiol. 94: 672-679.

[22] Kozak RG, Malanga G, Caro A, Puntarulo S (1997) Ascorbate free radical content in photosynthetic organisms after exposure to ultraviolet-B. Recent Res. Devel. Plant Physiol. 1: 233-239.

[23] Goldberg ED (1975) The mussel watch: a first step in global marine monitoring. Mar. Poll. Bull. 6: 111-132.

[24] Viarengo A, Burlando B, Cavaletto M, Marchi B, Ponzano E, Blasco J (1999) Role of metallothionein agaist oxidative stress in the mussel *Mytilus galloprovincialis*. Am. J. Physiol. Regul. Integr. Comp. Physiol. 277: 1612-1619.

[25] Alves de Almeida E, Dias Bainy AC, de Melo Loureiro AP, Martinez GR, Miyamoto S, Onuki J, Barbosa LF, Machado Garcia CC, Manso Prado F, Ronsein GE, Sigolo CA, Barbosa Brochini C, Gracioso Martins AM, Gennari de Medeirosa MH, Di Mascio P (2007) Oxidative stress in *Perna perna* and other bivalves as indicators of environmental stress in the Brazilian marine environment: Antioxidants, lipid peroxidation and DNA damage. Comp. Biochem. Physiol. 146(4)A: 588-600.

[26] González PM, Abele D, Puntarulo S (2010) Exposure to excess of iron in vivo affects oxidative status in the bivalve *Mya arenaria*. Comp. Biochem. Physiol. C 152: 167-174.

[27] Boveris A (1998) Biochemistry of free radicals: from electrons to tissues. Medicina 54: 350-356.

[28] Pierre JL, Fontecave M (1999) Iron and activated oxygen species in biology: the basic chemistry. Biometals 12: 195-199.

[29] Rauen U, Petrat F, Tongju L, De Groot H (2000) Hypothermia injury/cold induced apoptosis -evidence an increase of chelatable iron causing oxidative injury in spite of low O_2/H_2O_2 formation. FASEB J. 14: 1953-1964.

[30] Livingstone DR (2001) Contaminant-stimulated reactive oxygen species production and oxidative damage in aquatic organisms. Mar. Pollut. Bull. 42: 656-666.

[31] Baker RTM, Martin P, Davies SJ (1997) Ingestion of sub-lethal levels of iron sulphate by African catfish affects growth and tissue lipid peroxidation. Aquat. Toxicol. 40: 51-61.

[32] Desjardins LM, Hicks BD, Hilton JW (1987) Iron catalysed oxidation of trout diets and its elect on the growth and physiological response of rainbow trout. Fish Physiol. Biochem. 3: 173-182.

[33] Martin JH, Coale KH, Johnson KS, Fitzwater SE, Gordon RM, Tanner SJ, Hunter CN, Elrod VA, Nowicki JL, Coley TL, Barber RT, Lindley S, Watson AJ, Van Scoy K, Law CS, Liddicoat MI, Ling R, Stanton T, Stockel J, Collins C, Anderson A, Bidigare R, Ondrusek M, Latasa M, Millero FJ, Lee K, Yao W, Zhang JZ, Friederich G, Sakamoto C, Chavez F, Buck K, Kolber Z, Greene R, Falkowski P, Chisholm SW, Hoge F, Swift R, Yungel J, Turner S, Nightingale P, Hatton A, Liss P, Tindale NW (1994) Testing the iron hypothesis in ecosystems of the equatorial Pacific Ocean. Nature 371: 123-129.

[34] Gervais F, Riebesell U, Gorbunov MY (2002) Changes in primary productivity and chlorophyll a in response to iron fertilization in the Southern Polar Frontal Zone. Limnol. Oceanogr. 47(5): 1324-1335.

[35] Blain S, Quéguiner B, Armand L, Belviso S, Bombled B, Bopp L, Bowie A, Brunet C, Brussaard C, Carlotti F, Christaki U, Corbière A, Durand I, Ebersbach F, Fuda J-L, Garcia N, Gerringa L, Griffiths B, Guigue C, Guillerm C, Jacquet S, Jeandel C, Laan P, Lefèvre D, Lo Monaco C, Malits A, Mosseri J, Obernosterer I, Park Y-H, Picheral M, Pondaven P, Remenyi T, Sandroni V, Sarthou G, Savoye N, Scouarnec L, Souhaut M, Thuiller D, Timmermans K, Trull T, Uitz J, van Beek P, Veldhuis M, Vincent D, Viollier E, Vong L, Wagener T (2007) Effect of natural iron fertilization on carbon sequestration in the Southern Ocean. Nature 446: 1070-1074.

[36] Laulhère JP, Laboure AM, Briat JF (1990) Photoreduction and incorporation of iron into ferritins. Biochem. J. 269: 79-84.

[37] Seckbach, J. (1968) Studies on the deposition of plant ferritin as influenced by iron supply to iron-deficient beans. J. Ultrastruct. Res. 22: 413-423.

[38] Cataldo DA, McFadden KM, Garland TR, Wildung RE (1988) Organic constituents and complexation of nickel (II), iron (III), cadmium (II), and plutonium (IV) in soybean xylem exudates. Plant Physiol. 86: 734-739.

[39] Bienfait HF, Van Den Briel W, Mesland-Mul NT (1985) Free space iron pools in roots. Generation and mobilization. Plant Physiol. 78: 596-600.

[40] Caro A, Puntarulo S (1996) Effect of in vivo iron supplementation on oxygen radical production by soybean roots. Biochim. Biophys. Acta 1291: 245-251.

[41] Schaedle M, Basshamb JA (1977) Chloroplast Glutathione Reductase. Plant Physiol. 59(5): 1011-1012.

[42] Robello E, Galatro A, Puntarulo S (2007) Iron role in oxidative metabolism of soybean axes upon growth. Effect of iron overload. Plant Sci. 172: 939-947.

[43] Lobréaux S, Briat JF (1991) Ferritin accumulation and degradation in different organs of pea (*Pisum sativum*) during development. Biochem. J. 274: 601-606.

[44] Lescure AM, Massenet O, Briat JF (1990) Purification and characterization of an iron induced ferritin from soybean (*Glycine max*) cell suspensions. Biochem. J. 272: 147-150.

[45] Briat JF, Lobréaux S (1997) Iron transport and storage in plants. Trends Plant Sci. 2: 187-193.

[46] Lanquar V, Lelièvre F, Bolte S, Hamès C, Alcon C, Neumann D, Vansuyt G, Curie C, Schröder A, Krämer U, Barbie-Brygoo H, Thomine S (2005) Mobilization of vacuolar iron by AtNRAMP3 and AtNRAMP4 is essential for seed germination on low iron. EMBO J. 24: 4041-4051.

[47] Simontacchi M, Buet A, Puntarulo S (2011) The use of electron paramagnetic resonance (EPR) in the study of oxidative damage to lipids in plants. In: Catalá A editor. Lipid Peroxidation: Biological Implications. Kerala, India: Res. Signpost Transworld Res. Network. pp 141-160.

[48] Puntarulo S (2005) Iron, oxidative stress and human health. Molec. Asp. Med. 26: 299-312.

[49] Galleano M, Puntarulo S (1994) Mild iron overload effect on rat liver nuclei. Toxicology 93: 125-134.

[50] Galleano M, Simontacchi M, Puntarulo S (2004) Nitric oxide and iron. Effect of iron overload on nitric oxide production in endotoxemia. Molec. Asp. Med. 25: 141-154.

[51] Heinecke JW, Rosen H, Chait A (1984) Iron and copper promote modification of low density lipoprotein by human arterial smooth muscle cells in culture. J. Clin. Invest. 74: 1890-1894.

[52] Salonen JT, Nyyssonen K, Korpela H, Tuomilehto J, Seppanen R and Salonen R (1992) High stored iron levels are associated with excess risk of myocardial infarction in eastern Finnish men. Circulation 86: 803-811.

[53] Loeb LA, James EA, Waltersdorph AM, Klebanoff SJ (1988) Mutagenesis by the autoxidation of iron with isolated DNA. Proc. Natl. Acad. Sci. U.S.A. 85: 3918-3922.

[54] Aust SD, White BC (1985) Iron chelation prevents tissue injury following ischemia. Free Radic. Biol. Med. 1: 1-17.

[55] Katoh S, Toyama J, Kodama I, Akita T, Abe T (1992) Deferoxamine, an iron chelator, reduces myocardial injury and free radical generation in isolated neonatal rabbit hearts subjected to global ischaemia-reperfusion. J Mol Cell Cardiol. 24(11): 1267-1275.

[56] Burkitt M J and Mason R P (1991) Direct evidence for in vivo hydroxyl-radical generation in experimental iron overload: an ESR spin-trapping investigation. Proc. Natl. Acad. Sci. U.S.A. 88: 8440-8444.

[57] Bacon BR, Britton RS (1990) The pathology of hepatic iron overload: a free radical mediated process? Hepatology 11: 127-137.

[58] Dabbagh AJ, Mannion T, Lynch SM, Frei B (1994) The effect of iron overload on rat plasma and liver oxidant status in vivo. Biochem. J. 300: 799-803.

[59] Morrow JD, Harris TM, Roberts LJ (1990) Noncyclooxygenase oxidative formation of a series of novel prostaglandins: analytical ramifications for measurement of eicosanoids. Anal. Biochem. 184: 1-10.

[60] Morrow JD, Awad JA, Kato T, Takahashi K, Badr KF, Roberts LJ II , Burk RF (1992) Formation of novel non-cyclooxygenase-derived prostanoids (F2-isoprostanes) in carbon tetrachloride hepatotoxicity. An animal model of lipid peroxidation. J. Clin. Invest. 90: 2502-2507.

[61] Galleano M, Puntarulo S (1997) Dietary alpha-tocopherol supplementation on antioxidant defenses after in vivo iron overload in rats. Toxicology 124(1): 73-81.

[62] Cockell KA, Wotherspoon ATL, Belonje B, Fritz ME, Madère R, Hidiroglou N, Plouffe LJ, Nimal Ratnayake WM, Kubow S (2005) Limited effects of combined dietary copper deficiency/iron overload on oxidative stress parameters in rat liver and plasma. J. Nutr. Biochem. 16(12): 750-756.

[63] Fischer JG, Glauert HP, Yin T, Sweeney-Reeves ML, Larmonier N, Black MC (2002) Moderate iron overload enhances lipid peroxidation in livers of rats, but does not affect NF-kappaB activation induced by the peroxisome proliferator, Wy-14,643. J. Nutr. 132: 2525-2531.

[64] Tapia G, Troncoso P, Galleano M, Fernandez V, Puntarulo S, Videla LA (1998) Time course study of the influence of acute iron overload on Kupffer cell functioning and hepatotoxicity assessed in the isolated perfused rat liver. Hepatology 27: 1311-1316.

[65] Boisier X, Schon M, Sepulveda A, Cornejo P, Bosco C, Carrion Y, Galleano M, Tapia G, Puntarulo S, Fernandez V, Videla LA (1999) Derangement of Kuppfer cell functioning and hepatotoxicity in hyperthyroid rats subjected to acute iron overload. Redox Rep. 4: 243-250.

[66] Lucesoli F, Fraga CG (1995) Oxidative damage to lipids and DNA concurrent with decrease of antioxidants in rat testes after acute iron intoxication. Arch. Biochem. Biophys. 316: 567-571.

[67] Lucesoli F, Caliguri M, Roberti MF, Perazzo JC, Fraga CG (1999) Dose-dependent increase of oxidative damage in the testes of rats subjected to acute iron overload. Arch. Biochem. Biophys. 372: 37-43.

[68] Galleano M, Puntarulo S (1992) Hepatic chemiluminiscence and lipid peroxidation in mild iron overload. Toxicol. 76: 27-38.

[69] Chow CK (1992) Oxidative damage in the red cells of vitamin E-deficient rats. Free Radic. Res. Commun. 16: 247-258.

[70] Galleano M, Puntarulo S (1995) Role of antioxidants on the erythrocytes resistance to lipid peroxidation after acute iron overload in rats. Biochim. Biophys. Acta 1271: 321-326.

[71] Galleano M, Farre SM, Turrens JF, Puntarulo S (1994) Resistance of rat kidney mitochondrial membranes to oxidation induced by acute iron overload. Toxicology 88:141-149.

[72] Halliwell B (2006) Oxidative stress and neurodegeneration: where are we now? J. Neurochem. 97(6): 1634-1658.

[73] Maaroufi K, Ammari M, Jeljeli M, Roy V, Sakly M, Abdelmelek H (2009) Impairment of emotional behavior and spatial learning in adult Wistar rats by ferrous sulfate. Physiol. Behav. 96(2): 343-349.

[74] Maaroufi K, Save E, Poucet Sakly B, Abdelmelek H, Had-Aissouni L (2011) Oxidativestress and prevention of the adaptive response to chronic iron overload in the brain of young adult rats exposed to a 150 kilohertz electromagnetic field. Neuroscience 186: 39-47.

[75] Caro A, Puntarulo S (1995) Effect of iron-stress on antioxidant content of soybean embryonic axes. Plant Physiol. (Life Sci. Adv.) 14: 131-136.

Evaluation of Lipid Peroxidation Processes

Chapter 5

Automation of Methods for Determination of Lipid Peroxidation

Jiri Sochor, Branislav Ruttkay-Nedecky, Petr Babula, Vojtech Adam, Jaromir Hubalek and Rene Kizek

Additional information is available at the end of the chapter

1. Introduction

Free radicals are atoms or molecules having one (or rarely more) free electron(s). These compounds may attack most of the (bio)molecules in organisms, which leads to the oxidative stress, which belongs to the causes of pathological processes in organisms [1-6]. Oxidative stress occurs in a situation, when the imbalance between the production of free radicals and effectiveness of antioxidant defence system occurs in a healthy organism. Determination of antioxidant activity or eventually markers directly connected with this variable is one way how to monitor the damage of organisms by these compounds [7-14]. The negative effect of free oxygen radicals consists in the lipid peroxidation. This type of peroxidation is a chemical process, in which unsaturated fatty acids of lipids are damaged by free radicals and oxygen under lipoperoxides formation. Lipoperoxides are unstable and decompose to form a wide range of compounds including reactive carbonyl compounds, especially certain aldehydes (malondialdehyde (MDA), 4-hydroxy-2-nonenal (4-HNE)) [15-22] that damage cells by the binding the free amino groups of amino acids of proteins. Consequently, the proteins' aggregates become less susceptible to proteolytic degradation [23-25]. In tissues, the accumulation of age pigment spots appears. In addition, free radicals effects are connected with a formation of atherosclerotic lesions. In body fluids (blood, urine) the increased levels of peroxidation end-products (MDA, 4-HNE, isoprostanes) are present [26,27]. The lipid peroxidation by free radicals occurs in three stages: initiation, propagation and termination [2,26]. Reaction (1) represents initiation, in which a fatty acid molecule of lipid is attacked by free radicals leading to a detachment of the hydrogen atom under fatty acid radical formation. In its structure, a rearrangement of the double bond to form conjugated diene occurs. This diene structure subsequently reacts with oxygen molecule to form a lipoperoxyl radical, which leads to the initiation of the second phase called propagation (2). In another part of the promotion, lipoperoxyl radical further reacts with another molecule of fatty acid, from which a hydrogen atom is detached under formation of lipid hydroperoxide from original molecule (3). After pairing of all radicals, the last stage of the reaction called termination occurs. In addition to the above-mentioned chemical non-enzymatic peroxidation, enzymatic lipid peroxidation that is catalysed by the enzymes cyclooxygenase and lipoxygenase takes place. [26,28]. Both enzymes are involved in the formation of eicosanoids, which represent a group of biologically active lipid compounds derived from unsaturated fatty acids containing 20 carbon atoms. Cyclooxygenase is involved in the genesis of prostaglandins [29].

$$(1)\ LH + R^{\bullet} \rightarrow L^{\bullet} + RH\ (2)\ L^{\bullet} + O_2 \rightarrow LOO^{\bullet}\ (3)\ LOO^{\bullet} + L'H \rightarrow L'^{\bullet} + LOOH$$

Scheme 1. The scheme of lipid peroxidation. Initiation (1), the first part of the propagation (2), the second part of propagation (3).

For the monitoring of lipid peroxidation, spectrophotometric [30,31], chromatographic [32] and immunochemical [33] methods can be used. The analysis itself may be based on the analysis of the primary products of lipid peroxidation as conjugated dienes [34] and lipid hydroperoxides [35], or secondary products, such as malondialdehyde [36], alkanes [37] or isoprostanes [32,38-40]. Chromatographic methods represent the special group of methods, which are mostly based on the decrease of unsaturated fatty acids' concentration [41]. The scope of this review was to summarize the photometric analyses of lipid peroxidation. Less common method - FOX (ferrous oxidation in xylenol orange) was suggested to be automated.

1.1. Spectrophotometric methods in lipid peroxidation analysis

Spectrophotometric methods for the analysis of lipid peroxidation (see Table 1) are well reproducible and low cost. They usually consist of several steps that can be automated without much difficulty. Determination of conjugated dienes and TBARS belong to the one of the oldest and mostlz used methods for their rapidity and simplicity. On the other hand, they are criticized for their non-specificity [42,43]. Lipid hydroperoxides may be determined by the iodometric method and FOX test [44].

Determined analyte	Method	Type of analysed sample	Reference
Conjugated dienes	The structures of conjugated dienes absorb in the UV spectrum of 230-235 nm	Serum lipoproteins, tissue lipids	[34,45]
TBARS/MDA	TBA complex with MDA, Measurement at 532 nm	plasma, urine, tissues (liver), Cell lysates	[36,46-53]
Lipid hydroperoxides	Iodometric method FOX test	plasma, plant tissues plasma, serum lipoproteins, both animal and plant tissues	[44,54] [35,44,55,56]

Table 1. Summary of spectrophotometric methods used in lipid peroxidation determination. FOX –ferrous oxidation in xylenol orange, MDA – malondialdehyde, TBARS – thiobarbituric acid reactive substances

1.2. Conjugated dienes

The structures of conjugated dienes (Fig. 1) with alternating double and single bonds between carbon atoms (-C=C-C=C-) absorb wavelengths of 230-235 nm in the UV region. Therefore, it is possible to use UV absorption spectrometry for their determination [41,42]. The method is used for determination of a non-specific lipid peroxidation caused by free radicals in biological samples, and is successfully used in the study of peroxidation in isolated lipoprotein fractions (LDL lipoproteins) [45]. However, its use in the direct analysis of plasma is controversial because of the presence of interfering substances, such as heme proteins, purines or pyrimidines in the UV region measurement [42,57].

Figure 1. Structural formula of conjugated diene arising from the fatty acids by the free radicals effects during lipid peroxidation.

Increased sensitivity of the method can be achieved by an extraction of lipids into organic solvents in combination HPLC with UV detection [34,58]. However, the result of application the method to lipid extracts from human body fluids after HPLC separation was surprising, because the majority of pre-treated lipid fraction absorbs at wavelengths typical for conjugated dienes consisting of conjugated linoleic acid isomer (*cis*-9, *trans*-11-octadecadienoic acid) [59]. The main sources of conjugated isomer of linoleic acid (CLA) are dairy products and ruminant meat, especially beef [60]. They come into human serum and tissues probably from the diet [61], but can be also produced by bacteria [62,63]. Therefore, formation of large amounts of CLA by free radicals seems unlikely. In addition, the presence of CLA was not detected in the plasma of animals suffering from oxidative stress. *In vivo* induction of lipid peroxidation in rats treated with phenylhydrazin trichlorbrommethan did not cause an increase of CLA plasma values [64]. In the case of the use this method, it is necessary to take into account the above-mentioned shortcomings in the analysis of biological fluids or tissues.

1.3. TBARS, TBA-MDA adducts

TBARS (TBA-MDA) (Thiobarbituric Acid Reactive Substances) is the most widely used method for determination of lipid peroxidation method, especially due to its simplicity and cheapness. As the name of this method implies, it is based on the ability of malondialdehyde, which is one of the secondary products of lipid peroxidation, to react with thiobarbituric acid (TBA) [65]. The principle of this method consists in the reaction of MDA with thiobarbituric acid in acidic conditions and at a higher temperature to form a pink MDA-(TBA)$_2$ complex (Fig. 2), which can be quantified spectrophotometrically at 532 nm [17,66-70]. TBARS method measures the amount of MDA generated during lipid peroxidation, however, other aldehydes generated during lipid peroxidation, which also absorb at 532 nm, may react with TBA [71]. The results of the assay are expressed in μmol of MDA equivalents. TBARS method can be also used in the case of defined membrane systems, such as microsomes and liposomes, but its application in biological fluids and tissue extracts appears to be problematic [72-74]. The first problem is based on the fact that MDA can be formed by the decomposition of lipid peroxides under heating of the sample with TBA. This decomposition is accelerated by traces of iron in the reagents and is inhibited by the use of chelating agents [42]. At the decomposition of lipid peroxides in the analysis, the originating radicals can amplify the entire process and the amount of MDA could be overestimated [74]. To prevent the decomposition of lipid peroxides during the analysis, inhibitor of the lipid peroxidation called butylated hydroxytoluene is added to the sample [42]. One of the other problems of the TBARS method application has been found in the analysis of biological fluids. In this case, some substances, such as bile pigments and glycoproteins provide a false positive reaction with TBA [71,75]. Unspecificity TBARS test problems can be partially overcome by the using of HPLC techniques for the separation of "authentic", original MDA-(TBA)₂ adduct from other chromogens absorbing at 532 nm [76]. Nevertheless, this approach cannot solve all problems. In addition, next molecules, such as aldehydes originated from lipid peroxidation, can form with TBA a original MDA-TBA₂ adduct, which has been demonstrated in the deoxyribose [77]. Using of different techniques in the determination of lipid peroxides in plasma or serum of healthy people (spectrophotometric versus HPLC method) leads to significantly different results. When using spectrophotometric techniques, the content of TBARS in plasma (serum) reached values from 0.9 to 42.7 $\mu mol \cdot L^{-1}$ of MDA equivalents, when HPLC technique was used, the content of TBARS in human plasma (serum) reached values of 0.6 – 1.4 $\mu mol \cdot L^{-1}$ of MDA equivalents [78-84]. This was probably caused by the using different methods for modifying the preparation of plasma (serum) sample. Method for the non-specific index of lipid peroxidation determination in isolated purified lipid fractions seems to be most useful [42].

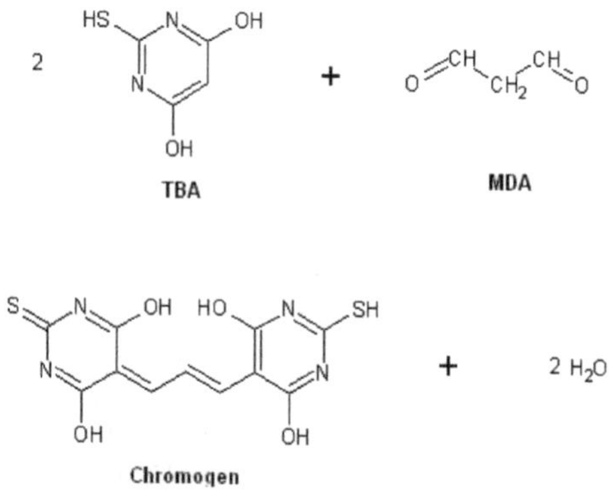

Figure 2. Chromophore produced by a condensation of MDA with TBA

1.4. Lipid hydroperoxides

1.4.1. Iodometric method

Iodometric method for lipid hydroperoxides determination is one of the oldest methods and is still used to determine lipid peroxide number [42,85]. Principle of this method is based on the ability of lipid hydroperoxides to oxidize iodide (I^-) to iodine (I_2), which further reacts with unreacted iodide (I^-) to triiodide anion (I_3^-) [86] and can be determined spectrophotometrically at 290 or 360 nm [87]. Modification of the iodometric method using commercially available reagent used for the determination of cholesterol can also be used to determine lipid (hydro)peroxides spectrophotometrically at 365 nm [54]. The method can be applied to extracts of biological samples without present the oxidizing agents. The possible interfering factors are especially the presence of oxygen, hydrogen peroxide and protein peroxides, which are able to oxidize iodide. Oxygen interference can be avoided by the using the anaerobic cuvettes and cadmium ions, which form a complex with unreacted iodide [86]. Values of lipid hydroperoxides in human plasma determined by iodometry are about 4 $\mu mol.L^{-1}$ [88,89].

1.4.2. Ferrous oxidation in xylenol orange

Total hydroperoxides can be determined using the oxidation of ferrous ions in the test with xylenol orange (FOX). The principle of the FOX method is based on the oxidation of ferrous ions to ferric by the hydroperoxide activity in the acidic environment [90-94]. The exact mechanism of the sequence of radical reactions is not known, but the mechanism has been designed by Gupta et al. [95] and is shown in reactions 1-4 (equation 2) [96]. The increase in the concentration of ferric ion is then detected using xylenol orange (Fig. 3), which forms a blue-violet complex with ferric ion (equation 2, reaction 5) with an absorption maximum at 560 nm [35]. However, the experimentally determined stoichiometry of 3 moles of Fe^{3+}-xylenol orange produced from 1 mol of peroxide [96,97] cannot be explained by the mechanism proposed by Gupta [95].

$$(1)\ Fe^{2+} + LOOH + H^+ \rightarrow Fe^{3+} + H_2O + LO^{\bullet}$$
$$(2)\ LO^{\bullet} + xylenol\ orange + H^+ \rightarrow LOH + xylenol\ orange^{\bullet}$$
$$(3)\ Xylenol\ orange^{\bullet} + Fe^{2+} \rightarrow xylenol\ orange + Fe^{3+}$$
$$(4)\ LO^{\bullet} + Fe^{2+} + H^+ \rightarrow Fe^{3+} + LOH$$
$$(5)\ Fe^{3+} + xylenol\ orange \rightarrow blue-violet\ complex\ (560\ nm)$$

Scheme 2. Equation of mechanism sequence of radical reactions.

Gay et al. [90] have found during comparison of the reactions of different peroxides with FOX reagents that the stoichiometry of the reaction ranged from 2.2 (H_2O_2) to 5.3 moles (Cu-OOH, t-BuOOH) Fe^{3+}-xylenol orange (Fe-XO) generated from 1 mol of peroxide, which was observed due to determination of molar absorption coefficients of Fe-XO complexes. Therefore, it is possible to compare only the results of FOX method analyses, in which the same type of peroxide was used in calibration. Hydrogen peroxide (H_2O_2) and Cumene hydroperoxide (Cu-OOH) are the most often peroxides used to calibrate the FOX method.

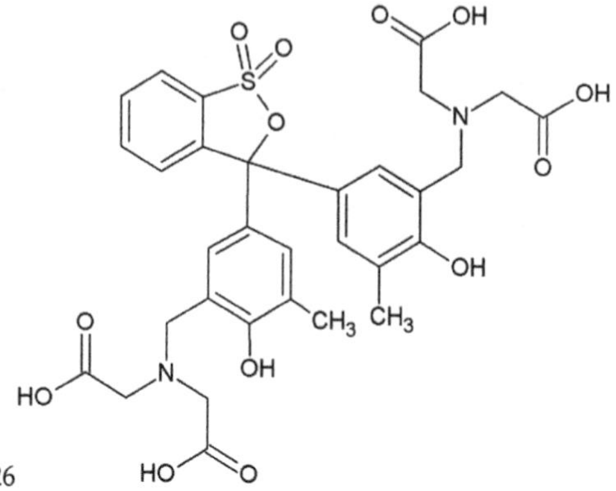

Figure 3. Structural formula of xylenol orange

The literature describes two versions of the FOX method called FOX1 and FOX2.. - FOX1 method can be used for the hydroperoxides determination in water phase and FOX2 method is suitable for the hydroperoxides of the lipid phase [30,35,98]. In the FOX1 method, chemicals used for a preparation of reagents (ferrous salt and sulphuric acid) are dissolved in water, whereas in FOX2 method methanol (90 % v/v) is the solvent [35]. FOX methods are not specific to hydroperoxides, the presence of oxidizing agent(s) in sample leads to the oxidization of ferrous ions to ferric ions. In the case of FOX2, the specificity of the method is achieved by the first FOX2 test performance in the presence of triphenylphosphine (TPP), which selectively reduces hydroperoxides to alcohols. The result of this test is used as a blank. After it, the FOX test without triphenylphosphine is performed and after deduction of blank values, we get the real value of lipid hydroperoxides. Improved specificity of the method using triphenylphosphine was later achieved also in FOX1 test [99]. Peroxidation chain reactions, which might occur during the analysis, are prevented by the addition of butylated hydroxytoluene prevented into the FOX1 agent. Plasma samples collected using ethylenediaminetetraacetic acid (EDTA) or diethylenetriaminepentaacetic acid pentasodium salt abbreviated as DETAPAC (anticoagulants or iron chelating agents) cannot be used due to interference with FOX reagents [30]. FOX1 method has been automated [100].

Measurement of lipid peroxidation in (blood) plasma

Banerjee et al. [99] enhanced sensitivity of FOX1 method by the addition of sorbitol into the FOX1 reagent in accordance with Wolff [98], and concurrently by the stabilization of pH of reagents at the values of 1.7 - 1.8. Improved specificity of method was obtained using triphenylphosphine and butylated hydroxytoluene. A comparison of both FOX1 and FOX2 methods on plasma samples of healthy individuals and diabetic patients was performed, where modified FOX1 method was more sensitive compared to the FOX2 method. Another advantage of the FOX1 method was based on the skip the centrifugation step that is necessary in FOX2 method. Nourooz-zadeh et al. [55] determined total lipid hydroperoxides in plasma by the use the FOX2 method and subsequently monitored content of lipid hydroperoxides in individual lipoprotein fractions (VLDL, LDL and HDL fractions). Content of total lipid hydroperoxides in plasma was 3.50 ± 2.05 µmol/L. The highest rate of hydroperoxides (67 %) was detected in LDL lipoprotein fractions. Södergren et al. [101] studied the impact of the storage of samples at low temperatures on the total lipid hydroperoxide content by the use the FOX2 method. They were focused on possible reduction of total lipid hydroperoxides content during the storage of samples under these conditions. Researchers found that storage of samples for 6 weeks at -70 °C leads to the 23 % average reduction of hydroperoxides content. The finding that the content of lipid hydroperoxides in short-term stored plasma samples (6 weeks) did not differ from the content of lipid hydroperoxides in the long-term stored samples (60 weeks) was interesting too.

Measurement of lipid peroxidation in animal tissues

Hermes-Lima et al. [96] proposed and elaborated methodology for application of FOX1 test in determination of lipid hydroperoxides in animal tissue extracts. They used methanol extracts of kidney, liver and heart from adult mice (*Mus musculus* Linnaeus), brain and lungs from adult Wistar rats (*Rattus norvegicus* Berkenhout var. *alba*), liver and adipose tissues from adult golden-mantled ground squirrels (*Spermophilus lateralis* Say), and liver and muscle tissues from adult red-eared slider turtles (*Trachemis scripta elegans* Wied-Neuwied). The highest values of lipid hydroperoxide content were detected in mice organs. The contents of lipid peroxides in animal tissues measured by the FOX1 method well correlated with results obtained by the TBARS. Grau et al. [102] adapted the FOX2 method for the determination of lipid hydroperoxides in raw and cooked dark chicken meat. Chickens were fed by a diet with different contents of α-tocopherol and fats from different sources. They determined the absolute values of lipid hydroperoxides in different experimental groups of chickens. Eymard et al. [56] modified the FOX1 method used by Hermes-Lima et al. [96] for the determination of lipid hydroperoxides in small pelagic fish. They used methanol extracts of ground tissues of the Atlantic horse mackerel (*Trachurus trachurus* Linnaeus). The original FOX1 reagent was replaced by the FOX2 reagent used by Wolff et al. [98] with the increased content of methanol to increase a solubility of extracts.

Measurement of lipid peroxidation in plant tissues

De Long et al. [44] applied the FOX2 method in the determination of hydroperoxides in plant tissues. They used ethanol extracts of pericarp of avocado (*Persea americana* P. Mill.), periderm of potatoes (*Solanum tuberosum* L.), leaves of red cabbage (*Brassica oleracea* convar. *capitata* var. *rubra* DC. Ranost), leaves of spinach (*Spinacia oleracea* L.), pericarp of the European Pear (*Pyrus communis* L.) and fruits of red pepper (*Capsicum annuum* L.) for analyses. The effect of UV radiation on lipid peroxidation was monitored. Parts of plants were exposed to UV radiation for 10-12 days prior the extraction due to induction of lipid peroxidation in plants. Lipid hydroperoxides were determined by the FOX2, the TBARS and the iodometric methods. UV radiation induced an increase in lipid peroxidation values in all samples of different plant tissues determined by the FOX method. The good correlation was found between the FOX and iodometric methods. However, the iodometric method had limitations in the determination of the low concentrations of lipid hydroperoxides. Similar results were obtained by the use the TBARS method. Griffiths et al. [103] applied the FOX2 method in determination of lipid peroxides in different types of plant tissues. They analysed plant tissues, such as extracts of bean hypocotyls (*Phaseolus* sp.) and microsomes, potato leaves (*Solanum tuberosum* L.), flowers of alstromeria (*Alstroemeria* spp.), broccoli (*Brassica oleracea* var. *italica* Plenck) and cells of green algae (*Chlamydomonas* sp.). Lipid hydroperoxide levels ranged from 26 to 602 nmol.g^{-1} of FW. The highest content of lipid hydroperoxides was detected in broccoli and green alga cells in their study.

2. Experimental section

2.1. Instruments

For dilution of stock solutions of standards an epMotion 5075 (Eppendorf, Germany) automated pipetting system was used (Fig. 4). The pipetting provides a robotic arm with adapters (TS 50, TS 300 and TS 1000) and Gripper (TG-T). The empty microtubes are placed in the position B3 (Fig. 4) in adapter Ep0.5/1.5/2 ml. Module Reservoir is located in the position B1, where stock solutions are available. The device is controlled by the epMotion control panel. The tips are located in the A4 (ePtips 50), A3 (ePtips 300) and A2 (ePtips 1000) positions. For preparation of the standards tips of sizes 300 µl and 1000 µl (Eppendorf – Germany) were used. For determination of antioxidant activity, a BS-400 automated spectrophotometer (Mindray, China) was used. It is composed of cuvette space tempered to 37±1 °C, reagent space with a carousel for reagents (tempered to 4±1 °C), sample space with a carousel for preparation of samples and an optical detector. Transfer of samples and reagents is provided by robotic arm equipped with a dosing needle (error of dosage up to 3 % of volume). Cuvette content is mixed by an automatic mixer including a stirrer immediately after addition of reagents or samples. Contamination is reduced due to its rinsing system, including rinsing of the dosing needle as well as the stirrer by MilliQ water. For detection itself, the following range of wave lengths can be used as 340, 380, 412, 450, 505, 546, 570, 605, 660, 700, 740 and 800 nm. In addition, a SPECOL 210 two beam UV-VIS spectrophotometer (Analytik Jena AG, Germany) with cooled semiconductor detector for measurement within range from 190 to 1,100 nm with control by an external PC with the programme WinASPECT was used as the manual instrument in this study. Laboratory scales (Sartorius, Germany) and pipettes (Eppendorf Research, Germany) were used.

2.2. Chemicals

Xylenol orange disodium salt, iron D-gluconate dihydrate, glycerol, *tert*-butylhydroperoxide (t-BHP) 70% in water, sodium chloride, sulphuric acid, formic acid and water ACS reagent were purchased from Sigma Aldrich (USA).

2.3. Preparation of reagents and standards

FOX1 reagents were prepared according Arab et al. [100]. The general acidic reagent (acidic reagent A) final concentrations were 0.9 % NaCl, 40 mM H_2SO_4, 20 mM formic acid and 1.37 M glycerol in ACS water. The pH of the reagent was adjusted to the value of 1.35. The reagent R1 contained 167 µM xylenol orange disodium salt, which was dissolved in acidic reagent A. The reagent R2 contained 833 µM iron D-gluconate dehydrate, which was also dissolved in acidic reagent A. Standards were prepared from the 70% water solution of *tert*-butylhydroperoxide, which was diluted by ACS water to the 20 mM pre-stock solution. From the pre-stock solution, five stock solutions: and 0.2, 3.9, 62.5, 375 and 1,000 µM were prepared daily by dilutions of pre-stock solution with 0.9 % NaCl. For further preparation of 20 standards from five stock solutions, an automated pipetting system epMotion 5075 was used to minimalize possible pipetting errors. The standards had following concentrations: 0.06, 0.12, 0.24, 0.48, 0.97, 1.9, 3.9, 7.8, 15.6, 31.2, 46.8, 62.5, 93.7, 125, 187, 250, 375, 500, 750 and 1000 µM. These standards were used for the preparation of calibration curves in both manual and automatic measurements.

2.4. Working procedure for manual spectrophotometric determination

A volume of 720 µl of the reagent R1 (167 µM xylenol orange in acidic reagent) was pipetted into plastic cuvettes. Subsequently, a volume of 100 µl of the sample was added. Absorbance was measured at λ = 591 nm. After it, a volume of 180 µl of the reagent R2 (833 µM iron D-gluconate in acidic reagent A) was pipetted to a reaction mixture and after 6 minutes of the incubation, absorbance was measured. Final value is calculated from the absorbance value of the mixture of the reagent R1 with sample and from the absorbance value after 6 minutes of incubation of the mixture with the reagent R2. The final concentrations in the cuvette of xylenol orange (R1) and iron D-gluconate (R2) were 120 and 150 µM, respectively.

2.5. Working procedure for automated spectrophotometric determination

A volume of 180 µL of the solution R1 (167 µM xylenol orange in acidic reagent) was pipetted into a plastic cuvette with subsequent addition of a 25 µL of sample. This mixture was incubated for 4.5 minutes. Subsequently, 45 µL of solution R2 (833 µM iron D-gluconate in acidic reagent) was added and the solution was incubated for next 6 minutes. Absorbance was measured at λ = 570 nm. Final value is calculated from the absorbance value of the mixture of reagent R1 with sample before the addition of the reagent 2 and from the absorbance value after 6 minutes of incubation of the mixture with the reagent 2. The final concentrations in the cuvette of xylenol orange (R1) and iron D-gluconate (R2) were 120 and 150 µM, respectively.

3. Results and discussion

Spectrophotometric methods for determination of lipid peroxidation have a relatively simple procedure of a measurement. In addition, they are relatively low-cost with easy applicability and they do not require specialized equipment or personnel. To maintain the sustainability of these methods, it is necessary to introduce these methods to automated operation, which has not been yet satisfactorily solved. Analyses of samples performed due to intensive work of personnel, which is expensive, slow, and, in addition, the human factor is responsible for a high percentage of errors. Requirement for laboratories, in which a large number of samples is analysed per day, consists in relatively simple and easy to apply method. Our aim was to automate the pre-analytical and analytical phase of the FOX1 method. For specification and comparison of this method, the method based on the use the manual spectrophotometer was also carried out.

3.1. Pre-analytical phase

Pre-analytical processing of biological samples in the laboratory is a necessary and important part of laboratory work. It represents a wide range of manual, often stereotyped operations that do not require special knowledge and skills, but require maintenance of the standard procedure(s) and prevent the possibility of errors connected with this analytical phase. Pre-analytical laboratory process is destined to automation and robotics. Automation and robotics of the pre-analytical phase brings many benefits and advantages to laboratory. It reduces the number of errors, the time necessary for sample manipulation, and the response time. It significantly increases the productivity, cost savings connected with productivity, and minimizes the exposure of personnel with biological material [104].

For automation of pre-analytical phase, the epMotion 5075 automated pipetting system was used. Stock solutions of *tert*-butylhydroperoxide (*t*-BHP) at the concentrations of 1000, 375, 62.5, 3.9 and 0.2 μM prepared in 0.9 % NaCl solution were applied into five vials. Sixth vial contained diluting solution (0.9% NaCl). Twenty empty Eppendorf tubes (1.5 ml) were placed into the metal holder. Scheme of the preparation of standards is shown in Table 2. Pipetting robot first pipetted different volumes of diluting solution (0.9% NaCl) into vials and after it, different volumes of stock solutions of various concentrations of *t*-BHP were pipetted. When pipetting the stock solution into the dilution buffer in micro test tube, robot three times mixed the solution by a pipetting.

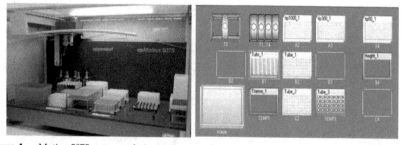

Figure 4. epMotion 5075 automated pipetting system from frontal part.

Tube nb.	Final concentration t-BHP (µM)	Pipetting volume (µl)					
		solution 0.9% NaCl	solution 1 1000 µM t-BHP	solution 2 375 µM t-BHP	solution 3 62.5 µM t-BHP	solution 4 3.906 µM t-BHP	solution 5 0.244 µM t-BHP
1	1000	-	1000	-	-	-	-
2	750.0	250	750	-	-	-	-
3	500.0	500	500	-	-	-	-
4	375.0	-	-	1000	-	-	-
5	250.0	750	250	-	-	-	-
6	187.5	500	-	500	-	-	-
7	125.0	875	125	-	-	-	-
8	93.75	750	-	250	-	-	-
9	62.50	-	-	-	1000	-	-
10	46.87	875	-	125	-	-	-
11	31.25	500	-	-	500	-	-
12	15.62	750	-	-	250	-	-
13	7.812	875	-	-	125	-	-
14	3.906	-	-	-	-	1000	-
15	1.953	500	-	-	-	500	-
16	0.977	750	-	-	-	250	-
17	0.488	875	-	-	-	125	-
18	0.244	-	-	-	-	-	1000
19	0.122	500	-	-	-	-	500
20	0.061	750	-	-	-	-	250

Table 2. Volume of the solution in the preparation of standards using epMotion 5075 automated pipetting system.

Using the epMotion 5075 automated pipetting system, work time of 20 minutes was saved (time, when laboratory staff was not needed). The only time-demanding operation consisted in replenishment of vials and initiation of the program. Potential errors that arise due to human activity were avoided. Accuracy of a pipetting was verified by weighing, the average error was approximately 1.8 %.

3.2. Analytical phase

Our goal was to introduce the FOX1 method to an automated operation and improve both analysis itself and conditions of analysis. The experiment was carried out using *tert*-butylhydroperoxide standard prepared at the concentrations from 0.06 to 1000 µM. Furthermore, the spectral curves of generated chromatic complexes were observed and the concentration dependence on temperature and time were determined. In addition, reaction kinetics during the reaction was established.

3.2.1. Monitoring the spectral courses at different concentrations and times

Spectral changes in the *t*-BHP concentration range from 0.06 to 1000 μM (Figures 5A and 4B) were observed. Two peaks at the wavelengths of 444 and 591 nm were detected in the formed complex at the recommended temperature of interaction of 37 °C.

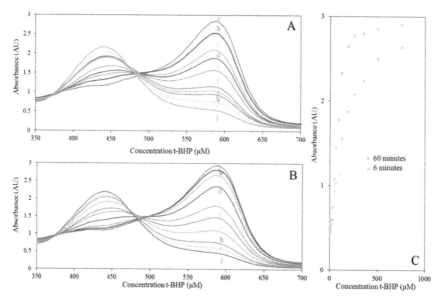

Figure 5. Courses of spectra of *t*-BHP in the concentrations from 0.06 to 1000 μM - a) 1000, b) 500, c) 250, d) 125, e) 62.50, f) 31., g) 7.8, h) 1.9, i) 0.4, j) 0.06 in the time of 6 (A) and 60 (B) minutes. (C) Comparison of values of absorption maximum at the wavelength of 591 nm and a time period of 6 and 60 minutes. The courses were measured in the interval form 350 to 700 nm using the SPECORD 210 apparatus. All analyses were carried out in triplicates.

Absorption maximum at low concentrations (up to the concentration of 0.122 μM) was at 444 nm, and with the increasing concentrations (higher than 0.122 μM) the absorption maximum was sifted and observed at 591 nm. Interaction of sample and reagents proceeded in six minutes, after this time, absorbance could be measured and the final value of lipid peroxidation calculated. We wanted to determine the changes in the absorbance during one hour. Comparison of absorbance values at the time of 6 and 60 min at $\lambda = 591$ nm is shown in Figure 5C. Absorbance values during the monitoring decreased for about 13 % on an average. When interlaying the trends points in the linear concentration part from 0.12 to 125 μM, the determination factor decreased from 0.996 (for the 6-minute reaction time) to 0.987 (for the 60-minute reaction time). This fact can be explained by unequal reaction kinetics during the analysis (see the reaction kinetics, Chapter 3.2.3) and oxidation of the sample during the analysis.

3.2.2. Monitoring the reaction under different temperature conditions

Dependences of representative concentration (62.5 μM) on the temperature conditions (17, 27, 37 and 47 °C) and the time from 0 to 30 minutes and absorption maximum of 591 nm is shown in Figure 6. The absorbance increased with the increasing temperature; after 6 minutes of reaction, the difference of absorbance value between the lowest (17 °C) and the highest (47 °C) temperature was about 0.64 AU. In other words, the value of absorbance at 47 °C was higher for 71 % compared to the absorbance determined at 17 °C. The highest values of absorbance and concurrently the most prominent difference was detected at 47 °C, therefore, this temperature was the most suitable for our purposes. On the other hand, this temperature may lead to degradation of biological samples. Due to this fact, the temperature of interaction of 37 °C was selected for further analyses.

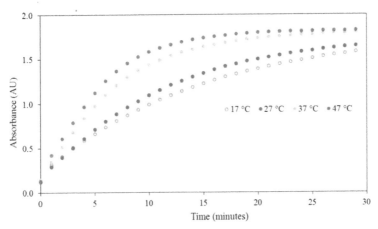

Figure 6. Dependences of representative concentration (62.5 μM) of applied *t*-BHP on temperature conditions (17, 27, 37 and 47 °C) and the time of interaction. Detected at 591 nm, interval of record is 1 minute, interval period 0 - 30 minutes. All analyses were carried out in triplicates.

3.2.3. Determination of reaction kinetics

Reaction kinetics at the temperature of 37 °C in the shortest time intervals in all concentrations (0.06 – 1000 μM) was monitored. Automated analyser BS-400 was used for this purpose. All samples could be studied at all once. This is not possible using the manual spectrophotometer, thus, use the automated analyser represents one of the most important steps in the analysis automation.

The curves were used for the calculating the reaction rate constants indicating the course and conception of the impact of the effect of *t*-BHP concentration on the reaction rate. The constant was calculated as the change in the absorbance per time unit (second, minute) according to the equation $x = A/t$, where x is the rate constant, A the value of absorbance after 6 minutes and t time for which the rate constant was related (second, minute). The effect of each of concentrations on the change in absorbance value was determined.

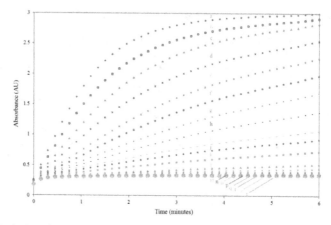

Figure 7. Monitoring of reaction curves of *t*-TBH in the concentrations from 0.06 to 1000 μM - **a)** 1000, **b)** 750, **c)** 500, **d)** 375, **e)** 250, **f)** 187, **g)** 125 **h)** 94, **i)** 63, **j)** 47., **k)** 31, **l)** 15.6, **m)** 7.8, **n)** 3.9, **o)** 1.9, **p)** 0.9, **q)** 0.4, **r)** 0.2, **s)** 0.1, and **t)** 0.06 μM in the time interval from 0 to 6 minutes. All analyses were carried out in triplicates.

Concentration	Logarithmic equation	Change in absorbance per second	Change in absorbance per minute	Change in abs. per minute recalculated to 1 μM t-BHP
1000	y = 3.7532ln(x) - 5.899	0.02304	1.383	0.0013
750.0	y = 3.6495ln(x) - 5.544	0.02211	1.345	0.0017
500.0	y = 3.4895ln(x) - 5.241	0.02168	1.301	0.0028
375.0	y = 3.1895ln(x) - 4.872	0.01987	1.258	0.0036
250.0	y = 3.2076ln(x) - 4.677	0.01853	1.112	0.0044
187.5	y = 2.7574ln(x) - 4.375	0.01534	0.924	0.0052
125.0	y = 2.2477ln(x) - 3.945	0.01298	0.779	0.0062
93.75	y = 1.7316ln(x) - 2.968	0.01000	0.600	0.0060
62.50	y = 1.2213ln(x) - 1.998	0.00705	0.423	0.0068
46.87	y = 1.0049ln(x) - 1.596	0.00580	0.348	0.0070
31.25	y = 0.7102ln(x) - 1.054	0.00410	0.246	0.0079
15.62	y = 0.3846ln(x) - 0.445	0.00222	0.133	0.0085
7.812	y = 0.2525ln(x) - 0.183	0.00146	0.088	0.0112
3.906	y = 0.1765ln(x) - 0.037	0.00102	0.061	0.0157
1.953	y = 0.1303ln(x) + 0.033	0.00075	0.045	0.0231
0.976	y = 0.1177ln(x) + 0.073	0.00068	0.041	0.0418
0.488	y = 0.1031ln(x) + 0.089	0.00060	0.036	0.0732
0.244	y = 0.0965ln(x) + 0.101	0.00057	0.034	0.1370
0.122	y = 0.0926ln(x) + 0.105	0.00055	0.033	0.2629
0.061	y = 0.0957ln(x) + 0.131	0.00053	0.032	0.5434

Table 3. Mathematical formularization of the course of reaction curves for *t*-TBH in the concentration range from 0.06 to 1000 μM by the use the logarithmic equation. Reaction rate constant is expressed as a

change in absorbance per second, and per minute. In addition, change in absorbance per minute recalculated to 1 μM t-BHP is introduced.

3.2.4. Dependence on concentration

By the using manual spectrophotometer and automated analyser, the dependence of t-TBH concentration (0.06 – 1000 μM) on the changes of coloured complex was determined. The calibration curves were calculated from final values.

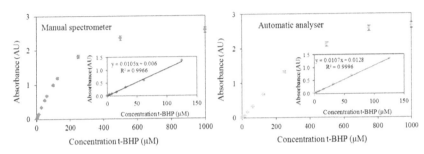

Figure 8. Dependence of absorbance on applied t-BHP concentration measured by manual spectrophotometer SPECOL 210 and automated analyser BS-400. All analyses were carried out in triplicates. For other experimental detail, see Fig. 7.

The analysis of 60 samples (20 samples in a standard three repetitions) took using the BS-400 automated analyser only 24 minutes. The analysis of 60 samples including delays for the pipetting, mixing and displacement of samples using the manual spectrophotometer took about 7 hours (6 minutes per sample + one minute of delay, 60 × 6 minutes of sample analysis). By using the fully automated analyser, results were obtained in more than 17 times less time compared to manual spectrophotometer. Shortening of the time of analysis contributes especially to higher quality of results due to reduction of possibility of chemical modification including degradation of the measured samples. This fact resulted in the preparation of calibration curves, where the determination factor for the calibration curve obtained using the automatic analyser was $R^2 = 0.9996$, while the determination factor for the results from manual spectrophotometer was $R^2 = 0.9966$. In addition, a limit of detection (LOD) and limit of quantification (LOQ) were determined. In the case of both automated and manual analyses, the LOD was determined as LOD = 0.06 μM of t-BHP, limit of quantification (LOQ) was also determined as LOQ = 0.2 μM of t-BHP (see Table 3). All measurements of all concentrations of t-BHP (concentration range from 0.06 to 1000 μM) were carried out in 3 repetitions and repeatability (RSD) was determined. In the case of automated method, the repeatability was RSD = 2.6 % compared to manual spectrophotometer, where RSD = 3.8 %.

Technical development is responsible for a tendency to increase the speed of analysis and analytical process itself. Automatic analysers allow analysing more samples at the same time, reducing the time required to analyse one sample and errors caused by incorrect pipetting and manipulation with sample, and generally provide higher data quality compared to manual analysis. Due to automation, the risk of sample confusion is significantly reduced. In addition, the whole process is much faster, the consumption of reagents and demands of personnel staff are reduced. The aim of automation is to eliminate stereotypical incompetent operation, eliminate the possibility of error, and to accelerate operations under significant increase of capacity while maintaining the precise performance of all necessary operations. The disadvantage, however, consists in still high acquisition costs and the need for compete service [105,106].

Apparatus	Wavelength (nm)	LOD	LOQ	Measuring range (μM)	Calibration equation	Confidence coefficient (R^2)	RSD	Time analysis of 60 samples (min)
SPECOL	591	0.06	0.2	0.012 - 125	y=0.0105x +0.006	0.9969	3.8	420
BS-400	570	0.06	0.2	0.012 - 125	y=0.0107x +0.0128	0.9996	2.6	24

Table 4. Analytic parameters for the FOX1 method for *t*-BHP analysis using manual SPECOL and automated BS-400 analysers.

4. Conclusion

This chapter brought a comprehensive overview of photometric methods used in the study of lipid peroxidation. Main attention was devoted to the detection of lipid peroxidation by using the less common FOX1 method. The proposal to automation the pre-analytical and analytical phases of the sample was introduced. In addition, conditions and parameters influencing the photometric reaction were studied and described. The comparison of results obtained using the manual and automated apparatuses (manual/automated operation) is introduced and discussed.

Author details

Jiri Sochor, Branislav Ruttkay-Nedecky, Vojtech Adam, Jaromir Hubalek and Rene Kizek[*]
Department of Chemistry and Biochemistry, Faculty of Agronomy, Mendel University in Brno, Brno, Czech Republic, European Union
Department of Microelectronics, Faculty of Electrical Engineering and Communication, Brno University of Technology, Brno, Czech Republic, European Union

Petr Babula
Department of Natural Drugs, Faculty of Pharmacy,
University of Veterinary and Pharmaceutical Sciences Brno, Brno, Czech Republic, European Union

---

[*] Corresponding Author

Acknowledgement

This project was supported by SIX research centre CZ.1.05/2.1.00/03.0072. The authors wish to express their thanks to Lukáš Nejdl for excellent technical assistance.

5. References

[1] Grinna LS (1977) Age-related changes in lipids of microsomal and mitochondrial-membranes of rat-liver and kidney Mech. Ageing Dev. 6: 197-205.

[2] De Zwart LL, Meerman JHN, Commandeur JNM, Vermeulen NPE (1999) Biomarkers of free radical damage applications in experimental animals and in humans. Free Radic. Biol. Med. 26: 202-226.

[3] Knight JA (1995) Diseases related to oxygen-derived free-radicals. Ann. Clin. Lab. Sci. 25: 111-121.

[4] Gutteridge JMC (1993) Free-radicals in disease processes - A compilation of cause and consequence. Free Rad. Res. Commun. 19: 141-158.

[5] Evans PH (1993) Free-radicals in brain metabolism and pathology. Br. Med. Bull. 49: 577-587.

[6] Adams JD, Odunze IN (1991) Oxygen free-radicals and Parkinsons-disease. Free Radic. Biol. Med. 10: 161-169.

[7] Pohanka M, Novotny L, Zdarova-Karasova J, Bandouchova H, Zemek F, Hrabinova M, Misik J, Kuca K, Bajgar J, Zitka O, Cernei N, Kizek R, Pikula J (2011) Asoxime (HI-6) impact on dogs after one and tenfold therapeutic doses: Assessment of adverse effects, distribution, and oxidative stress. Environ. Toxicol. Pharmacol. 32: 75-81.

[8] Diopan V, Babula P, Shestivska V, Adam V, Zemlicka M, Dvorska M, Hubalek J, Trnkova L, Havel L, Kizek R (2008) Electrochemical and spectrometric study of antioxidant activity of pomiferin, isopomiferin, osajin and catalposide. J. Pharm. Biomed. Anal. 48: 127-133.

[9] Babula P, Kohoutkova V, Opatrilova R, Dankova I, Masarik M, Kizek R (2010) Pharmaceutical importance of zinc and metallothionein in cell signalling. Chim. Oggi-Chem. Today. 28: 18-21.

[10] Jurikova T, Rop O, Mlcek J, Sochor J, Balla S, Szekeres L, Hegedusova A, Hubalek J, Adam V, Kizek R (2012) Phenolic Profile of Edible Honeysuckle Berries (Genus Lonicera) and Their Biological Effects. Molecules. 17: 61-79.

[11] Sochor J, Skutkova H, Babula P, Zitka O, Cernei N, Rop O, Krska B, Adam V, Provaznik I, Kizek R (2011) Mathematical evaluation of the amino acid and polyphenol content and antioxidant activities of fruits from different apricot cultivars. Molecules. 16: 7428-7457.

[12] Sochor J, Zitka O, Skutkova H, Pavlik D, Babula P, Krska B, Horna A, Adam V, Provaznik I, Kizek R (2010) Content of phenolic compounds and antioxidant capacity in fruits of apricot genotypes. Molecules. 15: 6285-6305.

[13] Rop O, Sochor J, Jurikova T, Zitka O, Skutkova H, Mlcek J, Salas P, Krska B, Babula P, Adam V, Kramarova D, Beklova M, Provaznik I, Kizek R (2011) Effect of five different

stages of ripening on chemical compounds in Medlar (Mespilus germanica L.). Molecules. 16: 74-91.

[14] Rop O, Jurikova T, Sochor J, Mlcek J, Kramarova D (2011) Antioxidant capacity, scavenging radical activity and selected chemical composition of native apple cultivars from Central Europe. J. Food Qual. 34: 187-194.

[15] Balestrieri ML, Dicitore A, Benevento R, Di Maio M, Santoriello A, Canonico S, Giordano A, Stiuso P (2012) Interplay between membrane lipid peroxidation, transglutaminase activity, and Cyclooxygenase 2 expression in the tissue adjoining to breast cancer. J. Cell. Physiol. 227: 1577-1582.

[16] Cai F, Dupertuis YM, Pichard C (2012) Role of polyunsaturated fatty acids and lipid peroxidation on colorectal cancer risk and treatments. Curr. Opin. Clin. Nutr. Metab. Care. 15: 99-106.

[17] Zalejska-Fiolka J, Wielkoszynski T, Kasperczyk S, Kasperczyk A, Birkner E (2012) Effects of Oxidized Cooking Oil and alpha-Lipoic Acid on Blood Antioxidants: Enzyme Activities and Lipid Peroxidation in Rats Fed a High-Fat Diet. Biol. Trace Elem. Res. 145: 217-221.

[18] Flohr L, Fuzinatto CF, Melegari SP, Matias WG (2012) Effects of exposure to soluble fraction of industrial solid waste on lipid peroxidation and DNA methylation in erythrocytes of Oreochromis niloticus, as assessed by quantification of MDA and m(5)dC rates. Ecotox. Environ. Safe. 76: 63-70.

[19] Janowska B, Kurpios-Piec D, Prorok P, Szparecki G, Komisarski M, Kowalczyk P, Janion C, Tudek B (2012) Role of damage-specific DNA polymerases in M13 phage mutagenesis induced by a major lipid peroxidation product trans-4-hydroxy-2-nonenal. Mutat. Res.-Fundam. Mol. Mech. Mutagen. 729: 41-51.

[20] Demir E, Kaya B, Soriano C, Creus A, Marcos R (2011) Genotoxic analysis of four lipid-peroxidation products in the mouse lymphoma assay. Mutat. Res. Genet. Toxicol. Environ. Mutagen. 726: 98-103.

[21] Guo LL, Chen ZY, Cox BE, Gragg S, Zhang YQ, Amarnath V, van Lenten B, Epand R, Davies SS (2011) Lipid Peroxidation Generates Aldehyde-Modified Phosphatidylethanolamines That Induce Inflammation. Free Radic. Biol. Med. 51: S103-S104.

[22] Guo J, Prokai L (2011) To tag or not to tag: A comparative evaluation of immunoaffinity-labeling and tandem mass spectrometry for the identification and localization of posttranslational protein carbonylation by 4-hydroxy-2-nonenal, an end-product of lipid peroxidation. J. Proteomics. 74: 2360-2369.

[23] Esterbauer H, Schaur RJ, Zollner H (1991) Chemistry and biochemistry of 4-hydroxynonenal, malonaldehyde and related aldehydes. Free Radic. Biol. Med. 11: 81-128.

[24] Benedetti A, Comporti M, Esterbauer H (1980) Identification of 4-hydroxynoneal as a cyto-toxic product originating from the peroxidation of liver microsomal lipids. Biochim. Biophys. Acta. 620: 281-296.

[25] Esterbauer H, Comporti M, Benedetti A (1980) Biochemical effects of 4-hydroxyalkenals, in particular 4-hydroxynonenal produced by microsomal lipid-peroxidation. J. Am. Oil Chem. Soc. 57: A144-A145.

[26] Gutteridge JMC (1995) Lipid-peroxidation and antioxidants as biomarkers of tissue-damage Clin. Chem. 41: 1819-1828.

[27] Tanrikulu AC, Abakay A, Evliyaoglu O, Palanci Y (2011) Coenzyme Q10, Copper, Zinc, and Lipid Peroxidation Levels in Serum of Patients with Chronic Obstructive Pulmonary Disease. Biol. Trace Elem. Res. 143: 659-667.

[28] Viita H, Pacholska A, Ahmad F, Tietavainen J, Naarala J, Hyvarinen A, Wirth T, Yla-Herttuala S (2012) 15-Lipoxygenase-1 Induces Lipid Peroxidation and Apoptosis, and Improves Survival in Rat Malignant Glioma. In Vivo. 26: 1-8.

[29] Kuehl FA, Egan RW (1980) Prostaglandins, arachidonic-acid, and inflammation. Science. 210: 978-984.

[30] Nourooz-Zadeh J Ferrous ion oxidation in presence of xylenol orange for detection of lipid hydroperoxides in plasma, in: L. Packer (Ed.), Oxidants and Antioxidants, Pt B, 1999, pp. 58-62.

[31] Rop O, Mlcek J, Jurikova T, Valsikova M, Sochor J, Reznicek V, Kramarova D (2010) Phenolic content, antioxidant capacity, radical oxygen species scavenging and lipid peroxidation inhibiting activities of extracts of five black chokeberry (Aronia melanocarpa (Michx.) Elliot) cultivars. J. Med. Plants Res. 4: 2431-2437.

[32] Nouroozzadeh J, Gopaul NK, Barrow S, Mallet AI, Anggard EE (1995) Analysis of F-2-isoprostanes as indicators of nonenzymatic lipid-peroxidation in-vivo by gas-chromatography mass-spectrometry - Development of a solid-phase extraction procedure. J. Chromatogr. B-Biomed. App. 667: 199-208.

[33] Bachi A, Zuccato E, Baraldi M, Fanelli R, Chiabrando C (1996) Measurement of urinary 8-epi-prostaglandin F-2 alpha, a novel index of lipid peroxidation in vivo, by immunoaffinity extraction gas chromatography mass spectrometry. Basal levels in smokers and nonsmokers. Free Radic. Biol. Med. 20: 619-624.

[34] Iversen SA, Cawood P, Dormandy TL (1985) A method for the measurement of a diene-conjugated derivative of linoleic-acid, 18-2(9,11), in serum phospholipid, and possible origins. Ann. Clin. Biochem. 22: 137-140.

[35] Nouroozzadeh J, Tajaddinisarmadi J, Wolff SP (1994) Measurement of plasma hydroperoxide concentrations by the ferrous oxidation xylenol orange assay in conjunction with triphenylphosphine. Anal. Biochem. 220: 403-409.

[36] Draper HH, Squires EJ, Mahmoodi H, Wu J, Agarwal S, Hadley M (1993) A comparative-evaluation of thiobarbituric acid methods for the determination of malondialdehyde in biological-materials. Free Radic. Biol. Med. 15: 353-363.

[37] Burk RF, Ludden TM (1989) Exhaled alkanes as indexes of invivo lipid-peroxidation. Biochem. Pharmacol. 38: 1029-1032.

[38] Nikolaidis MG, Kyparos A, Vrabas IS (2011) F(2)-isoprostane formation, measurement and interpretation: The role of exercise. Prog. Lipid Res. 50: 89-103.

[39] Janicka M, Kot-Wasik A, Kot J, Namiesnik J (2010) Isoprostanes-Biomarkers of Lipid Peroxidation: Their Utility in Evaluating Oxidative Stress and Analysis. Int. J. Mol. Sci. 11: 4631-4659.

[40] Soffler C, Campbell VL, Hassel DM (2010) Measurement of urinary F(2)-isoprostanes as markers of in vivo lipid peroxidation: a comparison of enzyme immunoassays with gas chromatography mass spectrometry in domestic animal species. J. Vet. Diagn. Invest. 22: 200-209.

[41] Gutteridge JMC, Halliwell B (1990) The measurement and mechanism of lipid-peroxidation in biological-systems. Trends Biochem.Sci. 15: 129-135.

[42] Halliwell B, Chirico S (1993) Lipid-peroxidation - its mechanism, measurement, and significance Am. J. Clin. Nutr. 57: S715-S725.

[43] Devasagayam TPA, Boloor KK, Ramasarma T (2003) Methods for estimating lipid peroxidation: An analysis of merits and demerits. Indian J. Biochem. Biophys. 40: 300-308.

[44] DeLong JM, Prange RK, Hodges DM, Forney CF, Bishop MC, Quilliam M (2002) Using a modified ferrous oxidation-xylenol orange (FOX) assay for detection of lipid hydroperoxides in plant tissue. J. Agric. Food Chem. 50: 248-254.

[45] Esterbauer H, Striegl G, Puhl H, Rotheneder M (1989) Continuous monitoring of invitro oxidation of human low-density lipoprotein. Free Rad. Res. Commun. 6: 67-75.

[46] Yagi K (1984) Assay for blood-plasma or serum Methods Enzymol. 105: 328-331.

[47] Hong R, Kang TY, Michels CA, Gadura N (2012) Membrane Lipid Peroxidation in Copper Alloy-Mediated Contact Killing of Escherichia coli. Appl. Environ. Microbiol. 78: 1776-1784.

[48] Martins DB, Mazzanti CM, Franca RT, Pagnoncelli M, Costa MM, de Souza EM, Goncalves J, Spanevello R, Schmatz R, da Costa P, Mazzanti A, Beckmann DV, Cecim MD, Schetinger MR, Lopes STD (2012) 17-beta estradiol in the acetylcholinesterase activity and lipid peroxidation in the brain and blood of ovariectomized adult and middle-aged rats. Life Sciences. 90: 351-359.

[49] Kwok CT, van de Merwe JP, Chiu JMY, Wu RSS (2012) Antioxidant responses and lipid peroxidation in gills and hepatopancreas of the mussel Perna viridis upon exposure to the red-tide organism Chattonella marina and hydrogen peroxide. Harmful Algae. 13: 40-46.

[50] Ahmad MK, Syma S, Mahmood R (2011) Cr(VI) Induces Lipid Peroxidation, Protein Oxidation and Alters the Activities of Antioxidant Enzymes in Human Erythrocytes. Biol. Trace Elem. Res. 144: 426-435.

[51] Naziroglu M, Akkus S, Celik H (2011) Levels of lipid peroxidation and antioxidant vitamins in plasma and erythrocytes of patients with ankylosing spondylitis. Clin. Biochem. 44: 1412-1415.

[52] Pimentel VC, Pinheiro FV, Kaefer M, Moresco RN, Moretto MB (2011) Assessment of uric acid and lipid peroxidation in serum and urine after hypoxia-ischemia neonatal in rats. Neurol. Sci. 32: 59-65.

[53] Tonin AA, Thome GR, Calgaroto N, Baldissarelli J, Azevedo MI, Escobar TP, dos Santos LG, Da Silva AS, Badke MRT, Schetinger MR, Mazzanti CM, Lopes STD (2011) Lipid

Peroxidation and Antioxidant Enzymes Activity of Wistar Rats Experimentally Infected with Leptospira interrogans. Acta Sci. Vet. 39: 966-976.

[54] Elsaadani M, Esterbauer H, Elsayed M, Goher M, Nassar AY, Jurgens G (1989) A spectrophotometric assay for lipid peroxides in serum-lipoproteins using a commercially available reagent. J. Lipid Res. 30: 627-630.

[55] NouroozZadeh J, TajaddiniSarmadi J, Ling KLE, Wolff SP (1996) Low-density lipoprotein is the major carrier of lipid hydroperoxides in plasma - Relevance to determination of total plasma lipid hydroperoxide concentrations. Biochem. J. 313: 781-786.

[56] Eymard S, Genot C (2003) A modified xylenol orange method to evaluate formation of lipid hydroperoxides during storage and processing of small pelagic fish. Eur. J. Lipid Sci. Technol. 105: 497-501.

[57] Corongiu FP, Banni S (1994) Detection of conjugated dienes by 2nd-derivative ultraviolet spectrophotometry. Oxygen Radicals in Biological Systems, Pt C. 233: 303-310.

[58] Banni S, Day BW, Evans RW, Corongiu FP, Lombardi B (1995) Detection of conjugated diene isomers of linoleic-acid in liver lipids of rats fed a choline-devoid diet indicates that the diet does not cause lipoperoxidation. J. Nutr. Biochem. 6: 281-289.

[59] Dormandy TL, Wickens DG (1987) The experimental and clinical pathology of diene conjugation. Chem. Phys. Lipids. 45: 353-364.

[60] Chin SF, Storkson JM, Liu W, Pariza MW (1991) Dietary sources of the anticarcinogen CLA (conjugated dienoic derivatives of linoleic-acid). Faseb J. 5: A1444-A1444.

[61] Britton M, Fong C, Wickens D, Yudkin J (1992) Diet as a source of phospholipid esterified 9,11-octadecadienoic acid in humans. Clin. Sci. 83: 97-101.

[62] Fairbank J, Ridgway L, Griffin J, Wickens D, Singer A, Dormandy TL (1988) Octadeca-9-11-dienoic acid in diagnosis of cervical intraepithelial neoplasia. Lancet. 2: 329-329.

[63] Jack CIA, Ridgway E, Jackson MJ, Hind CRK (1994) Serum octadeca-9,11 dienoic acid - an assay of free-radical activity or a result of bacterial production. Clin. Chim. Acta. 224: 139-146.

[64] Thompson S, Smith MT (1985) Measurement of the diene conjugated form of linoleic-acid in plasma by high-performance liquid-chromatography - A questionable noninvasive assay of free-radical activity Chem.-Biol. Interact. 55: 357-366.

[65] Janero DR (1990) Malondialdehyde and thiobarbituric acid-reactivity as diagnostic indexes of lipid-peroxidation and peroxidative tissue-injury. Free Radic. Biol. Med. 9: 515-540.

[66] Yu TC, Sinnhuber RO (1957) 2-Thiobarbituric acid method for the measurement of rancidity in fishery products. Food Technol. 11: 104-108.

[67] Mihara M, Uchiyama M (1981) Evaluation of thiobarbituric acid (TBA) value as an index of lipid-peroxidation in CCl4-intoxicated rat-liver. Yakugaku Zasshi-J. Pharm. Soc. Jpn. 101: 221-226.

[68] Mihara M, Uchiyama M (1983) Properties of thiobarbituric acid-reactive materials obtained from lipid peroxide and tissue homogenate. Chem. Pharm. Bull. 31: 605-611.

[69] Mohebbi-Fani M, Mirzaei A, Nazifi S, Shabbooie Z (2012) Changes of vitamins A, E, and C and lipid peroxidation status of breeding and pregnant sheep during dry seasons on medium-to-low quality forages. Trop. Anim. Health Prod. 44: 259-265.

[70] Keles H, Ince S, Kucukkurt I, Tatli, II, Akkol EK, Kahraman C, Demirel HH (2012) The effects of Feijoa sellowiana fruits on the antioxidant defense system, lipid peroxidation, and tissue morphology in rats. Pharm. Biol. 50: 318-325.

[71] Kosugi H, Kato T, Kikugawa K (1987) Formation of yellow, orange, and red pigments in the reaction of alk-2-enals with 2-thiobarbituric acid. Anal. Biochem. 165: 456-464.

[72] Frankel EN (1991) Recent advances in lipid oxidation. J. Sci. Food Agric. 54: 495-511.

[73] Bonnestaourel D, Guerin MC, Torreilles J (1992) Is malonaldehyde a valuable indicator of lipid-peroxidation Biochem. Pharmacol. 44: 985-988.

[74] Lapenna D, Cuccurullo F (1993) TBA test and free MDA assay in evaluation of lipid-peroxidation and oxidative stress in tissue systems Am. J. Physiol. 265: H1030-H1031.

[75] Gutteridge JMC, Tickner TR (1978) Thiobarbituric acid reactivity of bile-pigments. Biochem. Med. 19: 127-132.

[76] Largilliere C, Melancon SB (1988) Free malondialdehyde determination in human-plasma by high-performance liquid-chromatography. Anal. Biochem. 170: 123-126.

[77] Halliwell B, Gutteridge JMC, Aruoma OI (1987) The deoxyribose method - a simple test-tube assay for determination of rate constants for reactions of hydroxyl radicals. Anal. Biochem. 165: 215-219.

[78] Wade CR, Vanrij AM (1989) Plasma malondialdehyde, lipid peroxides, and the thiobarbituric acid reaction. Clin. Chem. 35: 336-336.

[79] Wade CR, Jackson PG, Vanrij AM (1985) Quantitation of malondialdehyde (MDA) in plasma, by ion-pairing reverse phase high-performance liquid-chromatography. Biochem. Med. 33: 291-296.

[80] Wade CR, Vanrij AM (1988) Plasma thiobarbituric acid reactivity - Reaction conditions and the role of iron, antioxidants and lipid peroxy-radicals on the quantitation of plasma-lipid peroxides. Life Sciences. 43: 1085-1093.

[81] Wong SHY, Knight JA, Hopfer SM, Zaharia O, Leach CN, Sunderman FW (1987) Lipoperoxides in plasma as measured by liquid-chromatographic separation of malondialdehyde thiobarbituric acid adduct. Clin. Chem. 33: 214-220.

[82] Santos MT, Valles J, Aznar J, Vilches J (1980) Determination of plasma malondialdehyde-like material and its clinical-application in stroke patients. J. Clin. Pathol. 33: 973-976.

[83] Mezes M, Bartosiewicz G (1983) Investigations on vitamin-E and lipid peroxide status in rheumatic diseases. Clin. Rheumatol. 2: 259-263.

[84] Kedziora J, Bartosz G, Gromadzinska J, Sklodowska M, Wesowicz W, Scianowski J (1986) Lipid peroxides in blood-plasma and enzymatic antioxidative defense of erythrocytes in Downs-syndrome. Clin. Chim. Acta. 154: 191-194.

[85] Takagi T, Mitsuno Y, Masumura M (1978) Determination of peroxide value by colorimetric iodine method with protection of iodide as cadmium complex. Lipids. 13: 147-151.

[86] Jessup W, Dean RT, Gebicki JM (1994) Iodometric determination of hydroperoxides in lipids and proteins. Oxygen Radicals in Biological Systems, Pt C. 233: 289-303.

[87] Hicks M, Gebicki JM (1979) Spectrophotometric method for the determination of lipid hydroperoxides. Anal. Biochem. 99: 249-253.

[88] Cramer GL, Miller JF, Pendleton RB, Lands WEM (1991) Iodometric measurement of lipid hydroperoxides in human plasma. Anal. Biochem. 193: 204-211.

[89] Chajes V, Sattler W, Stultschnig M, Kostner GM (1996) Photometric evaluation of lipid peroxidation products in human plasma and copper oxidized low density lipoproteins: Correlation of different oxidation parameters. Atherosclerosis. 121: 193-203.

[90] Gay C, Collins J, Gebicki JM (1999) Hydroperoxide assay with the ferric-xylenol orange complex. Anal. Biochem. 273: 149-155.

[91] Griffiths G, Leverentz M, Silkowski H, Gill N, Sanchez-Serrano JJ (2000) Lipid hydroperoxides in plants. Biochem. Soc. Trans. 28: 837-839.

[92] Nouroozzadeh J, Tajaddinisarmadi J, Birlouezaragon I, Wolff SP (1995) Measurement of hydroperoxides in edible oils using the ferrous oxidation in xylenol orange assay J. Agric. Food Chem. 43: 17-21.

[93] Burat KM, Bozkurt O (1996) Improvement of calibration curve for determining peroxide values of food lipids by the modified ferrous oxidation-xylenol orange method. J. AOAC Int. 79: 995-997.

[94] Bou R, Codony R, Tres A, Decker EA, Guardicila F (2008) Determination of hydroperoxides in foods and biological samples by the ferrous oxidation-xylenol orange method: A review of the factors that influence the method's performance. Anal. Biochem. 377: 1-15.

[95] Gupta BL (1973) Microdetermination techniques for H2O2 in irradiated solutions. Microchem J. 18: 363-374.

[96] Hermeslima M, Willmore WG, Storey KB (1995) Quantification of lipid-peroxidation in tissue-extracts based on Fe(III)xylenol orange complex-formation. Free Radic. Biol. Med. 19: 271-280.

[97] Jiang ZY, Hunt JV, Wolff SP (1992) Ferrous ion oxidation in the presence of xylenol orange for detection of lipid hydroperoxide in low-density-lipoprotein Anal. Biochem. 202: 384-389.

[98] Wolff SP Ferrous ion oxidation in presence of ferric ion indicator xylenol orange for measurement of hydroperoxides, in: L. Packer (Ed.), Oxygen Radicals in Biological Systems, Pt C, 1994, pp. 182-189.

[99] Banerjee D, Madhusoodanan UK, Sharanabasappa M, Ghosh S, Jacob J (2003) Measurement of plasma hydroperoxide concentration by FOX-1 assay in conjunction with triphenylphosphine. Clin. Chim. Acta. 337: 147-152.

[100] Arab K, Steghens JP (2004) Plasma lipid hydroperoxides measurement by an automated xylenol orange method. Anal. Biochem. 325: 158-163.

[101] Sodergren E, Nourooz-Zadeh J, Berglund L, Vessby B (1998) Re-evaluation of the ferrous oxidation in xylenol orange assay for the measurement of plasma lipid hydroperoxides. J. Biochem. Biophys. Methods. 37: 137-146.

[102] Grau A, Codony R, Rafecas M, Barroeta AC, Guardiola F (2000) Lipid hydroperoxide determination in dark chicken meat through a ferrous oxidation-xylenol orange method. J. Agric. Food Chem. 48: 4136-4143.

[103] Griffiths G, Leverentz M, Silkowski H, Gill N, Sanchez-Serrano JJ (2000) Lipid hydroperoxide levels in plant tissues. J. Exp. Bot. 51: 1363-1370.

[104] Huska D, Adam V, Babula P, Trnkova L, Hubalek J, Zehnalek J, Havel L, Kizek R (2011) Microfluidic robotic device coupled with electrochemical sensor field for handling of paramagnetic micro-particles as a tool for determination of plant mRNA. Microchim. Acta. 173: 189-197.

[105] Sochor J, Salas P, Zehnalek J, Krska B, Adam V, Havel L, Kizek R (2010) An assay for spectrometric determination of antioxidant activity of a biological extract. Lis. Cukrov. Repar. 126: 416-417.

[106] Sochor J, Ryvolova M, Krystofova O, Salas P, Hubalek J, Adam V, Trnkova L, Havel L, Beklova M, Zehnalek J, Provaznik I, Kizek R (2010) Fully automated spectrometric protocols for determination of antioxidant activity: Advantages and disadvantages. Molecules. 15: 8618-8640.

Chapter 6

Trends in the Evaluation of Lipid Peroxidation Processes

Mihaela Ilie and Denisa Margină

Additional information is available at the end of the chapter

1. Introduction

Oxidative stress occurs as a result of imbalance between the antioxidant and prooxidant systems acting at certain points in metabolic processes, in favor of the last. The oxidative stress, defined by H Sies following extensive research performed between 1981 and 1993, is the outcome of intense generation of reactive oxygen species (ROS), which are not counteracted by endogenous antioxidant molecules [Sies, 1985]. Current knowledge links many types of pathologies to oxidative damage; among them, most cited are atherosclerosis, diabetes mellitus, neurodegenerative disorders, cancers, rheumatic diseases, autoimmune disorders, etc. Figure 1, sometimes referred to as "oxidative stress wheel", presents the most important diseases in which oxidative stress is involved resulting in biochemical lessions

Free radicals are chemical species containing unpaired electrons, which can increase the reactivity of atoms or molecules. Free radicals are highly reactive and unstable, due to their impaired electrons; they can react locally, accepting or donating electrons, in order to become more stable. The reaction between a radical and a non-radical compound generally leads to the propagation of the radical chain reaction, and to an increasing generation of new free radicals. During biochemical processes that normally take place in living cells, many types of free radicals are generated: oxygen-, sulfur-, bromide- and chloride- centered species [Halliwell & Gutteridge, 2007]. The most common reported cellular free radicals are singlet oxygen ($^1\Sigma g^+ O_2$), hydroxyl (OH·), superoxide (O_2^-·) and nitric monoxide (NO·). Also, some other molecules like hydrogen peroxide (H_2O_2) and peroxynitrite (ONOO$^-$) (which are not free radicals from the chemical point of view, having all-paired electrons) are reported to generate free radicals in living organisms through various chemical reactions [Halliwell, 2006].

In this context, it is extremely important to evaluate the extent and rate of the lipid peroxidation process using different methods and experimental models, ranging from quantitative assay of lipoperoxides end products to the evaluation of changes in certain metabolic processes under the influence of pro-oxidative or antioxidative known substances. The present article aims at reviewing different techniques, methods and experimental models for the evaluation of lipid peroxidation that can be used in clinical research and in basic biochemical research as well. Simple, rapid, cost effective, and more elaborated, expensive methods are critically evaluated, presenting the advantages and limitations of each one. A special emphasis is given to fluorescent methods, which our team is frequently using to evaluate the lipid peroxidation processes.

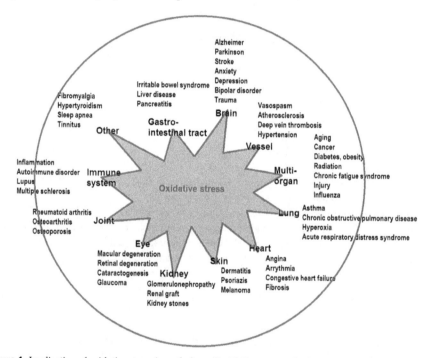

Figure 1. Implication of oxidative stress in pathology ("oxidative stress wheel")

2. Oxidative stress, ROS and implication in metabolic procesess

2.1. Oxygen centered reactive species (ROS)

Reactive oxigen species is a generic term that includes both oxygen radicals and certain non-radicals that are oxidizing agents and/or are easily converted into radicals, such as H_2O_2, ozone (O_3), singlet oxygen ($\Delta g^1 O_2$), peroxynitrite, hypochlorous acid (HOCl), etc. [Halliwell, 2006].

ROS are generated as a result of oxygen action on nutrients or on physiological components in living organisms.

The sources for oxidative stress are either endogenous (abnormal mithocondria and peroxisomes function, lipoxygenase, NADPH-oxidase, cytochrome P450 activity), endogenous antioxidant systems dysfunction (low amount of non-enzymatic antioxidants such as gluthatione, vitamins A, C and E, reduced enzimatic activity) or exogenous agents (ultraviolet or ionizing radiation, toxins, chemotherapy, bacteria, etc.) (Figure 2).

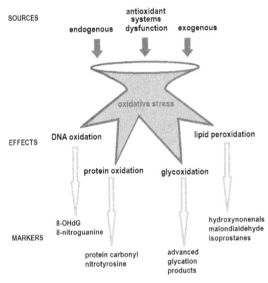

Figure 2. Sources, effects and main markers of oxidative stress

ROS can induce many damaging cellular processes, such as DNA oxidative lesions, loss of membrane integrity due to lipid peroxidation, protein and functional carbohydrate structural changes, etc. All these structural and functional changes have direct clinical consequences, leading to the acceleration of the general aging process, but also to some pathological phenomena, associated with the increase of the capillary permeability, impairment of the blood cell function, etc. ROS lesions are frequently associated with aging [Dröge & Schipper, 2007; Griffiths et al., 2011], atheroclerosis [Hulsmans & Holvoet, 2010], cardio-vascular disease [Dikalov & Nazarewicz, 2012; Puddu et al., 2009], type I or type II diabetes mellitus [Cai et al., 2004], autoimmune disorders, neurodegenerative disorders such as Parkinson [Yoritaka et al., 1996] or Alzheimer's disease [Sayre et al., 1997; Takeda et al., 2000], inflammatory diseases such as reumatoid arthritis [Griffiths et al., 2011] or different types of cancers [Lenaz, 2012; Li et al., 2009; Manda et al., 2009].

In order to counteract the damaging action of the physiologically generated ROS, the living organisms developed efficient antioxidant systems [Christofidou-Solomidou & Muzykantov 2006; Halliwell, 2006; Sies, 1997; Veskoukis et al., 2012]. Endogenous antioxidants in the human body act through different types of mechanisms:

- reducing the ROS generation – through chelating the metal ions (ferum, copper, etc.) by specific or non-specific proteins (ferritine, transferrine, albumin), so that these ions can no longer participate to redox reactions [Aruoma et al., 1989; Elroy-Stein et al., 1986; Freinbichler, 2011; Halliwell & Gutteridge, 1984; Velayutham, 2011]
- stopping the ROS formation chain reaction – generally via antioxidants with small molecular mass, such as reduced glutathion (GSH), vitamins E and C, uric acid, etc. [Gordon, 2012; Halliwell, 2006]
- scavenging ROS with antioxidant enzymes such as superoxyde dismutase (SOD), catalase (CAT), glutathion-peroxidase (GPx), glutathion-reductatse (GR), etc. [Halliwell, 2006; Sies, 1997]
- reparing lesions caused by ROS via specific enzymes such as endonucleases, peroxidases, lipases, etc. [Sies, 1997].

All these antioxidant systems act differently, depending on their structure and properties, their hydrophilic or lipophilic character, and also depending on their localization (intracellular or extracellular, in cell or organelles membrane, in the cytoplasm, etc.). All the aforementioned systems act sinergically and form a network which protects living cells from the destructive action of ROS (Figure 3).

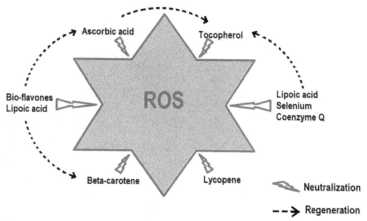

Figure 3. ROS neutralization by several biomolecules

2.2. Nitrogen centered radical species (RNS)

After the discovery of the physiological role of endogenously produced nitric oxide (NO), the capacity of this bio-molecule to react with other cellular components (such as proteins and lipids), specific nitrosative chemical changes have emerged as a key signaling mechanism in cell physiology. Several studies reported the involvement of excess generation of NO and its adducts in the etiology of multiple disease states, including insulin resistance and diabetes, atherosclerosis or Alzheimer's disease [Duplain et al., 2008; Parastatidis et al., 2007; Uehara, 2007; Yasukawa T et al., 2005; White et al., 2010].

The nitrosative modifications of proteins take two main forms: either S-nitrosylation of cysteine thiols or nitration of tyrosine residues. Both chemical processes may arise from protein interactions with NO or with secondary intermediates of NO, otherwise termed reactive nitrogen species (RNS) [White et al., 2010]. One of the very important members of the RNS group is represented by peroxynitrite (ONOO$^-$), produced from the reaction of NO with the superoxide anion (O^{2-}), which is considered as one of the major cellular nitrating agents [Hogg, 2002; White et al., 2010]. Other nitrating agents are the nitrosonium cation (derived from the action of myeloperoxidase), produced from the reaction of nitrite with hydrogen peroxide and nitroso-peroxocarbonate, which results from the reaction of carbon dioxide with peroxynitrite. Lipid peroxyl radicals have been recently shown to promote tyrosine nitration by inducing tyrosine oxidation and also by reacting with NO$_2^-$ to produce · NO$_2$ [Bartesaghi et al., 2010; Denicola et al., 1996; Lang et al., 2000].

2.3. Oxidative stress and lipid peroxidation

Among the targets of ROS and RNS, lipids are basically the most vulnerable, as their peroxidation products can result in further propagation of free radical reactions [Halliwell & Chirico, 1992]. The brain is a high oxygen-consuming organ and the nervous cell has also the greatest lipid-to-protein ratio; besides, the brain has a relatively week protection systems against ROS generation, therefore it is particularly vulnerable to oxidative stress. The age-related increase in oxidative brain damage results in intense generation of lipid peroxidation products, protein oxidation, oxidative modifications in nuclear and mitochondrial DNA [Grimsrud et al., 2008].

Polyunsaturated fatty acids (PUFAs) and their metabolites have many physiological roles such as energy generation, direct involvement in cellular and sub-cellular membrane structure and function, implication in cell signaling processes and in the regulation of gene expression as well. They constitute the main target of ROS action in the lipid peroxidation reactions.

The general process of lipid peroxidation consists of three stages: initiation, propagation, and termination [Auroma et al., 1989; Leopold & Loscalzo, 2009]. Many species can be responsible for the initiation of the chain reaction the radicals: hydroxyl, alkoxyl, peroxyl, superoxide or peroxynitrite. As a consequence, a free radical atracts a proton from a carbon of a fatty acyl side chain leaving the remaining carbon radical accessible to molecular oxygen to form a lipid peroxyl radical. This is also highly reactive and the chain reaction is propagated further. As a result, PUFA molecules are transformed into conjugated dienes, peroxy radicals and hydroperoxides, which will undergo a cleavage mainly to aldehydes. More than 20 lipoperoxidation end-products were identified [Niki, 2009]; among the components of PUFAs oxidative degradation products, the most frequently mentioned were acrolein, malondialdehyde (MDA), 4-hydroxyalkenals and isoprostanes [Esterbauer et al., 1991; Leopold & Loscalzo, 2009].

RNS also play a major role in lipid biology, by two major pathways: targeting some enzymes (COX-2 and cytochrome P-450) and thus influencing bioactive lipid synthesis and interacting with unsaturated fatty acids (such as oleate, linoleate and arachidonate) and generating novel nitro-fatty acids. Nitrated lipids, such as nitroalkenes, may undergo aqueous decay and release independent of thiols, isomerize to a nitrite ester with N-O bond cleavage, or generate an enol group and ·NO [Leopold & Loscalzo, 2009]. The nitro-fatty acids have distinct bioactivities from their precursor lipids [Baker et al., 2009; Freeman et al., 2008; Kim et al., 2005; Lee et al., 2008; White et al., 2010]. Studies identified a high number of nitro-fatty acid species and proved an elevated formation of RNS in hydrophobic environments, such as the lipid bilayer, suggesting that lipids might constitute candidates for nitrosative signal transduction [Jain et al., 2008; Moller et al., 2005; Moller et al., 2007; Thomas et al., 2001].

Nitroalkenes may also participate in reactions with cysteine and histidine residues in proteins and with the thiolate anion of glutathione (GSH) to initiate reversible modification(s) of proteins. Thiyl radicals may also initiate lipid peroxidation by extraction of a hydrogen atom from bis-allylic methylene groups of fatty acids generating pentadienyl radicals. These radicals, in turn, may react with oxygen to generate peroxyl radicals [Leopold & Loscalzo, 2009].

Lipid hydroperoxides (LOOHs) are intermediates of PUFAs lipid peroxidation and can be also found as minor constituents of cell membranes; these compounds are also final products of prostaglandin and leukotriene biosynthesis, and can be decomposed by transition metals to form alkoxyl and lipoperoxyl radicals. Furthermore, biomolecules such as proteins or amino-lipids, can be covalently modified by lipid decomposition products (i.e. by forming Schiff bases with aldehydes or/and by activating membrane-bound enzymes). In consequence, lipid peroxidation may alter the arrangement of proteins in bilayers and thereby interfere with their physiological role in the membrane function.

2.4. Pathological involvement of the lipid peroxidation process

Many lipid peroxidation products, either full chain or chain-shortened, have been reported to be harmful or to have pro-infl ammatory effects [Birukov 2006; Niki 2009; Salomon 2005].

Lipid peroxidation increases the permeability of cellular membranes, resulting in cell death.

The lipid peroxidation process located at the cell membrane level may lead to loss of integrity and viability, and also to altered cell signaling and finally to tissue dysfunction; the oxidation of plasma lipoproteins is probably a major contributor to the formation of lipid peroxidation products and is widely thought to be involved in atherosclerosis. Malondialdehyde and 4-hydroxy-2-nonenal (HNE), the well known products of lipid peroxidation, react with a variety of biomolecules, such as proteins, lipids and nucleic acids, and they are thought to contribute to the pathogenesis of human chronic diseases [Breusing et al., 2010].

A clear example of the pathological role of PUFAs peroxidation is the evolution of atherosclerotic lesions to cardio-vascular disease. Until 1970, it was considered that dyslipidemia was the main factor initiating the atherosclerotic lesions. Later on, researchers emphasized the involvement of inflammatory processes, growth factors, smooth muscle cells proliferation, as well as viruses, bacteria or tumor phenomenon in the atherosclerosis, besides lipids and lipoproteins.

RNS interaction with different types of proteins is directly involved in physio-pathological processes. Several studies proved that key enzymes involved in glycolysis, β-oxidation, the tricarboxylic acid cycle and electron transport chain are targets of tyrosine nitration or S-nitrosylation. Modifications by nitrosylation that reduce the activity or function of important tricarboxylic acid cycle and electron transport proteins have the potential to slow substrate oxidation and probably lead to the build up of metabolic intermediates (particularly lipids) that could impair signaling pathways to reduce insulin action. Also, key insulin-signaling intermediaries are S-nitrosylated, and this could constitute a potential mechanism of insulin resistance [Chouchani et al., 2010].

3. Evaluation of the end products of lipid peroxidation

Malonidialdehyde (MDA) is one of the most cited lipoperoxidation product originating from PUFAs. Several generation mechanisms have been proposed for MDA. Pryor & Stanley (1975) considered them as bicyclic endoperoxydes coming up from nonvolatiles MDA precursors, similar to prostaglandins. This mechanism was confirmed in 1983 by Frankel & Nef. Two other mechanisms were postulated by Esterbauer (1991) and consist in the successive generation of peroxydes and β-cleavage of the lipid chain or as a reaction of acroleine radical with a hydroxyl moiety (exemplification for arachidonic acid in Figure 4, adapted from Esterbauer, 1991). Hecker & Ulrich (1989) consider that MDA can be generated *in* vivo by means of enzymatic processes linked to prostaglandins.

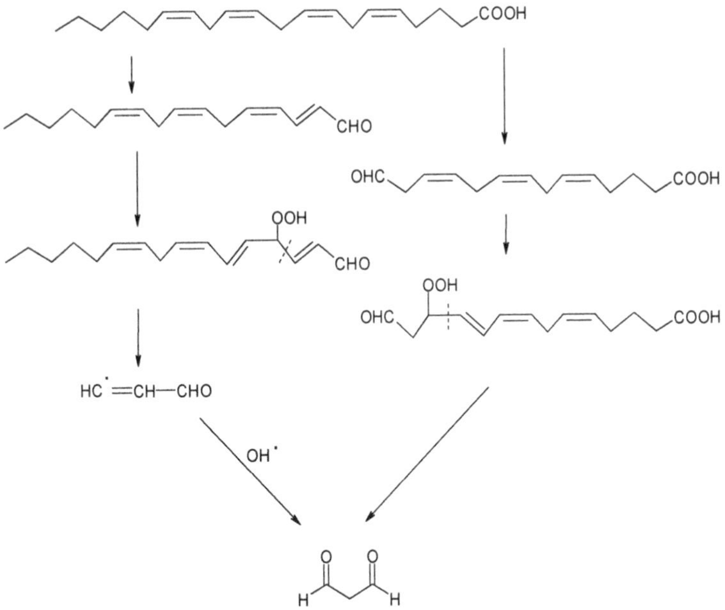

Figure 4. Generation of MDA as proposed by Esterbauer (1991)

Because the evaluation of the end products of lipid peroxidation at the tissue level is considered of maximum importance in both clinical and toxicological research, several methods to assay MDA have been proposed [Del Rio, 2005]. Since 1948, the most used method to assay lipoperoxidation end products is based on the studies of Bernheim et al., consisting in MDA condensation with the thiobarbituric acid (TBA) leading to a red complex which can be quantified by visible absorption spectrophotometry (in the range 500-600 nm, depending on the procedure used) or fluorescence spectroscopy (Figure 5). As TBA reacts also with several other aldehydes commonly present in the biological sample, the agents reacting with TBA are frequently denoted as thiobarbituric acid reacting species (TBARS). It must be said that TBARs and MDA are a rather imprecise measure of the lipid peroxidation process, since many substances that are present in human biological fluids can also react with TBA.

Moreover the reaction conditions (heating) can lead to the degradation of other molecules in the sample, increasing the amount of MDA that is available for the reaction with TBA. Therefore this assay usually gives an overestimation of free radical damage [Cherubini et al., 2005].

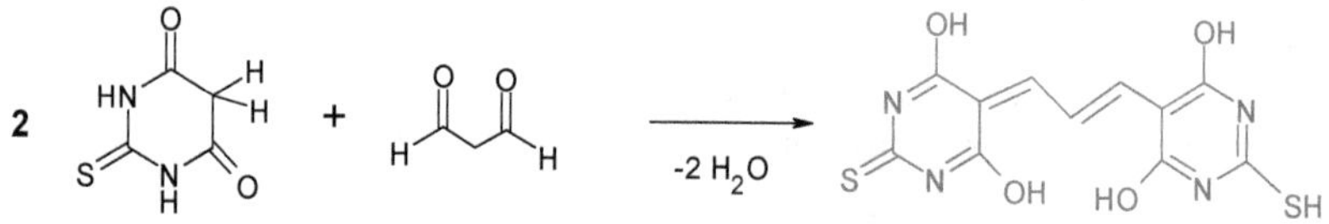

Figure 5. Mechanism of reaction for TBARS quantification

The authors generally avoid interferences by using different methods: Kwiecień et al., 2002, used BHT (butylated hydroxytoluene) to prevent further oxidation of the sample; deproteinisation was performed with trichloracetic acid [Cassini et al., 1986], forced the lipoperoxidation and stopped the reaction with SDS and acetic acid [Gautam et al., 2010; Ohkawa et al., 1979], added EDTA and refered the results to standards prepared from tetramethoxypropane [Houglum et al., 1990] or used the standard addition method [Sprinteroiu et al., 2010].

The specificity of the measurement is improved by HPLC to separate the MDA-TBA adduct from interfering chromogens [Agarwal & Chase, 2002; Del Rio et al., 2003; Lykkersfeldt, 2001; Templar et al., 1999].

The principles of TBARS assay is so popular, that a few companies even developed kits for clinical research to assay MDA spectrophotometrically from biological samples.

Apart from the above mentioned method, other methods were applied for the quantitative assay of lipoperoxidation end-products: direct HPLC [Karatas et al., 2002], capillary electrophoresis [Wilson et al., 1997], RP-HPLC, derivatisation with 2,4-diphenylhydrazine [Sim et al., 2003], pre-column derivatisation with diaminonaphtalene at acidic pH (for protein bound MDA) or alkaline pH (for non protein-bound MDA), followed by HPLC-UV analysis [Stegens, 2001], GC-MS analysis following derivatisation with phenylhydrazine [Cighetti et al., 2002], GC-ECD-MS after derivatisation with 2,4,6-trichlorophenylhydrazine [Stalikas & Konidari, 2001].

4. Evaluation of the antioxidant status

For the evaluation of the antioxidant status in biological samples several markers (either enzymatic or non-enzymatic) can be used. Generally, the results obtained from the evaluation of antioxidant status markers should be correlated with certain peroxidative parameters, in order to be able to draw conclusions from the experiments.

Among the enzymatic markers of the antioxidant defense mechanism of the biological samples, literature cites some specific enzymes, such as catalase, superoxide-dismutase, glutathione-reductase, glutathione-peroxidase, etc. Each of these enzymes can be assayed using specific kits.

Commonly used are also non-enzyimatic markers, such as reduced glutathione, some vitamins (ascorbic acid, tocopherols, carotenoides).

A series of commercial kits for measuring the antioxidant status in biological samples (blood, serum, but also food products) are also available - the so-called *total antioxidant status* kits. The assays can be colorimetric (one example is the kit using as a chromogen 2,2'-azino-di-[3-ethylbenzthiazole sulfonate], which reacts with methmyoglobin and hydrogen peroxide to give a coloured cation), chemiluminometric (using the reaction of luminol with hydrogen peroxyde), or physico-chemical (potentiometric).

These methods allow the evaluation of the total antioxidant capacity in the biological samples, thus accounting both for enzymatic and for non-enzymatic bio-molecules.

Our group developed a method enabling a distinct evaluation for the biological samples antioxidant capacity as resulting exclusively from redox hydrophilic biomolecules. The method is based on potentiometric evaluation of the status of oxidant and reducing species in samples of human serum. We used a micro Pt/AgCl combination redox electrode, with an internal reference, and a Tistand 727 Potentiometer (Metrohm AG, Switzerland). The baseline apparent redox potential of the human serum (ARP0) was measured; than a mild prooxidant chemical system (quinhydrone) was added to the biological samples. After incubating at 25°C for 1h respectively 3 hours, two final apparent redox potentials were recorded (ARPf). Quinhydrone mimics the prooxidant conditions developing *in vivo* and consumes the reducing species leading to an increase of the apparent redox potential in time. This dynamic recording of the data allowed the calculation of a difference between the final value of the ARP (ARPf) and the initial one (ARP0), thus defining the redox stability index (RSI). This parameter illustrates the serum sample capacity to counteract the prooxidant agent. The lower the RSI, the higher the activity of the hydrosoluble antioxidant protective systems in the serum [Margina et al., 2009].

5. Monitoring the induced-peroxidation process

J. Goldstein, M. Brown, and D, Steinberg emphasized more than a decade ago, that low density lipoproteins can be chemically modified, loosing their ability to be recognized by the classical LDL receptors [Brown & Jessup, 1999; Steinberg, 2009]. The oxidation of LDL does not take place in the blood stream, but inside the intimae, after lipoprotein complexes crossing through the endothelium. Therefore, in the oxidation process intracellular as well as extra cellular components are involved. This change of the LDL is realized mainly by oxidation with the free radicals that appear in large amounts in hypertension, diabetes mellitus, as a consequence of smoking, or in viral and bacterial infections, but LDL may be also modified by glycation (in type II diabetes mellitus), by association with proteoglycans or by incorporation in immune complexes [Ross, 1999].

Oxidised LDL particles (LDLox) have been proved to have different proatherogenic effects that can be predominantly attributed to their lipid components. Mainly, the uptake of LDLox by macrophages is enhanced through the scavenger receptor. Products generated from the decomposition of peroxidized lipids, such as aldehydes, modify the apolipoprotein B-100 (apoB-100) structure to a more electronegative form able to interact with the macrophage scavenger receptor. The modified LDL particles are taken up by scavenger receptors on the macrophages instead of the classical LDL receptors. This process is not regulated by feed back inhibition and allows the excessive build-up of cholesterol inside the cells, leading to their transformation into foam cells that are involved in the initiation and progression of atherosclerotic lesion [Parthasarathy et al., 1999; Shamir, 1996].

Alpha-tocopherol is an antioxidant from the LDL structure, and is the first one degraded during the radical attack on these lipoproteins. When this antioxidant protection is exhausted, PUFA are changed into lipid hydroperoxides. There is a great variability between subjects regarding the amount of PUFA and of antioxidants from LDL, which explains the variability concerning the susceptibility to oxidation of LDL particles. There are also a lot of other factors that influence *in vitro* evaluation of LDLox: some endogenous compounds, diet, some medicines, and probably genetic factors. Therefore, the assay of the *in vitro* LDL susceptibility to lipid peroxidation constitutes an important marker in the evaluation of atherogenic models/patients [Parthasarathy et al., 1999; Shamir, 1996].

The susceptibility of lipoprotein particles to lipid peroxidation can be assessed, after the isolation of LDL, either by treatment with copper salts, with mixtures of ferric compounds and ascorbic acid, or other prooxidant systems.

In order to evaluate this susceptibility to oxidation Esterbauer et al. proposed in 1989 a method based on the variation of absorbance of conjugated dienes in time at 234 nm. These dienes are relatively stable products resulted by a rearrangement of the double bonds from the PUFA molecules after the radical hydrogen abstraction. The increase in absorption at 234 nm is due to the formation of conjugated dienes during the peroxidation of polyunsaturated fatty acids. Absorbance at 234 nm shows an initial slower increase (lag phase) as antioxidants are destroyed and then increases more rapidly (propagation phase) reaching a plateau phase at which the absorbance is maximal as the rate of formation of dienes approaches their rate of decomposition [Esterbauer et al., 1990].

Another method for the assay of lipoprotein susceptibility to lipid peroxidation is based on the ability of lipid hydroperoxides to convert iodide (I^-) to iodine (I_2), which will than react with the iodide excess and form I_3^- that absorbs at 365 nm. There is a direct stoechiometric relationship between the amount of organic peroxides that resulted from the reaction and the concentration of I_3^- [Steinberg D., 1990, Esterbauer H. 1993]

One of the symplest methods for the assay of the susceptibility of LDL particles to lipid peroxidation is based on the selective precipitation of serum LDL with heparin at isoelectrical point (pH=5.4), method which is cheaper and quicker than the one based on ultracentrifugation. Personal results proved that the susceptibility of LDL to lipid peroxidation is correlated with the fasting plasma glucose level as well as with the lipid level [Margina et al., 2004].

The same type of assay can be used in order to evaluate the susceptibility to induced peroxidation for other kinds of biological samples (red blood cells, sub-cellular fractions such as mitochondria, or even tissue homogenates).

Previously published results [Margina D et al., 2011] proved that, for patients diagnosed with central obesity (BMI>30Kg/m2), adipose tissue susceptibility to lipid peroxidation correlated significantly with the total cholesterol (TC) level and with the LDL level. The susceptibility of adipose tissue to lipid peroxidation was assessed on white adipose tissue harvested from the abdominal area, homogenated in NaOH 0.015M, followed by TBARS evaluation. This parameter reflects the tendency to accumulation of free radicals in the adipose tissue of obese patients. In the same study, we pointed out that patients with impaired lipid profile (TC>220 mg/dl, LDL>150 mg/dl) had a significantly higher susceptibility of the adipose tissue to lipid peroxidation (p=0.036), associated with the decrease of the adiponectin level (Figure 6), compared to obese patients with physiologic lipid profile (TC<220mg/dl, LDL<150mg/dl).

Literature data also mention the assay of circulating LDLox, using different ELISA methods; one of these methods uses antibodies against a conformational epitope in the apolipoprotein B-100 (apoB-100) moiety of LDL that is generated as a consequence of substitution of at least 60 lysine residues of apoB-100 with aldehydes. This number of substituted lysines corresponds to the minimal number required for scavenger-mediated uptake of ox-LDL. The substituting aldehydes can be produced by peroxidation of lipids of LDL, leading to the generation of ox-LDL [Holvoet et al., 2006]. Another method might be the electrophoretic separation of LDL and LDLox particles from serum samples; studies proved that the electrophoretic mobility of oxidized LDL particles is increased compared to that of standard LDL [Lougheed et al., 1996].

Besides biochemical determination, noninvasive, real-time monitoring of lipid peroxidation using fluorescent probes has also been developed. The assays can be performed either on living cells (for example using cis-parinaric acid, fluoresceinated phosphoethanolamine, undecylamine-fluorescein, diphenyl-1-pyrenylphosphine – DPPP or other fluorescent

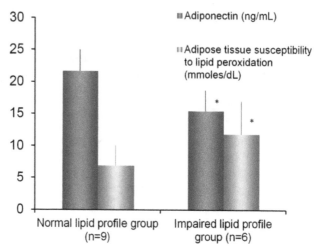

Figure 6. Obese patients with impaired plasma lipid profile (TC>220mg/dl, LDL>150mg/dl) are characterized by significantly different levels for the susceptibility of adipose tissue to induced lipid peroxidation as well as adiponectin level, compared to obese patients with normal plasma lipid profile (TC<220mg/dl, LDL<150mg/dl); * p<0.05 for cardio-vascular group compared to normal lipid profile group)

markers) or non-living samples (liposomes, tissue homogenates, plasma, serum, etc). In the cases of experiments that are performed on living cells, common limitation of use of fluorescent probes is that the probes often are cytotoxic or affect physiological activities of the cell [Drummen et al., 2004; Margina et al., 2012, Takahashi M. et al., 2001]. DPPP stoichiometrically reduces biologically generated hydroperoxides (such as fatty acid hydroperoxides, and triacylglycerol hydroperoxides) to their corresponding alcohols, and is transformed consequently into its oxide. DPPP is essentially non-fluorescent until oxidized to a phosphine-oxide by peroxides. Due to its solubility in lipids, DPPP intercalates into the membrane leaflets and reacts with lipid hydroperoxides, thus allowing the evaluation of peroxide formation in the membranes of live cells [Kawai et al.,, 2007]. Due to these chemical properties, the probe can be used in order to evaluate the extent of lipid peroxidation of biological materials such as cell membranes [Akasaka et al., 1993; Ohshima et al., 1996; Takahashi et al., 2001]. Because DPPP molecules are incorporated into the cell membranes, hydroperoxides located in the membrane are supposed to preferably react with DPPP.

Diphenyl-1-pyrenylphosphine (DPPP) is a synthetic compound with high reactivity against hydroperoxides, which has been used as a sensitive fluorescent probe for hydroperoxide analysis for HPLC methods. H_2O_2, which is least lipid-soluble, does not induce the peroxidation reaction of DPPP located in cell membranes. Although H_2O_2 is highly permeable to the membrane, it may not stay within the membrane long enough to react with DPPP effectively. Experimental lipid peroxidation can be induced by 10µM cumene hydroperoxide (CuOOH) which generates an effective reaction with DPPP in the membrane [Gomes et al., 2005; Takahashi et al., 2001].

We proved using DPPP in cell models (U937 human macrophage cell line as well as Jurkat lymphocytic cells) that the increase of certain polyphenol concentration (quercetin or epigallocatechin gallate) induces a decrease of the CuOOH induced lipid peroxidation of cell membranes. Fluorescence signals for the cells labeled with 5 μ M DPPP are presented in Figure 7 [Margina et al, 2012].

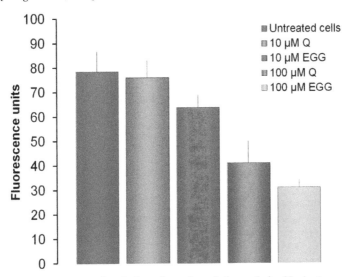

Figure 7. DPPP-oxide generation from Jurkat cell membrane before and after 20 minutes exposure to 10μM and 100 μM quercetin (Q) and epigallocatechin gallate (EGG) respectively; the fluorescence intensity was registered after 5 minutes of incubation of DPPP-labeled cells with CuOOH (for the induction of the lipid peroxidation)

6. Conclusions

Oxidative stress is among the most claimed causes of disease, as by its very definition indicates an abnormal biochemical function of the body. Among the targets of the oxidative stress, lipids are favorites, susceptible to structural changes that can decisively influence their normal function, and also generating hydroxyperoxides that further propagate the lipoperoxidation process.

Lipid peroxidation has been intensively studied in connection with normal and pathological metabolic processes; one of the main purposes was the understanding of toxicity triggered by lipoperoxidation end-products. That is why direct or indirect quantification of these products (TBARS, MDA, hydroxynonenals, prostaglandins, DNA and protein-adducts of the former, etc.) remains of interest for traditional and nowadays methods. The process of lipoperoxidation is often monitored in dynamics, even on living cells, using various techniques.

It is also of a great interest to seek for efficient antioxidants to prevent the excess lipoperoxidation, therefore one component of such kind of studies consist in finding out efficient markers for the oxidative stress and the antioxidant status. This can be fulfilled if a right and comprehensive understanding of the lipoperoxidation process is achieved. But this is still a faraway target.

Author details

Mihaela Ilie and Denisa Margină
Carol Davila University of Medicine and Pharmacy, Faculty of Pharmacy, Bucharest, Romania

Acknowledgments

The work was performed under the CNCSIS grant PD 132/30.07.2010 (PD 29/2010).

7. References

Agarwal R. & Chase S.D. (2002) Rapid, fluorimetric-liquid chromatographic determination of malondialdehyde in biological samples. *J Chromatogr B*, Vol. 775, No. 1, 121-126, ISSN 1570-0232.

Akasaka K., Ohrui H., Meguro H. & Tamura M. (1993). Determination of triacylglycerol and cholesterol ester hydroperoxides in human plasma by high-performance liquid chromatography with fluorometric postcolumn detection. *J. Chromatogr. B* Vol 617, No2, 205–211, ISSN 1570- 0232

Aruoma O.I., Halliwell B., Laughton M.J., Quinlan G.J. & Gutteridge J.M.C. (1989). The mechanism of initiation of lipid peroxidation. Evidence against a requirement for an iron(II)-iron(III) complex. *Biochem. J.* Vol. 258, No. 2, 617-620, ISSN 0264-6021

Baker P.R., Schopfer F.J., O'Donnell V.B. & Freeman B.A. (2009). Convergence of nitric oxide and lipid signaling: anti-inflammatory nitro-fatty acids. *Free Radic Biol Med* Vol 46, No 8, 989–1003, ISSN 0891-5849.

Bartesaghi S., Wenzel J., Trujillo M., Lopez M., Joseph J., Kalyanaraman B. & Radi R. (2010) Lipid peroxyl radicals mediate tyrosine dimerization and nitration in membranes. *Chem Res Toxicol* Vol 23, No 4, 821–835, ISSN 1520-5010.

Benkhai H., Lemanski S., Below H., Heiden J.U., Below E., Lademann J.,Bornewasser M., Balz T., Chudaske C. & Kramer A. (2010). Can physical stress be measured in urine using the parameter antioxidative potential? *GMS Krankenhaushyg Interdiszip.* Vol.5 No 2, Doc03.DOI: 10.3205/dgkh000146 ISSN 1863-5245

Bernheim F., Bernheim M.L.C. & Wilbur K.M. (1948). The reaction between thiobarbituric acid and the oxidation products of certain lipids. *J Biol Chem*, Vol. 174, 257-264, ISSN 0021-9258

Birukov KG. (2006) Oxidized lipids: the two faces of vascular inflammation. *Curr Atheroscler Rep* Vol 8, No 3, 223 – 231, ISSN 1534-6242.

Brown A.J. & Jessup W. (1999) Oxysterols and atherosclerosis. *Atherosclerosis* Vol. 142, 1–28, ISSN 0021-9150.

Cai W., He J.C., Zhu L., Peppa M., Lu C., Uribarri J. & Vlassara H. (2004). High Levels of Dietary Advanced Glycation End Products Transform Low-Density Lipoprotein Into a Potent Redox-Sensitive Mitogen-Activated Protein Kinase Stimulant in Diabetic Patients *Circulation*, Vol. 110, No. 3, 285-291, ISSN: 0009-7322.

Casini A.F, Ferrali M., Pompella A., Maellaro E. & Comporti M. (1986) Lipid Peroxidation and Cellular Damage in Extrahepatic Tissues of Bromobenzene-Intoxicated Mice. *Am J Pathol*, Vol. 123, 520-531, ISSN 0002-9440.

Cherubini A., Ruggiero C., Polidori M.C. & Mecocci P. (2005) Potential markers of oxidative stress in stroke. *Free Radic Biol Med* Vol. 39, No. 7, 841 – 852, ISSN 0891-5849

Chouchani E.T., Hurd T.R., Nadtochiy S.M., Brookes P.S., Fearnley I.M., Lilley K.S., Smith R.A. & Murphy MP. (2010) Identification of S-nitrosated mitochondrial proteins by S-nitrosothiol difference in gel electrophoresis (SNO-DIGE): implications for the regulation of mitochondrial function by reversible S-nitrosation. *Biochem J* Vol 430, No 1, 49–59, ISSN 0264-6021

Christofidou-Solomidou M. & Muzykantov V.R. (2006). Antioxidant Strategies in Respiratory Medicine, *Treat Respir Med* Vol. 5, No 1, 47-78, ISSN 1176-3450

Cighetti G., Allevi P., Anastasia L., Bortone L. & Paroni R. (2002). Use of methyl malondialdehyde as an internal standard for malondialdehyde detection: validation by isotope-dilution gas chromatography-mass spectrometry. *Clin Chem* Vol. 48, No. 12, 2266-2269, ISSN 1530-8561

Del Rio D., Pellegrini N., Colombi B., Bianchi M., Serafini M., Torta F, Tegoni M. & Musci M. (2003). Rapid fluorimetric method to detect total plasma malondialdehyde with mild derivatization conditions. *Clin Chem*, Vol. 49, No. 4, 690-692, ISSN 1530-8561

Denicola A., Freeman B.A., Trujillo M. & Radi R. (1996) Peroxynitrite reaction with carbon dioxide/bicarbonate: kinetics and influence on peroxynitrite mediated oxidations. *Arch Biochem Biophys* Vol 333, No 1, 49–58, ISSN 0003-9861.

Dikalov S.I. & Nazarewicz R.R. (2012). Angiotensin II-induced production of mitochondrial ROS: Potential mechanisms and relevance for cardiovascular disease. *Antioxid Redox Signal*. Mar 24. [Epub ahead of print], ISSN 1557-7716

Dröge W. & Schipper H.M. (2007). Oxidative stress and aberrant signaling in aging and cognitive decline. *Aging Cell* Vol. 6, No. 3, 361–370, ISSN 1474-9718.

Drummen G.P., Makkinje M., Verkleij A.J., Op den Kamp J.A., Post J.A. (2004) Attenuation of lipid peroxidation by antioxidants in rat-1 fibroblasts: comparison of the lipid peroxidation reporter molecules cis-parinaric acid and C11-BODIPY(581/591) in a biological setting. *Biochim Biophys Acta*. Vol. 1636, No. 2-3, 136-150, 0006-3002 .

Duplain H., Sartori C., Dessen P., Jayet P.Y., Schwab M., Bloch J., Nicod P. & Scherrer U. (2008) Stimulation of peroxynitrite catalysis improves insulin sensitivity in high fat diet-fed mice. *J Physiol* Vol 586, No 16, 4011–4016, ISSN 1469-7793

Eiserich J.P., Hristova M., Cross C.E., Jones A.D., Freeman B.A., Halliwell B. & van der Vliet A. (1998) Formation of nitric oxide-derived inflammatory oxidants by myeloperoxidase in neutrophils. *Nature* Vol 391, No 6665, 393–397, ISSN 1476-4687.

Elroy-Stein O., Bernstein Y. & Groner Y. (1986). Overproduction of human Cu/Zn-superoxide dismutase in transfected cells: extenuation of paraquat-mediated cytotoxicity and enhancement of lipid peroxidation. *EMBO J* Vol.5 No.3, 615-622, ISSN 0261-4189

Esterbauer H. (1993). Cytotoxicity and genotoxicity of lipid-oxydation products, *Am. J. Clin. Nutr.*, Vol 57, No 5, 779-785. ISSN 0002-9165.

Esterbauer H., Dieber-Rothender M., Waeg G., Striegl G. & Jurgens, G. (1990). Biochemical, structural, and functional properties of oxidized low-density lipoproteins. *Chem. Res. Toxicol.*, Vol. 3, No. 2, 77-92, ISSN 0893-228X .

Esterbauer H., Schaur R.J, & Zollner H. (1991). Chemistry and biochemistry of 4-hydroxynonenal, malonaldehyde and related aldehydes. *Free Radic Biol Med*, Vol. 11, No. 1, 81-128, ISSN 0891-5849.

Frankel E.N. & Neff W.E. (1983) Formation of malonaldehyde from lipid oxidation products, *Biochim Biophys Acta* Vol. 754, No. 3, 264-270, ISSN 0006-3002

Freeman B.A., Baker P.R., Schopfer F.J., Woodcock S.R., Napolitano A. & d'Ischia M. (2008) Nitro-fatty acid formation and signaling. *J Biol Chem* Vol 28, No 23, 15515–15519, ISSN 1083-351X.

Freinbichler W., Colivicchi M.A., Stefanini C., Bianchi L., Ballini C., Misini B., Weinberger P., Linert W., Varešlija D., Tipton K.F. & Della Corte L. 2011. Highly reactive oxygen species: detection, formation, and possible functions. *Cell Mol Life Sci.* Vol. 68 No.12, 2067-79, ISSN 1420-682X.

Gautam N., Das S., Mahapatra S.K., Chakraborty S.P., Kundu P.K. & Roy S. (2010). Age associated oxidative damage in lymphocytes. *Oxidative Medicine and Cellular Longevity* Vol. 3, No.4, 275-282, ISSN 1942-0900.

Gomes A., Fernandes E. & Lima J.L.F.C. (2005). Fluorescence probes used for detection of reactive oxygen species. *J. Biochem. Biophys. Methods* Vol 65, No 2-3, 45–80, ISSN 0165-022X

Gordon M.H. (2012). Significance of Dietary Antioxidants for Health. *Int. J. Mol. Sci.*, Vol. 13, No. , 173-179, ISSN 1422-0067.

Griffiths H.R., Dunston C.R., Bennett S.J., Grant M.M., Phillips D.C. & Kitas G.D. 2011. Free radicals and redox signalling in T-cells during chronic inflammation and ageing. *Biochem Soc Trans.* Vol. 39, No. 5, 1273-1278, ISSN 0300-5127.

Grimsrud P.A., Xie H., Griffin T.J & Bernlohr D.A. (2008). Oxidative Stress and Covalent Modification of Protein with Bioactive Aldehydes. *J Biol Chem* Vol. 283, No. 32, 21837–21841, ISSN 0021-9258.

Halliwell B. & Chirico S. (1993). Lipid peroxidation: its mechanism, measurement, and significance. *J Clin Nutr* Vol 57(suppl), No. 7, 155-255, ISSN 1938-3207.

Halliwell B. & Gutteridge J.M.C. (1984). Oxygen toxicity, oxygen radicals, transition metals and disease. *Biochem. J.* Vol. 219, No. 1, 1-14, ISSN 0264-6021

Halliwell B. (1992) Reactive oxygen species and the central nervous system. *J. Neurochem.* Vol. 59, No. 5, 1609-1623, ISSN 0022-3042.

Halliwell B.& Gutteridge J.M.C. (2007). *Free radicals in Biology and Medicine*, 4-th edition Oxford University Press, ISBN 978-0-19-856869-8, New York.

Halliwell, B. (2006). Reactive Species and Antioxidants. Redox Biology Is a Fundamental Theme of Aerobic Life. *Plant Physiology*, Vol. 141, No. 2, 312–322, ISSN 0981-9428.

Hecker M. & Ullrich V. (1989) On the mechanism of prostacyclin and thromboxane A2 biosynthesis. *J Biol Chem* Vol. 264, No. 1, 141-150, ISSN 1083-351X.

Hogg N. (2002). The biochemistry and physiology of S-nitrosothiols. *Annu Rev Pharmacol Toxicol* Vol 42, 585–600, ISSN 0362-1642.

Holvoet P., Macy E.,Landeloos M., Jones D., Nancy J.S.,Van de Werf F. & Tracy R.P. (2006). Analytical Performance and Diagnostic Accuracy of Immunometric Assays for the Measurement of Circulating Oxidized LDL. *Clinical Chemistry* Vol 52, No 4, 760-764, ISSN 1530-8561.

Houglum K., Filip M., Witztum J.L. & Chojkier M. (1990). Malondialdehyde and 4-Hydroxynonenal Protein Adducts in Plasma and Liver of Rats with Iron Overload. *J. Clin. Invest.* Vol. 86, 1991-1998, ISSN 0021-9738.

Hulsmans M. & Holvoet P. (2010). The vicious circle between oxidative stress and inflammation in atherosclerosis. *J Cell Mol Med* Vol. 14, No. 1-2, 70–78, ISSN 1422-0067.

Jain K., Siddam A., Marathi A., Roy U., Falck J.R. & Balazy M. (2008) The mechanism of oleic acid nitration by $\cdot NO_2$. *Free Radic Biol Med*, Vol 45, No 3, 269–283, ISSN 0891-5849.

Karatas F., Karatepe M. & Baysar A. (2002) Determination of free malondialdehyde in human serum by high-performance liquid chromatography, *Anal Biochem*, Vol 311, 76-79, ISSN 1096-0309.

Kawai Y., Miyoshi M., Moon J.H. & Terao J. (2007). Detection of cholesteryl ester hydroperoxide isomers using gas chromatography–mass spectrometry combined with thin-layer chromatography blotting. *Anal. Biochem.* Vol 360, No 1, 130–137, ISSN 1096-0309

Kim S.F., Huri D.A. & Snyder S.H. (2005) Inducible nitric oxide synthase binds, S-nitrosylates, and activates cyclooxygenase-2. *Science* Vol 310, No 5756, 1966–1970, ISSN 0036-8075.

Kwiecień S., Brzozowski T., Konturek P.Ch. and Konturek S.J. (2002). The role of reactive oxygen species in action of nitric oxide-donors on stress-induced gastric mucosal lesions. *Journal of Physiology and Pharmacology*, Vol. 53, No. 4, 761 -773, ISSN 0867-5910.

Lang J.D. Jr, Chumley P., Eiserich J.P., Estevez A., Bamberg T., Adhami A., Crow J. & Freeman B.A. (2000) Hypercapnia induces injury to alveolar epithelial cells via a nitric oxide-dependent pathway. *Am J Physiol Lung Cell Mol Physiol* Vol 279, No 5, L994–L1002, ISSN 1522-1504.

Lenaz G. (2012). Mitochondria and reactive oxygen species. Which role in physiology and pathology? *Adv Exp Med Biol.* Vol. 942, 93-136, ISSN 0065-2598.

Leopold J.A., & Loscalzo L. (2009). Oxidative Risk for Atherothrombotic Cardiovascular Disease. *Free Radic Biol Med.* Vol. 47, No. 12, 1673–1706, ISSN 0891-5849.

Liu Y & Schubert D.R. (2009). The specificity of neuroprotection by antioxidants. *J Biomed Sci* Vol. 16, 98-112, ISSN 1021-7770.

Lougheed M. & Steinbrecher U.P. (1996) Mechanism of Uptake of Copper-oxidized Low Density Lipoprotein in Macrophages Is Dependent on Its Extent of Oxidation, *J Biol Chem* Vol 271, No 20, 11798–11805, ISSN 1083-351X

Lykkersfeld, J. (2001). Determination of Malondialdehyde as Dithiobarbituric Acid Adduct in Biological samples by HPLC with fluorescence detection: Comparison with

Ultraviolet-Visible Spectrophotometry. *Clin Chem*, Vol. 47, No. 9, 1725-1727, ISSN 1530-8561.

Manda G., Nechifor M.T. & Neagu T.M. (2009). Reactive Oxygen Species, Cancer and Anti-Cancer Therapies. *Current Chemical Biology* Vol. 3, No. 1, 342-366, ISSN 1872-3136

Margina D, Gradinaru D., Panaite C., Cimponeriu D., Vladica M., Danciulescu R., Mitrea N. (2011) The association of adipose tissue markers for redox imbalance and the cardio-vascular risk in obese patients, *HealthMED* Vol. 5, No 1, 194-199, ISSN 1840-2291.

Margina D., Gradinaru D. & Mitrea N. (2004). Clinical study regarding the atherogenic properties of the oxidatively modified low-density lipoproteins. *Archives of the Balkan Medical Union*, Vol 39, No. 2, 78-82, ISSN 0041-6940

Margina D., Gradinaru D., Mitrea N. (2009) Development of a potentiometric method for the evaluation of redox status in human serum, *Revue Roumaine de Chimie*, Vol 54, No 1, 45-48, ISSN 0035-3930

Moller M., Botti H., Batthyany C., Rubbo H., Radi R. & Denicola A. (2005) Direct measurement of nitric oxide and oxygen partitioning into liposomes and low density lipoprotein. *J Biol Chem* Vol 280, No 10, 8850-8854, ISSN 1083-351X.

Moller M.N., Li Q., Lancaster J.R.Jr & Denicola A. (2007) Acceleration of nitric oxide autoxidation and nitrosation by membranes. *IUBMB Life*, Vol 59, No 4-5, 243-248, ISSN 1521-6551.

Niki E. (2009). Lipid peroxidation: physiological levels and dual biological effects. *Free Radic Biol Med*; Vol 47, No 5, 469 - 484. ISSN 0891-5849.

Ohkawa H., Ohishi N. & Yagi K. (1979). Assay for lipid peroxides in animal tissues by thiobarbituric acid reaction. *Anal Biochem* Vol. 95, No. 2, 351-358, ISSN1096-0309.

Ohshima T., Hopia A., German J.B. & Frankel E.N. (1996). Determination of hydroperoxides and structures by high-performance liquid chromatography with post-column detection with diphenyl-1-pyrenylphosphine. *Lipids* Vol 31, No 10, 1091-1096, ISSN: 1558-9307

Parastatidis I., Thomson L., Fries D.M., Moore R.E., Tohyama J., Fu X., Hazen S.L., Heijnen H.F., Dennehy M.K., Liebler D.C., Rader D.J. & Ischiropoulos H. (2007) Increased protein nitration burden in the atherosclerotic lesions and plasma of apolipoprotein A-I deficient mice. *Circ Res* Vol 101, No 4, 368-376, ISSN 0009-7330.

Parthasarathy S., Santanam N., Ramachandran S. & Meilhac O (1999) Oxidants and antioxidants in atherogenesis: an appraisal, *J. Lipid Res.* Vol 40, No 12, 2143-2157. ISSN 1539-7262.

Pryor W.A. & Stanley J.P. (1975). A suggested mechanism for the production of malonaldehyde during the autoxidation of polyunsaturated fatty acids. Nonenzymatic production of prostaglandin endoperoxides during autoxidation. *J Org Chem* Vol. 40, No. 24, 3615-3617, ISSN 1099-0690.

Puddu P., Puddu G.M., Cravero E., De Pascalis S & Muscari A. (2009). The emerging role of cardiovascular risk factor-induced mitochondrial dysfunction in atherogenesis *J. Biomed Sci* Vol. 16, 112, ISSN 1021-7770.

Rizzo M.A. & Piston D.W. (2003). Regulation of beta cell glucokinase by S-nitrosylation and association with nitric oxide synthase. *J Cell Biol* Vol 161, No 2, 243-248, ISSN 1540-8140.

Ross R. (1999) Atherosclerosis–an Inflammatory Disease, *NEJM,* Vol 340, No 2, 115-126, ISSN 1533-4406.

Salomon RG. (2005). Isolevuglandins, oxidatively truncated phospholipids, and atherosclerosis. *Ann NY Acad Sci* Vol 1043, 327 – 342, ISSN 1749- 6632.

Sayre L.M., Zelasko D.A., Harris P.L., Perry G., Salomon R.G. & Smith MA. (1997). 4-Hydroxynonenal-derived advanced lipid peroxidation end products are increased in Alzheimer's disease. *J. Neurochem.* Vol. 68, No. 5, 2092–2097, ISSN 0022-3042.

Shamir R., Johnson W., Morlock-Fitzpatrick K, Zolfaghari R., Ling Li, Mas E., Lombardo D., Morel D.W. & Fisher E.A. (1996). Pancreatic Carboxyl Ester Lipase: A Circulating Enzyme That Modifies Normal and Oxidized Lipoproteins In Vitro, *J Clin Invest* Vol 97, No 7, 1696-1704, ISSN 1365- 2362.

Sies H. (1997). Oxidative stress: oxidants and antioxidants. *Experim Phys* Vol. 82, No. 2, 291-295, ISSN 0144-8757

Sies, H. (1985). Oxidative stress: introductory remarks. In: *Oxidative Stress.* H. Sies, (Ed.). Academic Press, pp (1–7), ISBN 978-0-12-642760-8, London

Sim A.S., Salonikas C., Naidoo D. & Wilcken D.E. (2003). Improved method for plasma malondialdehyde measurement by high-performance liquid chromatography using methyl malondialdehyde as an internal standard. *J Chromatogr B* Vol. 785, No. 2, 337-344, ISSN 1570- 0232

Sprinteroiu M., Balalau D., Gutu C.M., Ilie M. & Petraru C. (2010). Quantitative analysis of malondialdehyde in normal human plasma using fluorescence and the standard addition method, *Farmacia* Vol. 58, No. 4, 509-514, ISSN 0014-8237

Stalikas C.D. & Konidari C.N. (2001). Analysis of malondialdehyde in biological matrices by capillary gas chromatography with electron-capture detection and mass spectrometry. *Anal Biochem* Vol. 290, No. 1, 108-115, ISSN 1096-0309

Steghens J.P., van Kappel A.L., Denis I. & Collombel C. (2001). Diaminonaphthalene, a new highly specific reagent for HPLC-UV measurement of total and free malondialdehyde in human plasma or serum. *Free Radic Biol Med,* Vol. 31, No. 2, 242-249, ISSN 0891-5849

Steinberg D. (1990). Arterial metabolism of lipoproteins in relation to atherogenesis. *Ann N Y Acad Sci,* Vol 598, 125-135, ISSN 1749-6632.

Steinberg, D. (2009). The LDL modification hypothesis of atherogenesis: an update. *J. Lipid Res.* Vol. 50 (S), S376–S381, ISSN 0022-2275.

Takahashi M., Shibata M. & Niki E. (2001). Estimation of lipid peroxidation of live cells using a fluorescent probe, Diphenyl-1-pyrenylphosphine. *Free Rad. Biol. Med.* Vol 31 No 2, 164-174, ISSN 1873-4596

Takeda A., Smith M.A., Avila J., Nunomura A., Siedlak S.L., Zhu X., Perry G., Sayre L.M. (2000). In Alzheimer's disease, heme oxygenase is coincident with Alz50, an epitope of tau induced by 4-hydroxy-2- nonenal modification. *J. Neurochem.* Vol. 75, No. 3, 1234–1241, ISSN 0022-3042.

Templar J., Kon S.P., Milligan T.P., Newman D.J. & Raftery M.J. (1999). Increased plasma malondialdehyde levels in glomerular disease as determined by a fully validated HPLC method. *Nephrol Dial Transplant,* Vol. 14, No. 4, 946-951, ISSN 0931-0509.

Thomas D.D., Liu X., Kantrow S.P. & Lancaster J.R. Jr. (2001) The biological lifetime of nitric oxide: implications for the perivascular dynamics of NO and O2. *Proc Natl Acad Sci USA* Vol 98, No 1, 355–360, ISSN 1091-6490.

Uehara T. (2007) Accumulation of misfolded protein through nitrosative stress linked to neurodegenerative disorders. *Antioxid Redox Signal* Vol 9, No 5, 597–601, ISSN: 1557-7716.

Velayutham M., Hemann C. & Zweier J.L. (2011). Removal of H_2O_2 and generation of superoxide radical: role of cytochrome c and NADH. *Free Radic Biol Med*. Vol. 51, No. 1, 160-170, ISSN 0891-5849.

Veskoukis A.S., Tsatsakis A.M. & Kouretas D. (2012). Dietary oxidative stress and antioxidant defense with an emphasis on plant extract administration. *Cell Stress and Chaperones* Vol. 17, No. 1, 11–21, ISSN 1355-8145

White P.J., Charbonneau A., Cooney G.J. & Marette A. (2010). Nitrosative modifications of protein and lipid signaling molecules by reactive nitrogen species, *Am J Physiol Endocrinol Metab*, Vol 299, No 6, E868-E878, ISSN: 0193-1849.

Wilson D.W., Metz H.N., Graver L.M. & Rao P.S. (1997). Direct method for quantification of free malondialdehyde with high-performance capillary electrophoresis in biological samples. *Clin Chem*. Vol. 43, No. 10, 1982-1984, ISSN 1530-8561

Yasukawa T., Tokunaga E., Ota H., Sugita H., Martyn J.A. & Kaneki M. (2005). S-nitrosylation-dependent inactivation of Akt/protein kinase B in insulin resistance. *J Biol Chem* Vol 280, No 9, 7511–7518, ISSN 1083-351X

Yoritaka A., Hattori N., Uchida K., Tanaka M., Stadtman E.R & Mizuno Y. (1996). Immunohistochemical detection of 4-hydroxynonenal protein adducts in Parkinson disease. *PNAS* Vol. 93, No. 7, 2696–2701, ISSN 1091-6490.

Chapter 7

Liposomes as a Tool to Study Lipid Peroxidation

Biljana Kaurinovic and Mira Popovic

Additional information is available at the end of the chapter

1. Introduction

Lipid peroxidation is used as a marker of cellular oxidative stress and contributes to the oxidative damage that occurs as a result of xenobiotics metabolism, inflammatory processes, ischemia, reperfusion injuries and chronic diseases such as atherosclerosis and cancer [1,2].

Cell membrane lipids (phospholipids, glycolipids and cholesterol) are the most common substrates of oxidative attack. Once initiated reaction autocatalytic continues, it has pro-gradient flow, and the ultimate consequence is the structural-functional changes of the substrate. Lipid peroxidation is one of the best studied processes of cell damage under conditions of oxidative stress [3-5]. In 1960s Hochstein et al. [6] found that the initiation of lipid peroxidation require the presence of iron ions. From that moment the mechanism of lipid peroxidation process has been studied in many *in vitro* systems. However, accurate and precise mechanism is still not fully understood. Peroxidation in liposomes is usually studied after adding iron ions (Fe^{2+} plus ascorbic acid). Although the mechanism is not fully understood, it is known that redox chemistry of iron plays an important role in the occurrence and the rate of lipid peroxidation. Many studies have shown that the iron-dependent lipid peroxidation in systems comprised initially of Fe^{2+} and liposomes requires Fe^{2+} oxidation. In their research work, Minotti and Aust [7] assumed that the complex is formed between Fe^{2+} and Fe^{3+} ions could be initiator of iron-dependet lipid peroxidation. However, the existence of this complex has never been proven. In contrast, Aruoma et al. [8] argue against the participation of a Fe^{2+}-Fe^{3+}-O_2 complex, or a critical 1:1 ratio of Fe^{2+} to Fe^{3+}, in the initiation of lipid peroxidation in liposomes. Study of Tang et al. [9] showed that whether adding 100 or 150 mM Fe^{2+} initially or adding 100 mM Fe^{2+} initially and then 50 mM Fe^{2+} later at various times during the latent period in the liposomal system, the concentration of the remaining Fe^{2+} at the end of the latent period was almost the same every time.

Since lipid peroxidation causes oxidative damage to cell membranes and all other systems that contain lipids, in investigation of total antioxidative activity of plant extracts it is necessary to investigate their effects on lipid peroxidation. However, the impact of various natural products (isolated compounds and extracts) on the intensity of lipid peroxidation is studied in a number of substrate (linoleic acid, liposomes, various fatty oils, liver homogenates or hepatocytes isolated from it). Some substrates (liposomes and linolenic acids) are used more frequently than others mainly because of the simpler ways of performing the method. Also, due to the complex composition, examining the process of lipid peroxidation in fatty oils and liver homogenates makes research more difficult.

2. Liposomes as a model system

Liposomes are microscopic structure consisting of the one or more lipid bilayer enclosing the same number of water compartments. First, they were produced in Great Britain in 1961 by Alex D. Bangham while he was studying blood clotting. It was discovered that when phospholipids were combined with water they immediately formed a sphere. This is due to the fact that one end of each molecule is water soluble, while the oposite end is water insoluble. Water-soluble medications added to the water were trapped inside the aggregation of the hydrophobic ends; fat-soluble medications were incorporated into phospholipids layer and then – an important delivery system was born! Generally, such a structure formed polar lipids (such as phospholipids) [10]. Liposomes could be characterized as particles, similar to the structure and composition of cell membrane (Figure 1.). They occur in nature and could be artificially prepared [11].

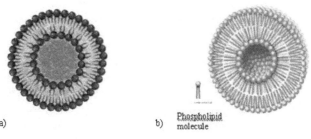

a)   b)   Phospholipid
                                                                         molecule

Figure 1. Example of a) empty liposome; b) liposome (2007 Encyclopadeia Britannica, Inc.)

The behaviour of liposomes in physical and biological systems is governed by the factors such as physical size, membrane permeability, percent entrapped solutes, chemical composition (estimation of phospholipids, phospholipids oxidation, and analysis of cholesterol), and quantity and purity of the starting material. Therefore, liposomes are characterized for physical attributes: shape, size, and its distribution; percentage drug capture; entrapped volume; lameliarity; percentage drug release. Based on the structure and size, we distinguish between different types of liposomes: Multilamellar Vesicles (MLV, size >0.5μm), Oligolamellar Vesicles (OLV, size 0.1-1μm), Unilamellar Vesicles (UV, all size ranges), Multivesicular Vesicle (MVV/MV, size >1μm). Unilamellar Vesicles are further divided into Small Unilamellar Vesicles (SUV, size 20-50nm), Medium Unilamellar Vesicles (MUV, size 50-100nm), Large Unilamellar Vesicles (LUV, size >100nm) and Giant Unilamellar Vesicles (GUV) (Figure 2.).

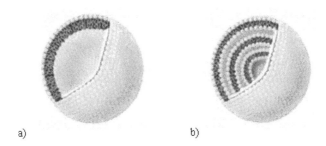

a)   b)

Figure 2. Example of a) unilamellar liposome; b) multilamellar liposome

Based on composition and applications, liposomes are divided into conventional liposomes (CL), fusogenic liposomes, pH sensitive liposomes, cationic liposomes, long circulatory (stealth) liposomes (LCL) and immuno-liposomes [12]. It is very difficult to measure dirrectly the phospholipid concentration, since dried lipids can often contain considerable quantities of residual solvent. Because of that, the method most widely used is an indirect one in which the phosphate content of the sample is first measured. The phospholipid concentration is measured using two methods - Bartlett and Stewart. In the Bartlett method the phospholipid phosphorous in the sample is first hydrolyzed to inorganic phosphate. This is converted to phospho-molybdic acid by the addition of ammonium molybdate and phospho-molybdic acid is quantitatively reduced to a blue colored compound by amino-naphthyl-sulfonic acid. The intensity of the blue color is measured spectrophotometrically and is compared with the curve of standards to give phosphorous and hence phospholipid content. This method is very sensitive. The problem is that test is easily upset by trace contamination with inorganic phosphate. In the other test, Stewart test, the phospholipid forms a complex with ammonium ferrothiocyanate in organic solution. The advantage of this method is that the presence of inorganic phosphate does not interfere with the test.

Until recently, liposomes are used as inert particles, carries of active principles, mostly for cosmetic purposes [13]. Today liposomes are used as very useful models, reagents and tools in various scientific disciplines, including biophysics (properties of cell membranes and channels), chemistry (catalysis, energy conversion, photosynthesis), biochemistry (the function of membrane proteins) and biology (excretion, cellular functions, transports and signaling, the transfer of genes and their functions). Liposomal formulation of several active molecules are currently in pre-clinical and clinical trials in different fields, with promising results. Two of the key problems in drug therapy (biodistribution throught the body and targeting to specific receptors) can be overcome by using liposomal formulations – liposomes protect encapsulated molecules from degradation and can passively target tissues or organs that have a discontinuous endothelium, such as liver, spleen, and bone marrow [14]. Comercial use of liposome was based on their colloidal, chemical and surface and microcapsuled proporties. These products include dosage formes of drugs (anti-cancer and antifugal agents, vaccines), cosmetic formulation (skin care products, shampoos), diagnostic products, a variety of applications in the food chemistry, as well as oral nutrient transport (liposomal vitamins, minerals and plants extracts for oral use). Liposome stability is an important aspect that must be met to be able to apply. By selecting the optimal value and size, pH and ionic strenght and the addition of complexing agents, liquid liposomical formulations could be stable for years.

Liposomal models have helped us to better understand the structure and dynamics of natural biomembrane systems. The concepts of structure and function of biomembranes, such as membrane fluidity, phase transition, the movement of lipids and proteins, triggering prosesses that affect metal ions or pH, have been very developed in this way. Modulatind effect of internal molecules (such as cholesterol) and insight into the mechanisms of membrane permeabillity for non-electrolytes and ions, are obtained by testing the model membranes. Liposomes that contains proteins as a components of membrane (reconstructes liposomes) were used in testing lipid-protein interactions in biological membranes, in examining the activities of active components such as membrane ionophores, anesthetics and divalent cations and mechanisms of antybody-antigen interactions [10].

Liposomes are very good models because they show the selectivity of the membrane to ions, osmotic swelling and response of range to agents that speed up or slow down the loss of ions and molecules from the particles in a way that at least, qualitatively mimic their activity in the natural membrane systems. Liposomes have also been successfully applied to „exclude the role" of membranes lipids and other components in biomembranes interact with the physical or chemical agents. Nevertheless, liposomal systems are useful because they allow the manipulation of membrane lipid composition, pH, temperature, content of different compounds in a limited way and provide the ability to determine the individual effect of the investigation product [15].

An example of the advantages of liposome in investigation of lipid peroxidation is that the influence of free radicals can be explored in the absence of chemical systems that produce free radicals, which may affect the test reaction. It is also possible to control the chemical composition of the liposome. This is particularly useful for the determination of lipid peroxidation induced by different systems for the generation of free radicals, and monitoring the overall effect of the combined system, or synergistic effects of combined systems that may arise. In addition to this it is possible to determine the antioxidant activity of tested compounds and determine which system works the best, by simple monitoring of lipid peroxidation. The tests used to determine the power of antioxidants to exert suppression of lipid peroxidation based on an assessment of the strength of oxidation of lipid substrate in the presence or absence of potential antioxidant molecules of plant extracts. There are four different strategies for assessing antioxidant capacity of molecules to the lipid substrate. They include the determination of oxygen consumed, the loss of substrate and formation of primary and secondary oxidation products [16]. The first method for determining the degree of lipid peroxidation, which includes the determination of oxygen consumed, based on following of initiation phase and its extension in the absence of antioxidants. The second method is based on measuring the loss of substrate in systems such as samples of food or biological samples and is very complicated, because they are full of potential oxiable substrates that are difficult to identify and characterize. The third method is based on monitoring the formation of primary oxidation products. It is a method that is well adapted to study such complex model systems and often involves the spectrophotometric determination of hydroperoxide, the dominant primary products of lipid peroxidation. Monitoring of secondary products of oxidation is the most commonly used method for the study of lipid model systems and lipid isolated from their natural environment. Both the *in vitro* and *in vivo* conditions are very often used TBA (thiobarbituric acid) test for detection of MDA (malondialdehyde), a secondary product of oxidation. This test is based on the reaction between TBA and MDA, which produces red chromophore with maximum of absorbance at 532 nm. This reaction is widely used and is performed by means of determination of many oxiable substrates (free fatty acids, LDL, body fluids). However, this method has some drawbacks. One is that the MDA is formed from free fatty acids which contain at least three double bonds. The next disadvantage is that the TBA is not specific for MDA because it can react with other aldehydes, such as occurs brown color that comes from the reactions of decomposition of sugar, amino acids, proteins and nucleic acids. Finally, MDA is not generated during the oxidation of many lipids and is often less important secondary oxidation product, and therefore not representative enough for the individual measurements. However, the TBA test was held for examination of lipid peroxidation, due to the simplicity of the method.

Despite all these advantage, liposomal systems remain different from the natural cellular systems. For this reason, regardless of the results obtained by testing the liposomes, they could not be reproduced on the natural membrane system, but they can provide useful information.

Therefore, the liposomes are still mostly used as a model system of biological membrane for testing the LP, especially when testing extracts and essential oils from plants on the intensity of LP. These studies are important because free radical oxidation of lipid components of food is a major strategic problem of food producers. The degree of oxidation of fatty acids and their esters in foods depends on the chemical structure of fatty acids, food processing technology, the temperature at which food is stored or prepared for eating and the presence of antioxidants. Synthetic antioxidants are widely used in many foods to retard undesirable changes as a result of oxidation. Chemicals, like tert-butyl-4-hydroxyanisole (BHA) and tert-butyl hydroxytoluene (BHT), can be used as antimicrobial and antioxidants agents. However, the use of some of these chemicals is restricted in several countries, as they may be dangerous to human health [17]. Therefore, the search for new natural antioxidant sources has been greatly intensified. For this reason, there is a growing interest in the studies of natural additives as potential antioxidants. The antioxidant properties of many herbs and spices are reported to be effective in retarding the process of lipid peroxidation in oils and fatty foods and have gained the interest of many research groups. A number of studies on the antioxidant activities of various aromatic plants have been reported over the last 20 years [18,19]. Their aroma is associated with essential oils, complex mixtures of volatile compounds, dominated by mono- and sesquiterpenes. It is known that essential oils exhibit significant biological and pharmacological activities such as anti-inflammatory, antimicrobial, spasmolytic, stimulant effect on the CNS and the like. New research shows that they possess significant antitumor activity [20], and act as inhibitors of growth of breast tumors [21]. It was confirmed that essential oils of some aromatic plants possess a high antioxidant potential [22]. Widely used in the food industry to improve the flavor of foods.

In addition to essential oils, aromatic plants and characterized by the presence of plant phenolic compounds, primarily phenylpropanoids and coumarins which are proven to have multiple pharmacological activities. Studies of these secondary biomolecules have become intensified when some commercial synthetic antioxidants found to be expressed toxic, mutagenic and carcinogenic activities [23]. In addition, it was found that excessive production of oxygen radicals in the body initiates oxidation and degradation of polyunsaturated fatty acids. It is known that free radicals attack the highly unsaturated fatty acid of membrane system and induce lipid peroxidation, which is a key process in many pathological conditions, and one of the reactions caused by oxidative stress. Particularly vulnerable are the biological membrane lipids in the spinal cord and brain because they contain high oxiable polyunsaturated fatty acids. These features facilitate the formation of oxygen radicals involved in the processes of aging, Alzheimer's and Parkinson's disease, ischemic damage, arthritis, myocardial infarction, arteriosclerosis and cancer. Phenolic antioxidants "stop" oxygen free radicals and free radicals formed from the substrate by giving hydrogen atom or an electron. Some flavonoids have strong inhibitory effect on lipid peroxidation processes. This action is based on their ability to chelate transition metal ions, thereby preventing the formation of radicals (initiators of LP), caught radicals initiators of LP (ROS), scavenge lipid-alkoxyl and lipid-peroxyl radicals and regenerate α-tocopherol by reduction of α-tocopheryl radicals. Flavonoids have the following characteristics: 3 ', 4'-dihydroxy group in ring B, or 4-keto and 3-hydroxy group in C ring, or 4'-keto group in C ring and 5-hydroxy group in A ring have the metal chelated properties (Figure 3.).

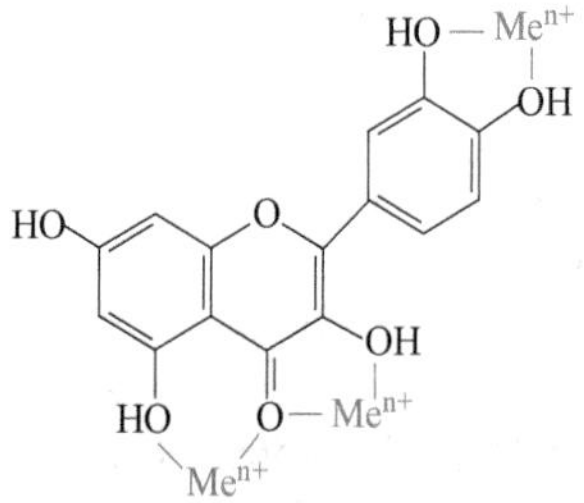

Figure 3. Possible places on flavonoids for chelating the transition metal ions in the process of lipid peroxidation.

Different metals have different binding affinity of the flavonoids [24]. Thus, for example, iron has the highest binding affinity for 3-OH group of ring C, then catechol group ring B and at the end of 5-OH group of ring A, while the copper ions bind to the first ring catechol group B [25]. Solubility of flavonoids in the lipid phase and the ability to penetrate the lipid membrane is small, since flavonoids in nature are mostly in the form of polar glycosides. Numerous tests of the inhibitory effects of flavonoids on lipid peroxidation were carried out on models of cell membranes. Based on these studies, it is assumed that quercetin and other flavonoids probably located on the surface membrane could easily capture radicals from the aqueous phase and thus prevent the initiation of LP. Thus located, flavonoids faster capture radicals initiators LP than α-tocopherol, which is located within phospholipid bilayer and that the switch is a typical chain reaction. Prevention of initial attacks radicals from the aqueous phase to membrane phospholipids is essential in the antioxidant protection of biomembranes because free radicals are constantly generated in the aqueous phase of cellular and sub cellular structure [25,26].

In the present chapter, lipid peroxidation in a liposomal system was initiated by Fe^{2+}-ascorbic acid system and the effects of four different Lamiaceae species (*Melittis melissophyllum, Marrubium peregrinum, Ocimum basilicum,* and *Origanum vulgare*) extracts and essential oils were investigated. Particular attention was paid to the chemical composition of extracts and essential oils and their capability to reduce lipid peroxidation. The plant leaves were dried in air and ground in a mixer. Finely powdered material (200 g) was macerated three times in 70% methanol (MeOH) with 4 L during a 24-h period. The macerates were collected, filtered, and evaporated to dryness under vacuum. The residues were dissolved in water and successively extracted with four solvents of increasing polarity: ether (Et$_2$O), chloroform (CHCl$_3$), ethyl acetate (EtOAc), and *n*-butanol (*n*-BuOH). The extraction was carried out until a colorless extract was obtained. The residue was the aqueous extract. All of five extracts (Et$_2$O, CHCl$_3$, EtOAc, *n*-BuOH, and H$_2$O) were evaporated to dryness and then dissolved in 50% ethanol to make 10% (w=v) solutions. Both, these and the diluted solutions, were further used for examination. Essential oil was made when air-dried plant material was submitted to hydrodistillation according to Eur. Pharm. 4 [27], using *n*-hexane as a collecting solvent. The solvent was removed under vacuum. The oils were dried over anhydrous sodium sulphate and kept at +4 ºC. The inhibition of LP was determined by measuring the formation of secondary components (malondialdehyde) of the oxidative process, using liposomes as an oxidizable substrate [28-30]. However, because the thiobarbituric acid test is not specific for MDA, other non-lipid substances present in plant extracts, or peroxidation products other then malondialdehyde, could react positively with TBA. These interfering compounds distort the results and therefore all the final results of investigated extracts have been corrected using the absorbances of the investigated extracts after the TBA-test (without liposomes) [31]. The commercial preparation of liposomes 'PRO-LIPO S' (Lucas-Meyer) pH = 5–7 was used as a model system of biological membranes. The liposomes, 225–250 nm in diameter, were obtained by dissolving the commercial preparation in demineralized water (1:10), in an ultrasonic bath.

3. Lamiaceae (Labiatae) family

The Lamiaceae family is one of the largest and most distinctive families of flowering plants, with about 220 genera and almost 4000 species worldwide [32]. Lamiaceae are best known for the essential oils common to many members of the family [33]. The family was established by De Jussieu in 1789 as the order Labiatae. This was the original family name, so given because the flowers typically have petals fused into an upper lip and a lower lip, the flower thus having an open mouth. Although this is still considered an acceptable alternative name, most botanists now use the name "Lamiaceae" in referring to this family. The main centre of diversity is the Mediterranean region to central Asia. Members are found in tropical and temperature regions [34]. All Lamiaceae are aromatic plants. The essential oil contains mainly monoterpenes, sesquiterpenes and phenylpropanoid compounds. Also, the plant species of Lamiaceae have been shown as rich sources of phenolic compounds mostly flavonoids and phenolic acids.

3.1. Balm (*Melittis melissophyllum* L.)

The name *melittis* of the genus derives from a Greek words *Melissa* or *Melitta*, meaning "honey bee" and refers to the properties of flowers of attracting these insects. The name *melissophyllum* of the species simply means "with leaves similar to melissa". This is a tall plant which likes shady places and is ideal for a sunny woodland edge or scrubby border, where it will be attractive to bees and other insects. Bastard balm is a strongly aromatic plant that smells like fresh mowed grass and has erect hairy stems. It blooms white with a large pinkish purple blotch on the lower lip. The flowers are hermaphrodite and get pollinated by bees and moths. It has oval, bluntly-toothed, leaves in opposite pairs up the stems. Bastard balm is a herb native to the Mediterranean region.

Main flavonoids in balm are glycosides of apigenin and luteolin. However, presence of some other flavonoids as kaempherol, quercetin (Figure 4.) and ramnocitrin have been also reported [35].

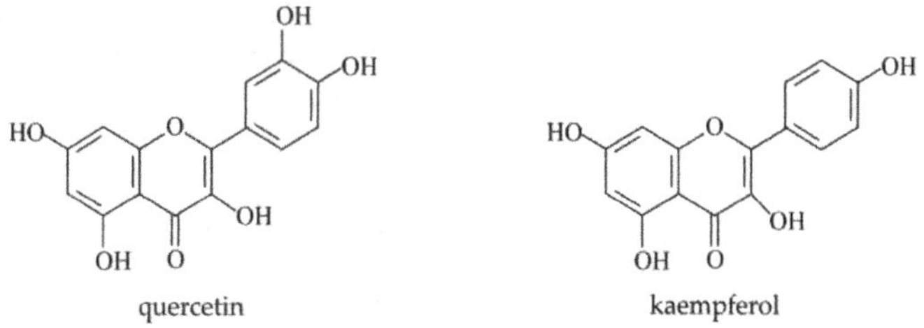

Figure 4. Structures of two flavonoids present in *M. melissophyllum*

Balm is characterized by the presence of the other important plant phenolic substances such as phenolic acids (caffeic, rosmarinic and chlorogenic acid) (Figure 5.).

Also, balm leaf is characterized by the presence of pentacyclic triterpenes (ursolic, pomolic and oleanolic acid) (Figure 6.). The main biopharmacological effects shared by ursolic and oleanolic acid are anti-inflammatory, hepatoprotective, antitumor, and antioxidative [36-39].

Essential oil is present in all parts of the plant. The largest amount of oil obtained from aerial parts of plants, harvested in late summer. Balm leaves contain no more than 0.13% of essential oil which is of complex and variable composition. Among the more than 50 compounds identified to date, citronellal (dominantly the (R) enantiomer), β-caryophyllene, ⊙-caryophyllene oxide, germacrene-D, nerol, geranial, citronellol, and geraniol amount to about 70% of the oil (Figure 7.) [40]. The composition is similar to that of lemongrass, but balm oil can be identified by its typical pattern of chiral compounds; for example, almost enantiomerically pure (R)-(+⊙)-methyl citronellate is a good indicator of true balm oil. For distinguish between two oils there is used the carbon isotopic ratio (IRMS-*isotope ratio mass spectrometry*) [41]. The essential oil exhibits spasmolitic action and acts as a muscle relaxant, sedative, narcotic, antibacterial, and antifungal [42,43].

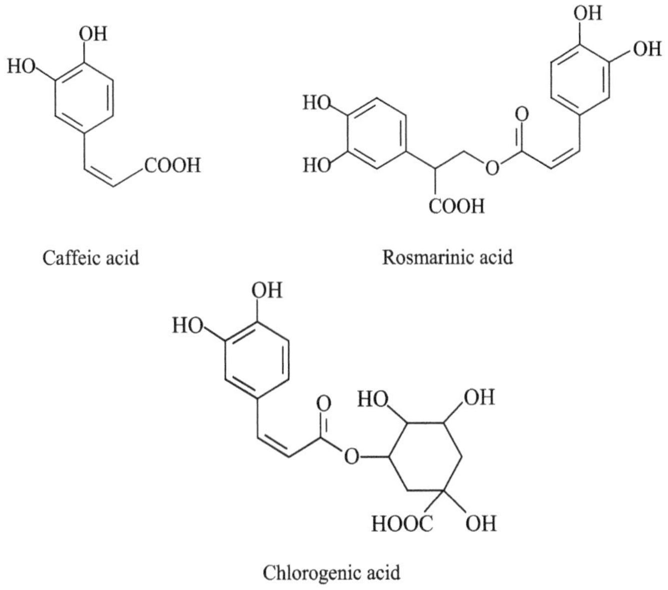

Caffeic acid

Rosmarinic acid

Chlorogenic acid

Figure 5. Structures of phenolic acids in *M. Melissophyllum*

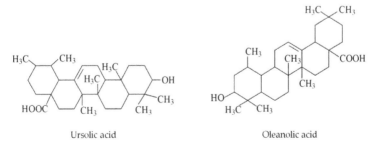

Ursolic acid

Oleanolic acid

Figure 6. Structures of triterpenoids compounds present in *M. melissophyllum* leaves.

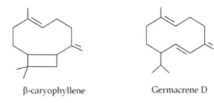

β-caryophyllene   Germacrene D

Figure 7. Sesquiterpenes in *M.melissophyllum* leaves

Beneficial effects of plants introduced by ancient Greeks and Romans.There is overlap with the use of plants in folk medicine and science. In relation to its complex composition it has multiple medicinal effects. Its herb has wide applications in the folk medicine. Due to the soothing action balm leaves enters into the composition of tea for calming, which is recommended for hysteria and neuralgia. Balm leaves mixed with bitter herbs are a great tool to enhance appetite. Various preparations containing extract or essential oil balm leaves are used as an addition to baths against rheumatism. In the folk medicine of Belarus alcoholic extract is drunk for stomach ulcer and duodenum, to calm the pain in the stomach, intestines, the liver, heart, and women's diseases. Terpenes found in essential oil of balm leaves, have a relaxing and antiviral effects. Eugenol calms muscle spasms and destroy bacteria [44]. It is also used as a carminative and sedative. Recent results indicate that the balm extract acts as depressants and have sedative effect on central nervous system of mice [45]. In the folk medicine of central Italy inflorescences of this plant, called "Erba Lupa", were used under infusion as antispasmodic, against insomnia and eyes inflammations [46,47].

Our research on balm was recently extended to the comprehensive *in vitro* and *in vivo* studies of antioxidant properties of balm essential oil and extracts measuring their capability to reduce lipid peroxidation in liposomes and effect on some enzymes of antioxidant defense systems [48]. Investigation of balm essential oil showed that with increasing concentration of essential oil reduces the intensity of lipid peroxidation compared to hexane-control (Table 1). Also, only the most diluted solution of essential oil of balm (0.213 and 0.535 μg/mL) has a weaker protective effect than the synthetic antioxidant BHT. The capability to reduce lipid peroxidation of essential oil was dose-dependent. This high inhibitory effect of balm essential oil was found to be in correlation with the content of monoterpene alcohols and ketones.

	Concentration (μg/mL)					
	BHT	**0.213**	**0.535**	**1.065**	**1.598**	**2.130**
LP	26.15	13.05	21.03	39.62	46.09	56.81

Table 1. Inhibition of LP (%) in Fe^{2+}/ascorbate system of induction by essential oil of balm leaves and BHT (as a positive control) in the TBA assay.

The protective effects on lipid peroxidation of balm extracts have been evaluated using the Fe^{2+}/ascorbate system of induction, by the TBA-assay (Table 2.). In general, all of the examined extracts (except n-BuOH extract) expressed strong antioxidant capacity and ability to reduce lipid peroxidation in liposomes. The largest inhibitory activity was exhibited by EtOAc and H$_2$O extracts because the 5% solutions show better protective effect than BHT. All extracts of the highest concentrations (10%) exhibited a better inhibitory effect than BHT. Protective activity of these extracts and its components towards Fe^{2+}-dependent LP of liposomes can be explained by present of phenolic acids and flavonoids and their influence on antioxidative capacity of ascorbic acid, which doesn't show a strong antioxidative effect in lipid phase, but different phenolic compounds can result increase of its antioxidant activity [49].

Concentration	Extracts					
	BHT	**Et$_2$O**	**CHCl$_3$**	**EtOAc**	**n-BuOH**	**H$_2$O**
1%	26.15	17.52	16.15	22.87	-10.59	24.24
5%	26.15	20.82	20.59	27.88	-13.19	39.36
10%	26.15	26.40	28.40	38.47	-18.52	41.32

Table 2. Inhibition of LP (%) in Fe^{2+}/ascorbate system of induction by extracts of balm leaves and BHT (as a positive control) in the TBA assay.

It is known that quercetin, like many other flavonoids, prevents oxidation of LDL cholesterol, and its anti-inflammatory activity comes from inhibition of the enzyme lipooxigenase and inhibition of inflammatory mediators [50]. Kaempferol acts synergistically with quercetin to reduce the proliferation of malignant cells, and treatments are a combination of quercetin and kaempherol efficient than their single use [51]. It is, also, known that rutin has strong antioxidant effects, as well as a feature to built chelates with metal ions (e.g. iron) and reduces the Fenton reaction in which the resulting harmful oxygen radicals. It is supposed to stabilize vitamin C. If rutin is taken together with vitamin C, increases the activity of ascorbic acid [52]. In addition, HPLC-DAD analysis showed that the aqueous extract, in large quantities, present phenolic acids (rosmarinic, chlorogenic and caffeic acid), which are known antioxidants. It was determined that rosmarinic acid has stronger antioxidant activity than vitamin E. Rosmarinic acid prevents cell damage caused by free radicals and reduce the risk of cancer and atherosclerosis. In contrast to the histamines, rosmarinic acid prevents activation of the immune system cells that cause swelling and fluid collection. Also, it is known that the caffeic acid by far surpassing other antioxidants because it reduces the production of α-toxin for more than 95% [35]. Furthermore, it can be supposed that the reduction process of lipid peroxidation is caused, besides flavonoids, also by triterpenoids acids (especially ursolic, oleanolic, and pomolic acid) since non-polar extracts (Et$_2$O and CHCl$_3$) also exhibited high antioxidant potential [39]. The n-BuOH extract shows a prooxidative effect that is increased by increasing concentration of added extract. It can be supposed that compounds with polar groups were extracted by n-BuOH, and are present in high concentration in the extract. It is notable that molecules which show antioxidant activity, when they are present in high concentration, might behave as prooxidants [53], so n-BuOH extract of balm leaves probably have this kind of activity. The antioxidant activities of all extracts of balm leaves were dose dependent.

The represented antioxidant activity results show that extracts of examined plant species, especially EtOAc and H2O extracts are efficient in the protection of tissues and cells from oxidative stress. Anyway, according to variations in regard to antioxidant activity of tested by different *in vitro* models, there are also requiste *in vivo* test that would confirm the capability of extracts to reduce the lipid peroxidation. *In vivo* tests are also necessary because a lot of plant phenols are biotransformed during their active metabolism. *In vivo* effects are evaluated on LP in the mice liver (Table 3.) and blood hemolysate (Table 4.) after treatment with examined balm extracts, or in combination with carbon tetrachloride (CCl4).

Parameter	Control	Et2O	CHCl3	EtOAc	n-BuOH	H2O
LP	7.19±0.23	7.36±0.21	7.91±0.19	6.71±0.16	7.12±0.23	6.19±0.27
LP + CCl4	8.91±0.29	7.12±0.21	7.06±0.24	6.92±0.17	6.98±0.24	6.81±0.24

Table 3. Effect of extracts of balm leaves on intensity of lipid peroxidation (nmol malondialdehyde/mg of proteines) in liver homogenate before and after treatment with CCl4

As compared with control, intensity of LP is statistically significant reduced during the treatment with ethylacetate and water extracts of balm leaves. The result derived by treatment with ethylacetate and water extracts is in according with amounts got *in vitro* experiment. Using CHCl3 extract leads to a significant increase of LP intensity, whereas the other two extracts had no effect on this parameter. All extracts of balm leaves combine with CCl4 have showed a statistically significant decrease of LP intensity, and this behavior of the extract probably results from the presence of secondary biomolecules like flavonoids and phenolic acids. Handa et al. [54] determined that secondary biomolecules such as flavonoids, xanthones and tannins in combination with CCl4 have protective effects on liver. Phenolic components present in balm leaves (rutin, luteolin, kaempherol) are known as strong inhibitors of CCl4-induced LP [55]. Flavonoids could affect the initiation phase of lipid peroxidation, where they influence the metabolism of CCl4, they scavenge the free radicals, or they decrease the microsomal enzyme systems that are claimed for CCl4 metabolism [56]. In continuation of this process, flavonoids can scavenge lipoperoxides and their radicals or they can act as chelating agents for Fe^{2+} ion, and in this way can stop Fenton reactions [57]. Furthermore, Afanas'ev et al. [28] found that quercetin and rutin exhibited a high inhibitory effect on the Fe^{2+}-induced liposomal LPx and NADPH-dependent CCl4-induction LPx in liver microsomes. Luteolin, one of the main active component in the balm, is responsible for the inhibitory effect on the former reaction. In addition to the above-mentioned mechanism (chelate formation with Fe^{2+}) it is possible that these compounds (of flavonoid type) act as scavengers of OH radicals, whereby they are transformed in the corresponding radical form which is stabilized by resonance. On the basis of these results, it can be concluded that all of extracts of balm leaves showed protection effect in relation to the CCl4-induced lipid peroxidation.

Similar results were obtained during examining the effects of extracts of bastard balm on LP in blood hemolysate in mice (Table 4.). Three extracts, CHCl3, EtOAc and H2O, induced a significant decrease of LP intensity, while Et2O and n-BuOH ones decreased the level of this enzyme insignificantly.

Parameter	Control	Et₂O	CHCl₃	EtOAc	*n*-BuOH	H₂O
LP	4.81±0.24	4.59±0.28	3.78±0.17	2.96±0.13	4.74±0.19	4.07±0.24
LP + CCl₄	5.11±0.24	5.31±0.17	4.92±0.21	3.02±0.24	5.17±0.25	2.98±0.12

Table 4. Effect of extracts of balm leaves on intensity of lipid peroxidation (nmol malondialdehyde/mL erythrocytes) in blood hemolysate before and after treatment with CCl₄

The LP value showed a statistically insignificant increase with CCl₄-treated animals compared with the untreated ones. A clear protective effect was seen in experimental animals administered H₂O extract and CCl₄ compared with untreated animals. Furthermore, EtOAc extract also significantly decreased the activity of LP, while Et₂O, CHCl₃ and *n*-BuOH extracts did not change notably the levels of lipid peroxidation. These results suggest that these two extracts (EtOAc and H₂O) had a protective effect. According to the literature data [58], the reduction of the serum LP might be the result of antioxidant activity of several classes of plant phenolic constituents, such as cinnamic acids (ferulic, caffeic, and chlorogenic), flavonoids and biflavonoids, 1,3,6,7-tetrahydroxyxynthones, and acylphoroglycinols such as hyperforin and adhyperforin. Cock and Samman [59] showed that quercetin and rutin and their glycosides show strong inhibitory effect in respect of LP. The observed differences in the action of particular balm extracts are probably due to the different contents of flavonoids, but the potential protective effects of some other groups of compounds can not be ruled out.

3.2. Horehound (*Marrubium peregrinum* L.)

Marrubium genus includes about 40 species. Species of this genus growing in dry pastures, abandoned the places along the roads in central and southern Europe, but also in North Africa, in parts of Asia and the Americas. Horehound is a perennial plant with a rectangular stem, branched in the upper part. Rhizomes of this species are ligneous, leaves oblong, flowers grouped in loose inflorescence [60]. A common plant blooms from July to September and harvested in that period. It has a bitter and pungent taste and smell. It is the drug of Herba *Marrubii albi*. This plant doesn`t require special conditions for growth.

In previous phytochemical investigations on *M. peregrinum*, different groups of chemicals were isolated: flavones (apigenin and luteolin) [61] (Figure 8.), flavonols (kaempferol) [62], glycosylated flavonoids (quercetin-3-O-β-D-rutinoside, naringenin-7-O-β-D-glucoside, kaempferol-3-O-β-D-rutinoside, quercetin-3-O-β-D-glucoside) [63], caffeic acid derivatives [64], and four diterpenoids (peregrinin, peregrinol, marrubiin and premarrubiin) [65]. T. Hennebelle et al. [66] have established the presence of acteoside, forsythoside B, arenarioside and terniflorine (apigenin-7-O-[6"-E-p-coumaroyl]β-D-glucopyranoside) in the MeOH extract of *M. peregrinum*.

Marrubium peregrinum essential oil yield between 0.02-0.07% [67]. Dominant monoterpenes are: α-pinene, sabinene, limonene, camphene and α-terpinolene. In a Greek sample, β-phellandrene, epi-bicyclosesquiphellandrene and bicyclogermacrene proved to be the major compounds [68], whereas the essential oil of a sample from Central Europe was rich in β-caryophyllene and its oxide, bicyclogermacrene and germacrene D [69]. The main sesquiterpene compounds are Z- and E- β-farnesene (~12%), β-caryophyllene (~8.5%), heksahidrofarnesil acetone (~ 6.5%), spathullenol (~5%) i germacrene D (~4.5%) (Figure 9.) [68].

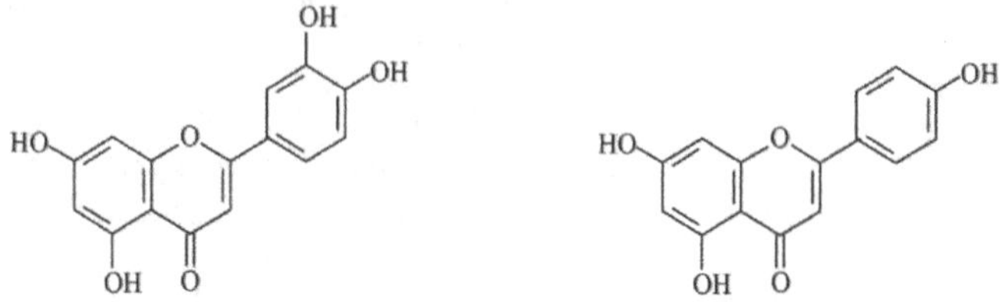

luteolin   apigenin

Figure 8. Figure 8. Structures of two main flavonoids in *M. peregrinum*.

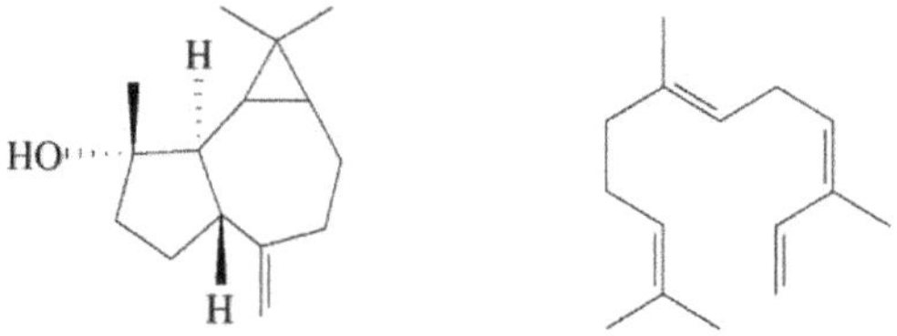

spathullenol   Z- and E-β-farnesene

Figure 9. Main constituents of *M. peregrinum* essential oil

Some species of *Marrubium* are used in traditional and modern medicine. Many studies have shown various activities in this genus, such as hypoglycemic effect, anti-schistosoma, antioxidant, calcium channel blocker and hypotensive activity [70]. As a medicinal plant, *M. peregrinum* have been employed against vascular diseases (antihypertensive, antispasmolitic) [61].

In our comprehensive study of chemical and biochemical investigation of *M. peregrinum* from three different locations (Backo Gradiste-Rimski Sanac (No 1.); Novi Knezevac (No 2.) and Senta (No 3.)), we have identified more than 40 compounds in essential oil of *M. peregrinum* (44 for *M. peregrinum* from Senta locality, 42 for *M. peregrinum* from Novi Knezevac locality and 41 for *M. peregrinum* from Rimski Sanac locality, representing 96.15%, 87.60% and 83.66% of the total oil contents, respectively), in which dominant compounds were β-caryophyllene (13.20-17.99%), bicyclogermacrene (6.42-9.80%) and germacrene-D (6.79-9.05%). Besides sesquiterpene hydrocarbons, oxygenated sesquiterpenes, spathulenol (3.76-5.78%) and caryophyllene oxide (3.73-4.78%) are also present in relevant quantities. However, we must point out that the amounts of these components in essential oil from different localities are very different. Essential oil obtain from plant collected in Senta is the richest of sesquiterpene hydrocarbons (62.71%), while oxygenated sesquiterpenes are most represented (11.84%) in essential oil from plants collected in the Rimski Sanac area.

	Concentration (µg/mL)					
	BHT	0.213	0.535	1.065	1.598	2.130
M. peregrinum (No 1.)	26.15	12.24	26.00	35.36	44.26	56.18
M. peregrinum (No 2.)	26.15	19.15	20.05	39.37	52.14	61.18
M. peregrinum (No 3.)	26.15	21.17	37.02	55.81	65.16	71.32

Table 5. Inhibition of LP (%) in Fe^{2+}/ascorbate system of induction by essential oil of *M. peregrinum* from three different location, and BHT (as a positive control) in the TBA assay.

Also, our study showed that all of the examined essential oils express strong antioxidant activity and capability to reduce lipid peroxidation (Table 5.). The largest inhibitory activity was exhibited by essential oil from plant collected at Senta locality (No. 3.). Solution of all concentrations, except the most diluted (0.213 µg/mL), have exhibited a stronger protective effect (from 37.02 to 71.32% of inhibition of LP) than BHT (26.15%). The other two essential oils (from Rimski sanac and Novi Knezevac), at higher concentration (from 1.065 to 2.130 µg/mL), have also exhibited more intense protective effect than BHT [71].

The effect of crude MeOH extracts of *M. peregrinum* was preliminarily determined from the three localities. There were taken three concentrations of MeOH extracts (1, 5, and 10% extracts). All of the examined extracts expressed stronger antioxidant capacity as compared to the 50% solution of MeOH. In particular, the largest inhibitory activity was established by the MeOH extracts of *M. peregrinum* collected from Senta locality. Also, the best results were obtained using solutions of the highest concentrations [72]. Because of all this there was carried out successive extractions of *M. peregrinum* from all three localities, and for further work 10% extracts are prepared. Successive extraction was performed as the extraction of antioxidant substances of different chemical structure, was achieved using solvents of different polarity. Numerous investigations of qualitative composition of plant extracts revealed the presence of high concentration of phenols in the extracts obtained using polar solvents [73]. The extracts that perform the highest antioxidant activity have the highest concentration of phenols. Phenols are very important plant constituents because of their scavenging ability on free radicals due to their antioxidant action [74]. The examination of capability to reduce intensity of lipid peroxidation of plant extracts from *M. peregrinum* showed different values (Table 6.). The first two extracts (Et₂O and CHCl₃) obtained from plants from all three locality are exhibited weaker protective effect than BHT, while the other three extracts (EtOAc, n-BuOH and H₂O) showed better protective properties than synthetic antioxidant.The largest inhibitory activity, again, was exhibited by the EtOAc and H₂O extracts of *M. peregrinum* collected from Senta locality.

Obtained results can be related to the experiments in which the total amount of flavonoids was determined, which show that EtOAc and H₂O extracts from Senta locality contains the largest amounts of total flavonoids, namely of luteolin, either being present as free or in the form of its glucosides. The suggested mechanism of flavonoid antioxidative action is as follows: the double bond in position 2, 3 is conjugated with C_4-carbonyl group, and free OH groups (C_5, C_3 and C_7) can form chelates with ions of d-elements. Once formed, complex with Fe^{2+} ion prevents formation of $OH^{\bullet}$ radicals in Fenton's reaction [59]. Also, luteolin is thought to play an important role in the human body as an antioxidant, a free radical scavenger, an agent in the prevention of inflammation, a promoter of carbohydrate metabolism, and an immune system modulator. These characteristics of luteolin are also believed to play an important part in the prevention of cancer. Multiple research experiments describe luteolin as a biochemical agent that can dramatically reduce inflammation and the symptoms of septic shock [75]. Furthermore, it is well known that some other flavonoids isolated from *M. peregrinum* possess certain biological and pharmacological activity. For example, apigenin, one of the flavonoids present in *M. peregrinum*, was shown to express strong antioxidant effects, increasing the activities of antioxidant enzymes and, related to that, decreasing the oxidative damage to tissues [61].

	Extracts					
	BHT	**Et₂O**	**CHCl₃**	**EtOAc**	**n-BuOH**	**H₂O**
M. peregrinum (No 1.)	26.15	9.38	14.22	29.41	27.39	32.35
M. peregrinum (No 2.)	26.15	14.27	21.19	29.54	26.83	37.55
M. peregrinum (No 3.)	26.15	17.11	23.52	38.83	28.73	41.18

Table 6. Inhibition of LP (%) in Fe^{2+}/ascorbate system of induction by extracts of *M. peregrinum* and BHT (as a positive control) in the TBA assay.

3.3. Basil (*Ocimum basilicum* L.)

Basil is originally native to India and other tropical regions of Asia, having been cultivated there for more then 5.000 years. Ocimum genus includes about 150 species [76]. There are many varieties of *Ocimum basilicum*, as well as several related species or species hybrids also called basil. These varieties differ in morphological and general structure, and also in the content and composition of essential oil. The chemotype is determined by chemical composition of essential oil and it is basic for chemotaxonomy within the genus Ocimum and species *Ocimum basilicum* [77].

The word market has several types of essential oils that differ in chemical structure, composition and fragrance. The dominant compounds of basil essential oil occur in two different biochemical pathways: phenylpropanoids (methyl chavicole, eugenol, methyl eugenol, and methyl cinnamate) through shicimic acid, and terpenoids (linalool and geraniol) through mevalonic acid. Based on chemical content, basils can be divided into four groups: European (French) *O. basilicum* (contains lower amounts of phenols); Exotic (contains methyl chavicol (40-80%)); Reunion and Javanean. European type of essential oil is the finest quality, has the finest fragrance and the highest price in the market. Other components that can be found in higher concentrations in this type of oil are: linalool, methyl chavicol (estragole) (Figure 10.), 1,8-cineole, eugenol, geraniol, germacrene D, α-terpinolene, β-caryophyllene, ocimene, sabinene, thujone, and γ-terpinene [78].

Among phenolic constituents flavonoids and their glucosides are dominant. The major flavonoids are: quercetin, kaempferol, apigenin, luteolin and rutin. Quercetin-3-Odiglucoside and kaempferol-3-O-β-rutinoside have been also identified. Beside, basil is rich in triterpenoid acids (ursolic and oleanolic), cinnamic acid (caffeic and rosmarinic), vitamin C and β-carotene, as well with calcium, copper, magnesium, sodium and potassium [79].

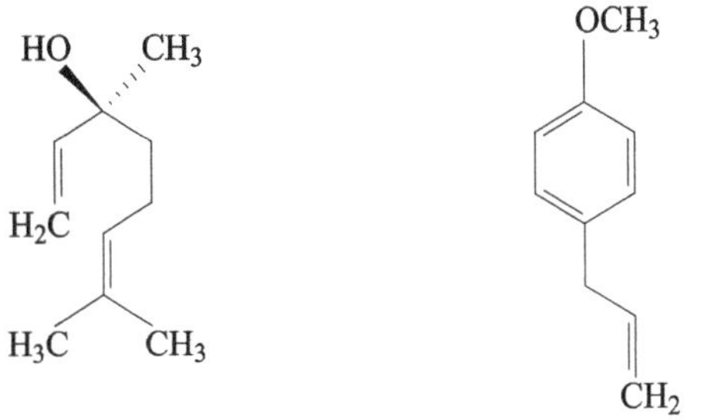

linalool

methyl chavicol

Figure 10. Main constituents of *O. basilicum* essential oil

Basilici herba has been used in traditional and homeopathic medicine to treat number of diseases. Essential oil (*Basilici aetheroleum*) extracted from fresh leaves and flowers can be used as aroma additives in foods, pharmaceuticals, and cosmetics [80]. Traditionally, basil has been used as a medicinal plant in the treatment of headaches, coughs, diarrhea, constipation, warts, worms, and kidney malfunction. Major aroma compounds from volatile extracts of basil present anti-oxidative activity [81]. Among the many studies to determine the antioxidant activities of basil, most have focused mainly on the antioxidant activities of crude extracts, using methanol, acetone, or water as a solvent [82,83].

	Concentration (μg/mL)				
BHT	0.213	0.535	1.065	1.598	2.130
LP	26.15	24.12	35.17	48.41	64.13	79.14

Table 7. Inhibition of LP (%) in Fe^{2+}/ascorbate system of induction by essential oil of basil leaves and BHT (as a positive control) in the TBA assay.

In our investigation, the examined essential oil expressed strong antioxidant activity (Table 7.). Solutions of all concentrations, except the most diluted (0.213 μg/mL), have exhibited a stronger protective effect (from 35.17 to 79.14% of inhibition of LP) than BHT (26.15%). The largest inhibitory activity was achieved by using the solution of the highest concentration. For the inhibition of LP, the most responsible compounds were the oxygenated phenolic monoterpens (methyl chavicole) and the mixture of mono- and sesquiterpene hydrocarbons. These findings are in correlation with the earlier published data on the antioxidant activities of the investigated essential oil and selected oil components [84,85].

Concentration	Extracts					
	BHT	Et₂O	CHCl₃	EtOAc	n-BuOH	H₂O
1%	26.15	-0.87	-0.86	37.42	26.31	31.74
5%	26.15	-0.94	-0.89	38.91	27.06	35.29
10%	26.15	-1.01	-1.04	41.56	28.83	36.54

Table 8. Inhibition of LP (%) in Fe^{2+}/ascorbate system of induction by extracts of basil leaves and BHT (as a positive control) in the TBA assay.

The data presented in Table 8. show that the last three extracts of *O. basilicum* (EtOAc, *n*-BuOH and H₂O) reduced the intensity of lipid peroxidation, while the first two extracts (Et₂O and CHCl₃) increased the intensity of LP, but statistically insignificant. The largest inhibitory activity was exhibited by ethyl acetate extract. High inhibitory effect of this three extracts can be related to the presence of the amount of total phenolic compounds and content of total flavonoids in the extracts, because a considerably content of total phenolic compounds and total flavonoids was determined in EtOAc and H₂O extract of *O. basilicum*. Preliminary 2D-TLC (Two Dimensional - Thin Layer Chromatography) analysis showed that the dominant flavonoid in the EtOAc extract of *O. basilicum* is derivative of quercetin. It is known that quercetin shows high antioxidant activity because of present OH groups in position 3' ring B (includes a 3', 4'-dihydroxy group). In the same experiment we established the presence of caffeic acid and its derivatives in the H₂O extract, which has two hydroxyl groups in ortho position. This was confirmed once again that the antioxidant capacity depends not only on quantity, but also depends on of the type of phenols and flavonoids present in the extracts. However, in extracts of *O. basilicum* higher content of total phenols and flavonoids from the EtOAc extract had an H₂O extract. From all this it can be assumed that the polarity of flavonoid components affects their ability to inhibit the process of LP. Specifically, in this test we used liposomes as a model system of biological membranes, and the least polar flavonoids present in the EtOAc extract could help to approach the scene and engage in the process of defense from the LP, compared to more polar compounds that are found in H₂O extract. A little less of total flavonoids was determined in *n*-BuOH extracts, while the smallest quantity of these compounds was found in Et₂O and CHCl₃ extracts. Differences in the amount of total phenolic compounds and flavonoid content between extracts can be explained by different number of secretory structures in various plant tissues [86]. Furthermore, the obtained results could be related to the protective role of phenolics, especially the flavonoid aglycones, in plants collected on the outskirts of big cities. One of the functions of these biomolecules, which are produced in response to ecological stress factors like pollution, is to serve as UV-B filters in plants [87]. It was established that flavonoids act as mighty scavengers of free radicals [88]. Different flavonoids inhibit LP *in vitro* and the most pronounced effect is exhibited by quercetin whose presence is found in extracts of *O. basilicum* using 2D-TLC [89]. More investigation is required to explain the enhanced production of phenolics in certain geographic areas [90]. Also, from the presented results we can conclude that the increase in concentration of the extracts does not significantly affect the inhibition of lipid peroxidation.

3.4. Oregano (*Origanum vulgare* L.)

Origanum is one of the most variable genera of Lamiaceae family. Originates from Europe, but is now cultivated throughout the world including USA, India and South America. This is an extremely variable species with several subspecies and named cultivars grown for ornamental, culinary and medicinal uses. Oregano is a bushy, semi-woody sub-shrub with upright or spreading stems and branches. Some varieties grow in mound like mats, spreading by underground stems (called rhizomes), and others with a more upright habit. The aromatic leaves are oval-shaped. Oregano will grow in a pH range between 6.0 (mildly acid) and 9.0 (strongly alkaline) with a preferred range between 6.0 and 8.0. The flowers are purple, 3–4 mm long, produced in erect spikes.

As the other three Lamiaceae species oregano is characterized by the presence of essential oil, flavonoids, phenolic acids (caffeic, chlorogenic and rosmarinic), triterpenoid acids (oleanolic and ursolic) and tannins. The oregano essential oil yield between 0.35-0.55% [91]. According to Arnold et al. [92], the content of essential oil in *Origanum ssp.* may come up even to 8.8%. Essential oils obtained from different parts of plant have a similar chemical profile. The dominant components are oxygenated phenolic monoterpenes thymol and carvacrol (Figure 11.), as well as sabinene, linalool, terpine-4-ol, α-pinene, caryophyllene, caryophyllene-oxide and 1,8-cineole.

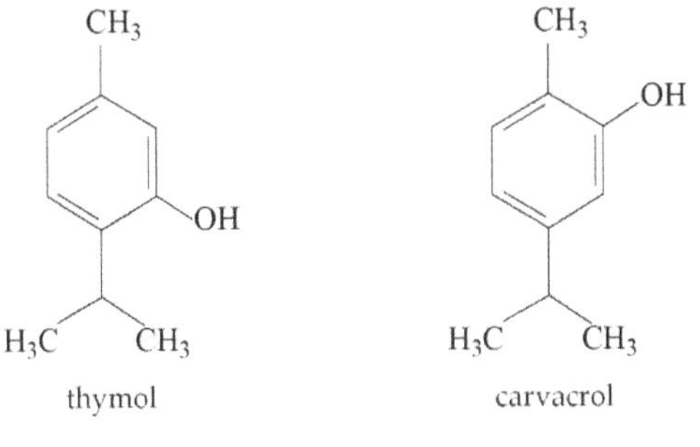

Figure 11. Oxygenated phenolic monoterpens from *O. vulgare* essential oil

According to Duke [93], flavonoids are found in the leaves and whole plant, mostly as kaempferol, quercetin, apigenin, luteolin and rutin. Beside, oregano is rich in apigenin-7-O-β-D-glucoside and luteolin-7-O-β-D-glucuronide. In oregano flavanon naringenin and flavanon glucoside (naringin), have also been identified (Figure 12.).

Most of the healing properties are attributed to the essential oil and flavonoids. It has been widely used in agricultural, pharmaceutical and cosmetic industries as a culinary herb, flavoring substances in food products, alcoholic beverages and perfumery for its spicy fragrance [94]. Regarding the nonvolatile components, the extracts of oregano have the most effective antioxidant activity among aromatic herbs [95]. Oregano family, is widely known as possessing therapeutic properties (diaphoretic, carminative, antispasmodic, antiseptic, tonic) being used in traditional medicine systems in many countries. Different groups of researchers [96,97] studied oregano alcohol extracts. The antioxidant effect of the mentioned extracts is generally due to the presence of rosmarinic and caffeic acid [98].

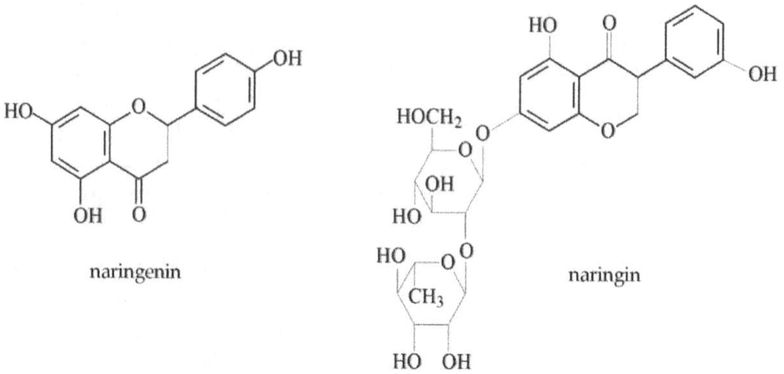

Figure 12. Structures of main flavonoids of *O. vulgare*

Concentration (µg/mL)					
BHT	0.213	0.535	1.065	1.598	2.130
LP	26.15	17.31	24.35	37.17	49.58	51.13

Table 9. Inhibition of LP (%) in Fe^{2+}/ascorbate system of induction by essential oil of oregano leaves and BHT (as a positive control) in the TBA assay.

Our tests showed that only concentrated solutions of essential oil exhibit a greater ability to inhibit LP in liposomes of synthetic antioxidant BHT. The antioxidant activities were dose dependent, but it is noticeable that the values obtained using two most concentrated solution of essential oils (1.598 and 2.130 µg/mL) are very close (49.58 and 51.13% of inhibition of LP). For the inhibition of LP, the most responsible compounds were the oxygenated phenolic monoterpens (thymol and carvacrol) and the mixture of mono- and sesquiterpene hydrocarbons [98].

Concentration	Extracts					
	BHT	**Et₂O**	**CHCl₃**	**EtOAc**	***n*-BuOH**	**H₂O**
1%	26.15	-0.46	-0.92	24.17	11.31	13.58
5%	26.15	-0.77	-0.94	26.04	14.57	16.49
10%	26.15	-0.91	-0.97	30.28	19.78	23.24

Table 10. Inhibition of LP (%) in Fe^{2+}/ascorbate system of induction by extracts of oregano leaves and BHT (as a positive control) in the TBA assay.

The data presented in Table 10. show that the last three extracts of *O. vulgare* (EtOAc, *n*-BuOH and H₂O) reduced the intensity of lipid peroxidation while the first two extracts (Et₂O and CHCl₃) have prooxidative effect (but not statistically significant). The largest inhibitory activity was exhibited by ethyl acetate extract. High inhibitory effect of this extract and its components towards Fe^{2+}-dependent LP of liposomes can be related to the presence of flavonoids in the extract. It was established that flavonoids that antiradical potential of flavonoids are the most pronounced towards OH, peroxy- and alkoxy radicals, which are formed in the process of lipid peroxidation [99]. Also, these results are consistent with 2D-TLC analysis which showed that the dominant component of the EtOAc extract was kaempferol monoglycoside, while the H_2O extract contains multiple kaempferol diglycosides. From the literature it is known that additional glycosylation reduces the antioxidant activity and capability to reduce lipid peroxidation [100].The antioxidant and prooxidant activities of all extracts of oregano leaves were dose dependent.

4. Conclusions

It was found that excessive production of oxygen radicals in the body initiates oxidation and degradation of polyunsaturated fatty acids. It is known that free radicals attack the highly unsaturated fatty acid of membrane system and induce lipid peroxidation. Since lipid peroxidation causes oxidative damage to cell membranes and all other systems that contain lipids, in any investigation of total antioxidative activity of extracts and essential oils it is necessary to investigate their effects on lipid peroxidation. Some substrates (for example liposomes) are used more frequently than others, mainly because of the simplicity of the methods involved. In this way we get very useful information to direct future research. The results of our *in vitro* assays of examined four different Lamiaceae species extracts expressed significant protective effects on LP, which was found to be correlated to different compounds. It can be concluded that ethyl acetate and water proved to be the best solvent for extraction of plant material. Also, a very strong protective activity of the EtOAc and H_2O extracts in lipid peroxidation processes was recorded, which means that they may have a protective role in oxidative stress. Experimental results indicate that the essential oil of *M. peregrinum* collected from the Senta locality (No.3) exhibited the strongest inhibitory effect on lipid peroxidation. Furthermore, the present chapter on the chemistry and biological activity of four well known Lamiaceae species explicitly prove that these plants may be an important sources of pharmalogically active substances, and thus can be used in the preparation of various herbal medicine.

Author details

Biljana Kaurinovic and Mira Popovic

Department of Chemistry, Biochemistry and Environmental Protection, University of Novi Sad, Novi Sad, Republic of Serbia

Acknowledgement

This work was supported by the Ministry of Science and Environmental Protection of the Republic of Serbia (Project No. 172058) and by the Provincial Secretariat for Science and Technological Development, Autonomous Province of Vojvodina, Republic of Serbia.

5. References

[1] Dargel R (1992) Lipid Peroxidation – A Common Pathogenic Mechanism? Exp. toxicol. pathol. 44: 169-181.

[2] Ursini F, Maiorino M, Sevanian A (1991) Membrane Hydroperoxides. In: Sies H, editor. Oxidative Stress: Oxidants and Antioxidants. London: Academic Press. pp. 319-336.

[3] Kagan VE (1988) Lipid Peroxidation in Biomembranes. Boca Raton: CRC Press. 1 p.

[4] Farber JL, Kyle ME, Coleman JB (1990) Mechanisms of Cell Injury by Activated Oxygen Species. Lab. invest. 62: 670-679.

[5] Halliwell B, Gutteridge JMC. (1984) Lipid Peroxidation, Oxygen Radicals, Cell Damage and Antioxidant Therapy. Lancet. 23: 1396-1397.

[6] Holchestein P, Nordenbrand K, Ernster L (1964) Evidence for the Involvement of Iron in the ADP-Activated Peroxidation of Lipids in Microsomes and Mitochondria. Biochem. biophys. res. commun. 14: 323-328.

[7] Minotti G, Aust SD (1987) The Role of Iron in the Initiation of Lipid Peroxidation. Chem. phys. lipids. 44: 191-208.

[8] Aruoma OI, Halliwell B, Laughton MJ, Quinlan J, Gutteridge JMC (1989) The Mechanism of Initiation of Lipid Peorxidation. Evidence Against a Requirement for an Iron(II)-Iron(III) Complex. Biochem. j. 258: 617-620.

[9] Tang L, Zhang Y, Qian Z, Shen X (2000) The Mechanism of Fe^{2+}-Initiated Lipid Peroxidation in Liposomes: The Dual Function of Ferrous Ions, the Roles of the Pre-Existing Lipid Peroxides and the Lipid Peroxyl Radical. Biochem. j. 352: 27-36.

[10] Chatterjee SN, Agarwal S (1988) Liposomes as Membrane Model for Study of Lipid Peroxidation. Free rad. biol. med. 4: 51-72.

[11] Bombardelli E, Cristoni A, Morazzobi P (1994) PHYTOSOME® in Functional Cosmetics. Fitoterapia. 65: 397-401.

[12] http://www.pharmaxchange.info

[13] Foldvari M (2000) Non-Invasive Administration of Drugs Through the Skin. Challenges in Delivery System Design. PSTT. 3: 417-425.

[14] Immordino ML, Dosio F, Cattel L (2006) Stealth Liposomes: Review of the Basic Science, Rationale, and Clinical Applications, Existing and Potential. IJN 1(3): 297-315.

[15] Kujundžić S (2002) Biochemical Investigaion of Plant Species from the Apiaceae Family (in Serbian). Master Thesis. Novi Sad: PMF. 48 p.

[16] Laguerre M, Lecomte J, Villeneuve P (2007) Evaluation of the Ability of Antioxidants to Counteract Lipid Oxidation: Existing Methods, New Trends and Challenges. Prog. lipid res. 46: 244-282.

[17] Safer AM, Al-Nughamish AJ (1999) Hepatotoxicity Induced by the Antioxidant Food Additive Butylated Hydroxytoluene (BHT) in Rats: An Electron Microscopical Study. Histol. histopathol. 197: 391-406.

[18] Brraco U, Loliger J., Viret J (1981) Production and Use of Natural Antioxidants. JAOCS. 58: 686-690.

[19] Lagouri V, Blekas G, Tsimidou M, Kokkini S, Boskou D (1993) Composition and Antioxidant Activity of Essential Oil from Oregano Plants Grown in Greece. Z. lebensmitt.- unters. forsch. 197: 20-23.

[20] Crowell PL, Ayoubi S, Burke YD (1996) Antitumorigenic Effects of Limopnene and Perillyl Alcohol Against Pancreatic and Breast Cancer. Adv. exp. med. biol. 401: 131-136.

[21] Carlson J, Borg AK, Unelius R, Shoshan MC, Wilking N, Ringborg U, Linder S (1996) Inhibition of Tumor Cell Growth by Monoterpenes *in Vitro*: Evidence of a Ras-Independent Mechanism of Action. Anti-cancer drugs. 74(4): 422-429.

[22] Ruberto G, Baratta M.T, Deans SG, Dorman HJD (2000) Antioxidant and Antimicrobal Activity of *Foeniculum vulgare* and *Crithmum maritimum* Essential Oils. Planta med. 66: 687-693.

[23] Namiki M (1990) Antioxidants/Antimutagens in Food. Food sci. nutr. 29: 273-300.

[24] Rice-Evans CA, Miller NJ, Paganga G (1996) Structure-Antioxidant Activity Relationships of Flavonoids and Phenolic Acids. Free radic. biol. med. 20: 933-956.

[25] Cotelle N (2001) Role of Flavonoids in Oxidative Stress. Cur. topics med. chem. 1: 569-590.

[26] Rice-Evans CA, Miller NJ, Paganga G (1996) Antioxidant Properties of Phenolic Compounds. Trends plant sci. 2: 152-159.

[27] 1. & 2. European Directorate for the Quality of Medicines (2002) European Pharmacopeia, 4th edition. Council of Europe. France: Strasbourg Cedex. pp. 183-184.

[28] Afanas'ev A, Dorozhko I, Brodski V, Kostyk A, Potapovich I (1989) Chelating and Free Radical Scavenging Mechanism of Inhibitory Action of Rutin and Quercetin in Lipid Peroxidation. Biochem. pharmacol. 38: 1763-1769.

[29] Fukuzawa K, Seko T, Minami K, Terao J (1993) Dynamics of Iron-Ascorbate-Induced Lipid Peroxidation in Charged and Uncharged Phospholipid Vesicles. Lipids. 28: 497-503.

[30] Buege AJ, Aust DS (1988) Methods in Enzymology. New York: Academic Press. 302 p.

[31] Janero DR (1990) Malondialdehyde and Thiobarbituric Acid-Reactivity as Diagnostic Indices of Lipid Peroxidation and Peroxidative Tissue Injury. Free rad. biol. med. 9: 515-540.

[32] Naghibi F, Mosaddegh M, Motamed MS, Ghorbani A (2005) Labiatae Family in Folk Medicine in Iran: From Ethnobotany to Pharmacology. IJPR. 2: 63-79.

[33] Hammer KA, Carson CF, Riley TV (1999) Antimicrobial Activity of Essential Oils and Other Plant Extracts. J. appl. microbiol. 86(6): 986-990.

[34] Cantino PD, Harley RM, Wagstaff SJ (1992) Genera of Labiatae: Status and Classification. In: Harley RM, Reynolds T, editors. Advances in Labiate Science. London: Royal Botanic Gardens. pp. 511-522.

[35] Kaurinović B (2008) Antioxidant Activities of *Laurus nobilis* L. and *Melittis melissophyllum* L. Extracts (in Serbian). PhD Thesis. Novi Sad: PMF. 68 p.

[36] Honda T, Rounds BV, Bore L, Finlay HJ, Favaloro FG, Suh N, Wang Y, Sporn MB, Gribble GW (2000) Synthetic Oleanane and Ursane Triterpenoids with Modified Rings A and C: A Series of Highly Active Inhibitors of Nitric Oxide Production in Mouse Macrophages. J. med. chem. 43: 4233-4246.

[37] Kuzuhara H, Nishiyama S, Minova N, Sasaki K, Omoto S (2000) Protective Effects of Soyasapogenol a on Liver Injury Mediated by Immune Response in a Concavalin A-Induced Hepatitis Model. EJP. 391: 175-181.

[38] Shibata SS (2001) Chemistry and Cancer Preventing Activities of Ginseng Saponins and Some Related Triterpenoid Compounds. J. Korean med. sci. 16: 28-37.

[39] Zhang Z, Chang Q, Zhu M, Huang Y, Ho WK, Chen Z (2001) Characterization of Antioxidants Present in Hawthorn Fruits. J. nutr. biochem. 12: 144-152.

[40] Velasco-Negueruela A, Sanz J, Perez-Alonso MJ, Pala-Paul J (2004) The Volatile Components of the Aerial Parts of *Melittis melissophyllum* L. Bot. complutensis. 28: 133-136.

[41] http://www.ang.kfunigraz.ac.at/~katzer/engl.html

[42] Skrzypczak-Pietraszek E, Hensel A (2000) Polysaccharides from *Melittis melissophyllum* L. Herb and Callus. Pharmazie. 55: 768-771.

[43] Skrzypczak, E.; Skrzypczak, L. (1993) The Tissue Culture and Chemical Analysis of *Melittis melissophyllum* L. Acta hort. 330: 263-267.

[44] http://www.umm.edu/altmed.html

[45] http://www.hort.purdue.edu/newcrop/med-aro.html

[46] Guarrera PM (2005) Traditional Phytotherapy in Central Italy (Marche, Abruzzo, and Latium). Fitoterapia. 76: 1–25.

[47] Maggi F, Bílek T, Lucarini D, Papa F, Sagratini G, Vittori S (2009) *M. melissophyllum* L. subsp. *melissophyllum* (Lamiaceae) from Central Italy: A New Source of a Mushroom-Like Flavour. Food chem. 113: 216-221.

[48] Kaurinovic B, Popovic M, Vlaisavljevic S, Raseta M (2011) Antioxidant Activities of *Melittis melissophyllum* L. (Lamiaceae). Molecules. 16: 3152-3167.

[49] Doba T, Burton GW, Ingold KU (1985) Antioxidant and Co-Oxidant Activity of Vitamin C. The Effects of Vitamin C, Either Alone or in the Presence of Vitamin E or a Water-Soluble Vitamin E Analogue, Upon the Peroxidation of Aqueous Multimalleral Phospholipid Liposomes. Biochim. biophys. acta. 835: 298-303.

[50] http://www.phytochemicals.info/phytochemicals/quercetin.php

[51] Acland ML, van de Waarsenburg S, Jones R (2005) Synergistic Antiproliferative Action of the Flavonols Quercetin and Kaempferol in Cultured Human Cancer Cell Lines. In vivo. 19(1): 69-76.

[52] http://www.phytochemicals.info/phytochemicals/rutin.php

[53] Decker EA (1997) Phenolics: Prooxidants or Antioxidants? Nutr. rev. 55: 396-407.

[54] Handa SS, Sharma A, Chakraborti KK (1986) Natural Products in Plants as Liver Protecting Drugs. Fitoterapia. 57: 307-310.

[55] Cholbi MR, Paya M, Alcaraz MJ (1991) Inhibitory Effect of Phenolic Compounds on CCl₄-Induced Microsomal Lipid Peroxidation. Experimentia. 47: 195-198.

[56] Sousa RL, Marletta MA (1985) Inhibition of Cytochrome P-450 Activity in Rat Liver Microsomes by the Natural Occuring Flavonoid, Quercetin. Arch. biochem. biophys. 240: 345-348.

[57] Rekka E, Kourounakis PN (1991) Effect of Hydroxyethyl Rutoside and Related Compounds on Lipid Peroxidation and Free Radical Scavening Activity. Some Structural Aspects. J. pharmac. pharmacol. 43: 486-490.

[58] Larson RA (1988) The Antioxidant of Higher Plants. Phytochemistry. 27: 969-978.

[59] Cock C, Samman S (1996) Flavonoids Chemistry Metabolism, Cardioprotectioeffects and Dietary Sources. Nutrit. biochem. 7: 66-76.

[60] Stanković SM (2011) Total Phenolic Content, Flavonoid Concentration and Antioxidant Activity of *Marrubium peregrinum* L. Extracts. Kragujevac j. sci. 33: 63-72.

[61 Vlaisavljevic S (2007) Antioxidant Activities of *Marrubium peregrinum* L. Essential Oil and Extracts. Master Thesis. Novi Sad: PMF. 36 p.

[62] Nagy M, Gergel D, Grancai D, Novomesky P, Ubik K (1996) Antilipoperoxidative Activity of Some Phenolic Constituents from *Marrubium peregrinum* L. Farmaceuticky-Obzor. 65: 283-285.

[63] Sahpaz S, Hennebelle T, Bailleul F (2002) Marruboside, a New Phenylethanoid Glycoside from *Marrubium vulgare* L. Nat. prod. lett. 16: 195-199.

[64] Gruenwald J, Brendler T, Jaenicke C (2000) PDRc for Herbal Medicines. NewYork: Medical Economics Co. 271 p.

[65] Salei LA, Popa DP, Lazurēvskii GV (1967) Diterpenoids from *Marrubium peregrinum* L. Chem. nat. comp. 2: 200-201.

[66] Hennebelle T, Sahpaz S, Skaltsounis AL, Bailleul F (2007) Phenolic Compounds and Diterpenoids from *Marrubium peregrinum*. Biochem. systemat. ecol. 35: 624-626.

[67] Demirci B, Baser HC, Kirimer N (2004) Composition of the Essential Oil of *Marrubium bourgaei* ssp. Caricum P.H. Davis. J. essent. oil res. 16: 133-134.

[68] Lazari DM, Skaltsa HD, Constantinidis T (1999) Essential Oils of *Marrubium velutinum* Sm. and *Marrubium peregrinum* L., Growing Wild in Greece. Flavour frag. j. 14: 290-292.

[69] Nagy M, Svajdlenka E (1998) Comparison of Essential Oils from *Marrubium vulgare* L. and *M. peregrinum* L. J. essent. oils res. 10: 585-587.

[70] Khanavi M, Ghasemian L, Motlagh EH, Hadjiakhoondi A, Shafiee A (2005) Chemical Composition of the Essential Oils of *Marrubium parviflorum* Fisch. & C.A. Mey. and *Marrubium vulgare* L. from Iran. Flavour frag. j. 20: 324-326.

[71] Kaurinovic B, Vlaisavljevic S, Popovic M, Vastag Dj, Djurendic-Brenesel M (2010) Antioxidant Properties of *Marrubium peregrinum* L. (Lamiaceae) Essential Oil. Molecules. 15: 5943-5955.

[72] Kaurinovic B, Popovic M, Vlaisavljevic S, Zlinska J, Trivic S (2011) *In Vitro* Effect of *Marrubium Peregrinum* L. (Lamiaceae) Leaves Extracts. FEB. 20(12): 3152-3157.

[73] Čanadanović-Brunet J, Ćetković G, Đilas S, Tumbas V, Bogdanović G, Mandić A, Markov S, Cvetković D, Čanadanović V (2008) Radical Scavenging, Antibacterial, and Antiproliferative Activities of *Melissa officinalis* L. Extracts. J. med. food. 11: 133-143.

[74] Tosun M, Ercisli S, Sengul M, Ozer H, Polat T (2009) Antioxidant Properties and Total Phenolic Content of Eight *Salvia* Species from Turkey. Biol. res. 41: 175-181.

[75] Mann J (1992) Secondary Metabolism, 2nd edition. Oxford: Oxford University Press. 279 p.

[76] Jelačić ČS, Beatović VD, Prodanović SA, Tasić RS, Moravčević ŽĐ, Vujošević MA, Vučković MS (2011) Chemical Composition of the Essential Oil of Basil (*Ocimum basilicum* L. Lamiacea). Chem. ind. 65(4): 465-471.

[77] Grayer RI, Kite GC, Goldstone J, Bryan SE, Paton A, Putievsky E (1996) Intraspecific Taxonomy and Essential Oil Chemotypes in Sweet Basil, *Ocimum basilicum*. Phytochemistry. 43: 1033-1039.

[78] Marotti M, Piccaglia E, Giovanelli E (1996) Differences in Essential Oil Composition of Basil (*Ocimum basilicum* L.) Italian Cultivars Related to Morphological Characteristics. J. agric. food chem. 44: 3926-3929.

[79] http://www.ars-grin.gov/cgi-bin/duke/farmacy2.pl

[80] Telci I, Elmastas M, Sahin A (2009) Chemical Composition and Antioxidant Activity of *Ocimum minimum* Essential Oils. Chem. nat. comp. 45: 568-571.

[81] Soran ML, Cobzac-Codruta S, Varodi C, Lung I, Surducan E, Surducan V (2009) The Extraction and Chromatographic Determination of the Essentials Oils from *Ocimum basilicum* L. by Different Techniques. J. phys. conf. ser. 182: 012016.

[82] Gülçin I, Elmastaş M, Aboul-Enein YH (2007) Determination of Antioxidant and Radical Scavenging Activity of Basil (*Ocimum basilicum* L. Family Lamiaceae) Assayed by Different Methodologies. Phytother. res. 21: 354-361.

[83] Lee SJ, Umano K, Shibamoto T, Lee KG (2005) Identification of Volatile Components in Basil (*Ocimum basilicum* L.) and Thyme Leaves (*Thymus vulgaris* L.) and Their Antioxidant Properties. Food chem. 91: 131-137.

[84] Ruberto G, Baratta MT (2000) Antioxidant Activity of Selected Essential Oil Components in Two Lipid Model Systems. Food chem. 68: 167-174.

[85] Lewinsohn E, Ziv-Raz I, Dudai N, Tadmor Y, Lastochkin E, Larkov O, Chaimovitsh D, Ravid U, Putievsky E, Pichersky E, Shoham Y (2000) Biosynthesis of Estragole and Methyl Eugenol in Sweet Basil (*Ocimum basilicum* L.). Developmental and Chemotypic Association of Allaphenol O-Methyltransferase Activities. Plant sci. 160: 27-35.

[86] Tucakov J (1997) Healing With Herbs (Phytotherapy) (in Serbian). Beograd: Rad. 34 p.

[87] Kootstra A (1994) Protection from UV-B-Induced DNA Damage by Flavonoids. Plant mol. biol. 26: 771-774.

[88] Robak J, Gryglewski J (1988) Flavonoids are Scavengers of Superoxide Anion. Biochem. pharmacol. 37: 837-841.

[89] Kaurinovic B, Popovic M, Vlaisavljevic S, Trivic S (2011) Antioxidant Capacity of *Ocimum basilicum* L. and *Origanum vulgare* L. Extracts. Molecules. 16: 7401-7416.

[90] Lakic N, Mimica-Dukic N, Isak J, Božin B (2010) Antioxidant Properties of *Galium verum* L. (Rubiaceae) Extracts. Cent. eur. j. biol. 5: 331-337.

[91] W'glarz Z, Osidska E, Geszprych A, Przybyb J (2006) Intraspecific Variability of Wild Marjoram (*Origanum vulgare* L.) Naturally Occurring in Poland. Rev. bras. pl. med. botucatu. 8: 23-26.

[92] Arnold N, Bellomaria B, Valentini G (2000) Composition of the Essential Oil of Different Species of *Origanum* in the Eastern Mediterrenean. JEOR. 12: 192-196.

[93] Duke J (1997) The Green Pharmacy, the Ultimate Compendium of Natural Remedies from the World's Foremost Authority on Healing and Herbs. New York: St. Martin's Press.

[94] Souza EL, Stamford TLM, Lima EO, Trajano VN (2007) Effectiveness of *Origanum vulgare* L. Essential Oil to Inhibit the Growth of Food Spoiling Yeasts. Food Control. 18: 409-413.

[95] Vekiari SA, Oreopoulou V, Tzia C, Thomopoulos CD (1993) Oregano Flavonoids as Lipid Antioxidants. J. am. oil chem. soc. 70: 483-487.

[96] Chevolleau S, Mallet JF, Ucciani E, Gamisans J, Gruber M (1992) Effects of Rosemary Extracts and Major Constituents on Lipid Oxidation and Soybean Lipoxygenase Activity. J. am. oil chem. soc. 69: 1269-1271.

[97] Kramer RE (1985) Antioxidants in Clove. J. am. oil chem. soc. 62: 111-113.

[98] Milos M, Mastelic J, Jerkovic I (2000) Chemical Composition and Antioxidant Effect of Glycosidically Bound Volatile Compounds from Oregano (*Origanum vulgare* L. ssp. *Hirtum*). Food chem. 71: 79-83.

[99] Sichel G, Corsaro C, Scalia M, Dibilio J, Bonomo R (1991) *In Vitro* Scavenger Activity of Some Flavonoids and Melanins Against O_2. Free rad. biol. med. 11: 1-8.

[100] Rice-Evans C, Miller NJ, Paganga G (1997) Antioxidant Properties of Phenolic Compounds. Trends plant sci. 2: 152-159.

Chapter 8

Liposomes as a Tool to Study Lipid Peroxidation in Retina

Natalia Fagali and Angel Catalá

Additional information is available at the end of the chapter

1. Introduction

In living organisms, the oxidative stress is associated with several physio- pathological affections (e.g. atherosclerosis, cancer, aging, neurodegenerative diseases). The oxidative stress is generally initiated by generation of reactive oxygen (ROS) and nitrogen species (RNS) (Halliwell & Gutteridge, 1990). ROS are continuously formed during cellular metabolism and are removed by antioxidants defences. ROS from endogenous and exogenous sources results in continuous and accumulative oxidative damage to cellular components and alters many cellular functions. The most vulnerable molecules to oxidative damage are proteins, lipids and DNA (Kohen & Nyska, 2002; Catalá, 2009, 2011a, 2011b).

In mammalian retina, free radicals and lipoperoxides seem to play important roles in the evolution of different retinopathies including glaucoma, cataractogenesis, diabetic retinopathy, ocular inflammation and retinal degeneration (Ueda et al., 1996; De La Paz & Anderson, 1992). Because of free radicals production induces the lipid peroxyl radical formation, known as secondary free radicals products; this chain reaction of lipid peroxidation can damage the retina, especially the membranes that play important roles in visual function (Catalá, 2006). The retina is the neurosensorial tissue of the eye. It is very rich in membranes and therefore in polyunsaturated fatty acids (PUFAs) such as docosahexenoic acid (22:6 n-3), that are quite vulnerable to lipid peroxidation. Also, the human retina is a well oxygenated tissue. High-energy short-wavelength visible light promotes the formation of ROS which can initiate lipid peroxidation in the macula and elsewhere. The macular carotenoids are thought to combat light-induced damage mediated by ROS by absorbing the most damaging incoming wavelengths of light prior to the formation of ROS and by chemically quenching ROS once they are formed.

Although peroxidation in model membranes may be very different from peroxidation in biological membranes, the results obtained in model membranes may be used to progress our understanding of subjects that cannot be studied in biological membranes. Nevertheless, in spite of the relative simplicity of peroxidation of liposomal lipids model, these reactions are still relatively complex because they depend in a complex fashion on liposome type, reaction initiator and reaction medium (Fagali & Catalá, 2009). This complexity is the most likely cause of the apparent contradictions of literature results.

Biological membranes are complex systems. In view of this complexity and in order to avoid collateral effects that may arise during lipid peroxidation process of whole retinal membranes, we have attempted to gain understanding of the mechanisms responsible for peroxidation in a simple model system, made by dispersing retinal lipids in the form of liposomes.

This chapter describes a very useful method to prepare liposomes with natural phospholipids and the necessary methodology to follow the lipid peroxidation of these liposomes.

2. Materials and methods

2.1. Materials

Chloroform, methanol, trizma base, butylated hydroxytoluene (BHT), NaCl, FeSO$_4$ heptahydrate and 2-thiobarbituric acid (TBA) were purchased from Sigma Chemical Co. Suitable plastic lab ware was used throughout this study to avoid effects of adventitious metals. Other reagents were of the highest quality commercially available. All solutions were prepared using distilled water treated with a Millipore Q system.

2.2. Isolation of bovine retina

Eyes were enucleated at slaughter (Frigorífico Gorina), transported in ice to laboratory where retinas were taken out within 1–2 h. Under red light and with all tubes and solutions in ice buckets, corneas were excised; lenses and vitreous were subsequently removed. Eye cups were inverted and retinas were carefully peeled from the eyes. Retinas were briefly homogenized in 0.15 M NaCl (1 ml/retina) 120 s (20on-20off) at 4 ºC in an Ultraturrax X25 homogenizer at 7000 rpm.

2.3. Lipid extraction

Total lipids were extracted from retinal homogenates with chloroform/methanol (2:1 v/v) (Folch et al, 1957) at 4 ºC (sample:Folch = 1:5). A volume of water corresponding to 20 % of total volume was added. This mixture was shaken and kept in rest in cold to allow phases separation. Chloroformic phase was kept at -22 °C.

2.4. Preparation of liposomes made of retinal lipids

Total lipids obtained from retinal homogenates dissolved in chloroform were evaporated under nitrogen until constant weight and submitted to vacuum to remove traces of chloroform. Resultant films were dispersed at room temperature in a saline solution (0.15 M NaCl). Dispersed lipids were mixed to homogeneity using a vortex-mixer to obtain non-sonicated liposomes (NSL). Sonicated liposomes (SL) were prepared by sonication of NSL under nitrogen and ice cooling (Huang, 1969), using a Sonics vibra cell, probe-sonicator Model VCX 750 (750 W, 20kHz) at 75% of maximal output. Preparation of liposomes required about 2.5 min of sonication to reach apparently minimal optical density values.

2.5. Determination of liposomes size by Dynamic Light-Scattering (DLS)

The time correlation $G(q,t)$ of the light-scattering intensity was measured at 90° with a goniometer, ALV/CGS-5022F, with a multiple-τ digital correlator, ALV-5000/EPP, covering a 10^{-6}-10^3 s time range. The light source was a helium/neon laser with a wavelength of 633 nm operating at 22 mW. Each correlation function was analyzed by the well known cumulant fit yielding the apparent mean diffusion coefficient and the distribution δD of this value (Koppel, 1972). The measurements were carried out with 80 µl of SL and NSL (lipid concentration= 2 mg/ml) in water, 0.15 M NaCl and Tris-HCl buffer 20 mM (final volume= 2 mL).

2.6. Measurements of lipid peroxidation by detection of conjugated dienes and trienes production

In order to determine conjugated dienes and trienes production, absorption spectra were recorded by means of a Shimadzu UV-1800 spectrophotometer, in the range 200 to 300 nm, at 22 °C, with 1 cm path length quartz cell. Liposomes made of retinal lipids (80 µl, 2 g/l of lipids) were diluted to 2 ml with water, 0.15 M NaCl or 20 mM Tris-HCl pH 7.4, and oxidation was initiated by the addition of FeSO₄ (final concentration = 25 µM). Lipid peroxidation was assessed continuously by measuring the increase in absorbance at 234 nm (formation of conjugated dienes) and 270 nm (formation of conjugated trienes) taken at 1 min intervals. Oxidation rates were determined as the slope of a regression line drawn through linear range of absorbance versus time curve. Lag times were determined as time corresponding to intersection of oxidation rate regression line with a regression line drawn through initial phase of oxidation (Sargis & Subbaiah, 2003).

2.7. Measurements of lipid peroxidation by detection of Thiobarbituric Reactive Substances (TBARS)

During Fe^{2+}- initiated reactions, extent of liposomal lipid peroxidation was assessed using a TBA assay. In this procedure, 850 µL of TBA (0.375% w/v TBA, 0.25 N HCl) were added to aliquots of 150 µl of reaction mixture containing BHT (0.1 % w/v in ethanol) to prevent possible peroxidation of liposomes during incubation. The aliquots were taken at different intervals of time. Samples were heated for 30 min at 75 °C. Absorbance was measured at 532 nm for determination of aldehydic breakdown products of lipid peroxidation.

2.8. Preparation of Fatty Acids Methyl Esters (FAME)

Lipids from retina, liposomes or liposomes exposed to peroxidation initiated by Fe^{2+}, in absence or presence of BHT, were extracted according to the method of Folch et al (1957). A similar reaction mixture to that used in the analysis of conjugated dienes but scaled up 7.5 times was used to analyze the fatty acid composition of the samples

After one hour of incubation of liposomes with or without Fe^{2+} in the presence or absence of BHT, the samples were mixed with 15 ml of chloroform:methanol (2:1 v/v) containing 0.01 % BHT to stop the reaction. The mixture was stirred, gassed with nitrogen and kept in refrigerator overnight to achieve separation of phases. The lower chloroform phase was filtered through paper filter containing anhydrous sodium sulphate. The solvent was evaporated to dryness under nitrogen. Dry lipids of retina and/or liposomes were transmethylated with 300 μl of 1.3 M BF_3 in methanol at 65°C during 180 min. After incubation 1 ml of 0.15 M NaCl was added and the fatty acid methyl esters were extracted with 1 ml of hexane. This phase was injected onto the chromatograph.

2.9. Gas chromatography – Mass spectrometry analyses

GC–MS analyses were done using a Perkin Elmer Clarus 560D MS - gas chromatograph equipped with a mass selective detector with quadrupole analyzer and photomultiplier detector and a split/splitless injector. In the gas chromatographic system, a Elite 5MS (Perkin Elmer) capillary column (30 m, 0.25 mm ID, 0.25 μm df) was used. Column temperature was programmed from 130 to 250 °C at a rate of 5 °C/min and 250 °C for 6 min. Injector temperature was set to 260 ºC and inlet temperature was kept at 250 °C. Split injections were performed with a 10:1 split ratio. Helium carrier gas was used at a constant flow rate of 1 ml/min. In the mass spectrometer, electron ionization (EI+) mass spectra was recorded at 70 eV ionization energy, in full scan mode (50-400) unit mass range. The ionization source temperature was set at 180 °C. The fatty acid composition of the lipid extracts was determined by comparing their methyl derivatives mass fragmentation patterns with those of mass spectra from the NIST databases.

3. Results

3.1. Size of sonicated and non-sonicated liposomes made of retinal lipids in different aqueous media

Average hydrodynamic radii of liposomes determined by DLS studies are presented in Table 1. We noted that NSL display a multimodal size distribution when analyzed by inverse Laplace transform (CONTIN), a result that is compatible with the high polydispersity index (PI > 0.4) from cumulants fit. Thus, hydrodynamic radii for NSL, at room temperature in different aqueous media, cover a broad range with intensity weighted maxima centered between 190 and 320 nm. On the other hand, results for liposomes formed by sonication gave, through cumulant method, hydrodynamic radii in the order of 76.4-83.3

nm, showing as expected significant influence of sonication on size and distribution. It is clear that NSL possessed higher hydrodynamic radii than SL. Either NSL or SL in water were slightly smaller than that in 0.15 M NaCl and Tris-buffer.

Type of liposome	Aqueous media	Hydrodynamic radii (nm)	Polydispersity index
Sonicated	Water	76.4	0.31
	0.15 M NaCl	83.3	0.27
	20 mM Tris–HCl	83.3	0.27
Non-sonicated	Water	190–225	0.44
	0.15 M NaCl	260–320	0.45
	20 mM Tris–HCl	200–240	0.43

Table 1. Summary of values obtained by dynamic light scattering of SL and NSL made of retinal lipids in different aqueous media. Hydrodynamic radii values are the average of at least three representative determinations in each media.

3.2. Evolution of UV spectra as a function of time for Fe^{2+} initiated lipid peroxidation of SL and NSL in different aqueous media

Figure 1 shows evolution of UV spectra as a function of time, for Fe^{2+} initiated lipid peroxidation of SL and NSL, in different aqueous media. This figure showed increases in UV absorption with a maximum at 234 nm and at 270 nm, due to conjugated dienes and trienes respectively, and a decrease of absorbance at 200-215 nm, due to loss of methylene interrupted double bonds (unoxidized lipids). When lipid peroxidation was carried out in water or 0.15 M NaCl decreases at 200-215 nm were more notorious than in reactions carried out in Tris-buffer.

3.3. Conjugated dienes, trienes and TBARS are excellent markers of lipid peroxidation of liposomes made of retinal lipids

Figure 2 shows changes in TBARS production and variation of absorbance at 234 nm and 270 nm as a function of time.

When SL were peroxidized in water (**Figure 2A**) a lag phase of 30 min, followed by a fast rate, was observed in TBARS production. Absorbance final value at 532 nm reached was 0.24. Increase of absorbance at 234 nm showed a small lag phase followed by a fast initial phase until 40 min, since then speed of reaction became slighter. This behaviour was also observed in measured absorbance at 270 nm, but all absorbance values were lower than that at 234 nm in the range of time studied.

Lipid peroxidation of SL in 0.15 M NaCl (**Figure 2B**). showed an immediate and fast production of TBARS without lag phase, reaching a final value (Abs$_f\approx$ 0.23) similar to that obtained in water. The absorbance at 234 nm increased with an initial speed greater than

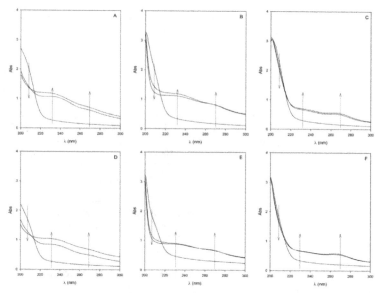

Figure 1. Time evolution (0, 90 and 180 min) of UV spectra of liposomes peroxidized with Fe²⁺ as an initiator of the reaction. SL in A) water, B) 0.15 M NaCl, C) buffer Tris. NSL in D) water and) 0.15 M NaCl, F) buffer Tris

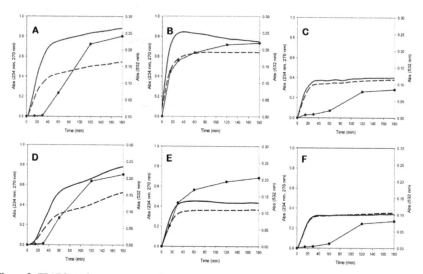

Figure 2. TBARS production (—•—) and variation of absorbance at 234 nm (—) and 270 nm (—) as a function of time, during Fe²⁺-catalyzed lipid peroxidation of SL (top) and NSL (bottom). TBARS were determined at 0, 15, 30, 60, 120 and 180 min after addition of Fe²⁺. Aqueous media where reactions were carried out: A, D: water; B, E: 0.15 M NaCl; C, F: 20 mM Tris-HCl pH 7.4.

that observed in water, became the highest to 30 minutes and, then, diminished slowly. The absorbance at 270 nm increased with an initial speed greater than that observed in water, became the highest around the 30 min and then remained constant.

Lipid peroxidation of SL in Tris- buffer (**Figure 2C**) showed the largest lag phase and the lowest final value of absorbance (Abs$_f$= 0.09) for TBARS formation. The initial speed of TBARS production was also the lowest. Initial speed of reaction observed, by increase of absorbance at 234 nm, was lower than that measured on water and 0.15 M NaCl. Absorbance reached the maximum at 30 minutes and then remained constant. Conjugated trienes production was very similar to that of conjugated dienes.

Lipid peroxidation of NSL in water (**Figure 2D**) showed a lag phase of 30 min for the TBARS production and a final value of absorbance of 0.21. Changes of absorbance at 234 nm displayed a lag of 16 min, increased quickly from this time to 60 min and since then continued increasing with lower speed. Values of absorbance at 270 nm were below than those observed at 234 nm, although the behavior was similar.

Lipid peroxidation of NSL in 0.15 M NaCl (**Figure 2E**) showed an initial speed of TBARS production greater than that observed in water, but with a very similar final value (Abs$_f$= 0.21). Changes in absorbance at 270 nm and 234 nm showed greater initial speeds than the corresponding ones in water. These speeds stayed until 30 minutes and since then, absorbance values did not change. Final values of absorbance in 0.15 M NaCl were smaller than the water ones.

Lipid peroxidation of NSL in Tris-buffer (**Figure 2F**) showed the greatest lag phase (60 min) in TBARS production and the smallest initial reaction rate. The final value was 0.08, a result much smaller than those obtained in water and 0.15 M NaCl. Values of change of absorbance determined at 270 nm and 234 nm were practically the same. Initial speeds were similar to those obtained in water and slower to those observed in 0.15 M NaCl. The reached final values were below to those obtained in water and 0.15 M NaCl.

SL were more susceptible to lipid peroxidation than NSL both in water as in 0.15 M NaCl. Nevertheless, both types of liposomes were equally peroxidized in Tris-buffer.

3.4. Fatty acid composition of retinal lipids and liposomes made of these retinal lipids

Figure 3 shows the fatty acid composition (area %) of retinal lipids and of liposomes made of these retinal lipids (SL-Fe, control). This table also compares fatty acid profiles of control with liposomes incubated with Fe^{2+} for 1 h, in absence and in presence of BHT. Retinal lipids show a high percent (25.8 ± 0.6 %) of docosahexaenoic acid (22:6 n-3), characteristic of this tissue. The retina has approximately 40 percent of PUFAs and 60 percent of saturated and monounsaturated fatty acids. SL prepared with these lipids show a decrease of 22:6 n-3. The PUFAs diminished significantly after incubation with Fe^{2+}. This produce a relative increase of saturated and monounsaturated fatty acids. 5 μM BHT protected PUFAs avoiding lipid peroxidation effects and the fatty acid profile there was not significant differences with control.

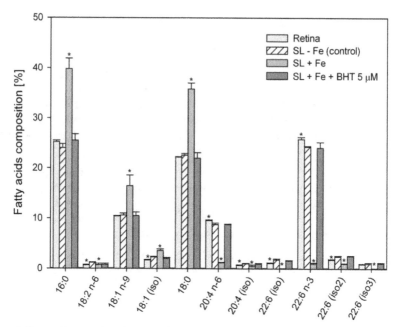

Figure 3. Fatty acid composition (area %) of retinal lipids, liposomes made of retinal lipids (SL -Fe, control), liposomes incubated with Fe^{2+} (SL + Fe) and liposomes incubated with Fe^{2+} in the presence of BHT. Results are expressed as $\tilde{x} \pm SD$. $\tilde{x}$: Average of area % of 3 assays, SD: standard deviation. Significant differences analyzed by ANOVA with control are marked with (*).

4. Conclusion

In summary, the presented results are indicative that liposomes made of retinal lipids by their structural similarities with the biomembranes constitute a very useful analytical system and can mimic the cellular membranes, providing additional information to that obtained with the whole retina. In addition, SL prepared with phospholipids obtained from selected tissues should be used in order to measure lipid peroxidation and the effect of different antioxidants. Additionally, we presented some simple techniques of many possibles that can be applied to study the lipid peroxidation process, different reaction initiators and the antioxidant effect of new compounds.

Abbreviations

16:0: palmitic acid, 18:1 n-9: oleic acid, 18:2 n-6: linoleic acid, 20:4 n-6: arachidonic acid, 22:6 n-3: docosahexaenoic acid, BHT: butylated hydroxitoluene, GC-MS: gas chromatography–mass spectrometry, PUFAs: polyunsaturated fatty acids, TBARS: thiobarbituric acid reactive substances, RNS: reactive nitrogen species, ROS: reactive oxygen species, SL: sonicated liposomes.

Author details

Natalia Fagali* and Angel Catalá
*Instituto de Investigaciones Fisicoquímicas Teóricas y Aplicadas,
(INIFTA-CCT La Plata-CONICET), Facultad de Ciencias Exactas,
Universidad Nacional de La Plata. Casilla de Correo 16, La Plata, Argentina*

Acknowledgement

The financial support of Consejo Nacional de Investigaciones Científicas y Técnicas (CONICET), Argentina, Grant PIP-0157, is gratefully acknowledged. The authors thank at the Departamento de Física, Facultad de Ciencias Exactas, Universidad Nacional de La Plata, Argentina for providing light scattering measurements facilities. Special thanks to Prof. J.L. Alessandrini for fitting experimental data.

5. References

Halliwell, B. and Gutteridge, J.M. (1990). Role of free radicals and catalytic metal ions in humandisease: an overview, *Methods Enzymol*. 186, 1–85

Huang, C. (1969). Studies on phosphatidylcholine vesicles, formation and physical characteristics. *Biochemistry* 8, 344–352.

Kohen, R., Nyska, A. (2002). Oxidation of biological systems: oxidative stress phenomena, antioxidants, redox reactions, and methods for their quantification. *Toxicol. Pathol*. 30, 620–650.

Catalá, A. (2009). Lipid peroxidation of membrane phospholipids generates hydroxy-alkenals and oxidized phospholipids active in physiological and/or pathological conditions, *Chem Phys Lipids*, 157, 1-11.

Catalá, A. (2011a). In: Catalá, A. (Ed.), Lipid Peroxidation: Biological Implications, Transworld Research Network Kerala, India, pp. 1-173.

Catalá, A. (2011b). Lipid peroxidation of membrane phospholipids in the vertebrate retina, *Front. Biosci.* (Schol Ed.) 3, 52–60, Review.

Fagali, N., Catalá, A. (2009) Fe^{2+} and Fe^{3+} initiated peroxidation of sonicated and non-sonicated liposomes made of retinal lipids in different aqueous media, *Chem Phys Lipids*, 159, 88-94.

Folch, J., Lees, N., Sloane Stanley, G.A. (1957). A simple method for the isolation and purification of total lipids from animal tissues. *J Biol Chem*, 226, 497–509.

Koppel, D.E. (1972). Analysis ofmacromolecular polydispersity in intensity correlation spectroscopy: the methods of cumulants, *J Chem Phys*, 57, 4814–4820.

Sargis, R.M., Subbaiah, P.V. (2003). Trans unsaturated fatty acids are less oxidizable than cis unsaturated fatty acids and protect endogenous lipids from oxidation in lipoproteins and lipid bilayers. *Biochemistry*, 42, 11533–11543.

---

* Corresponding Author

Ueda, T., Ueda, T., Armstrong, D. (1996). Preventive effect of natural and synthetic antioxidants on lipid peroxidation in the mammalian eye. *Ophthalmic Res*, 28, 184–192.

Lipid Peroxidation in Vegetables, Oils, Plants and Meats

Chapter 9

Effects of Hydroperoxide in Lipid Peroxidation on Dough Fermentation

Toshiyuki Toyosaki

Additional information is available at the end of the chapter

1. Introduction

The oxidation of lipids in foods is responsible for the formation of off-flavors and chemical compounds that may be detrimental to health; it is a well known problem in the food chemistry and biochemistry fields (Ames et al., 1994; Gardnaer, 1996; Grosch, 1987; Pokorny, 1999; Shewfelt & Del Rosario, 2000; Mercier & Gelinas, 2001; Toyosaki & Sakane, 2002; Toyosaki & Koketsu, 2004). Currently, the various lipid peroxides produced by such lipid peroxidation are treated only as a nuisance. However, among longstanding traditional foods there are foods with fine flavors that are brought out by inducing lipid peroxidation; such foods include fine, thin noodles and certain dried foods. Thus, lipid peroxides produced by lipid peroxidation can also be advantageous. The properties of foods can be improved by better use of the properties of lipid peroxides. The current work provided an interesting finding: when lipoxygenase was added during the fermentation of bread dough, the fermentation of dough was promoted. During this event, hydroperoxides produced by lipid peroxidation triggered the promotion of fermentation and promoted the fermentation of dough. The phenomenon by which hydroperoxides are produced by lipid peroxidation and promote the fermentation of bread dough is decidedly not beneficial when assessed from a nutritional standpoint, but this phenomenon is extremely desirable when assessed from a food science standpoint. The objective of the current study was to investigate the bread dough fermentation-promoting action of hydroperoxides produced by lipid peroxidation in bread dough and the mechanism that is involved.

2. Methods

Preparation of dough

The wheat flour (strong flour) that was used to make adjustments in the bread dough was the type that is readily commercially available. The lipid added was linoleic acid (more than

95% pure), which was added at a 3% level. Other ingredients used to make the bread were all commercially available. Lipoxygenase was added at the end of bread dough adjustment and underwent primary fermentation in an incubator at 37°C with 75-80% humidity. After fermentation, gas was released; after a bench time of 10 min, the dough underwent final fermentation for 90 min and was then baked for 12 min at 200°C.

Preparation of the model system of gluten and linoleic acid

By mixing a fixed amount of commercially available gluten and linoleic acid (more than 95% pure) of 3% level, which served as the test sample, a model system was created. This sample underwent fermentation by lipoxygenase induction.

Measurement of the rate of dough expansion

To determine the rate of dough expansion with fermentation, a fixed amount of dough was placed in a graduated cylinder and fermented in an incubator (temperature 30°C, humidity 75%). The rate of expansion over a fixed period of time was then measured.

Measurement of hydroperoxide

Hydroperoxide concentration was calculated in terms of 2',7'-dichlorofluorescein (DCF). To de-emulsify, the 5.0 ml samples were centrifuged (10,000 x g, 30 min). The linoleic acid of the supernatant was then measured to determine the hydroperoxide level using the method of Cathcart et al. (1984). First, 1.0 ml of a 1.0 mM solution of DCF in ethanol and 2.0 ml of 0.01N NaOH were mixed and stirred for 30 min before being neutralized with 10 ml of 25 mM phosphate buffer (pH 7.2). Then 2.0 ml of the neutralized DCF solution were added to a solution of hematin (10 mM) in 25 mM phosphate buffer (pH 7.2; 0.01 mg DCF/ml); subsequently, 2.0 ml of this hematin-DCF solution and 10 ml of the linoleic acid sample were mixed and left at 50°C for 50 min, before fluorometry treatment (excitation. 400 nm; emission. 470 nm) to measure DCF. This method measures hydroperoxide with more sensitivity than the iron rhodanide method that is usually used.

Diethylaminoethyl (DEAE) column chromatography

The extracted dough was separated by Tris-HCl buffer (pH 8.0). The extracted sample fractions were separated by DEAE-cellulose (DE52, Whatman, Ltd., Tokyo, Japan) column chromatography as follows. A DEAE-cellulose column (3.8 x 54 cm) was equilibrated with 50 mM Tris-HCl buffer (pH 8.0), washed with the same buffer, and developed in a linear gradient made with 300 ml of this buffer and 300 ml of the same buffer containing 0.6 M NaCl. The flow rate was 30 ml/hr, and 3.0 ml fractions were collected.

Measurement of the amount of protein

The amount of protein was measured using the Lowry method (1951).

SDS-polyacrylamide gel electrophoresis (PAGE)

The measurement was done according to the method of Laemmli (1979). Electrophoresis was performed using the Mini-Protean II Electrophoresis Cell (Bio-Rad Laboratories, Inc.,

Tokyo, Japan) at 18 mA/gel with Ready Gel J of differing gel concentrations. After electrophoresis, the gels were stained using Coomassie brilliant blue-R250. In addition, automated electrophoresis (Phast System; Pharmacia LKB, Biotechnology AB, Uppsala, Sweden) equipment was used.

Statistical analysis

Analysis of variance (ANOVA) was performed and mean comparisons were obtained by Duncan's multiple range test (Steel & Torrie, 1980). Significance was established at $P<0.05$.

3. Results and discussion

Lipid peroxidation during fermentation and accompanying changes in the rate of expansion

After 3% linoleic acid was added to the other bread ingredients, the doughs with and without lipoxygenase were individually mixed with a mixer for a fixed time. Next, the dough underwent primary fermentation in an incubator for 90 min, and the amount of hydroperoxide produced over this time period was measured. These results are shown in Fig. 1. Lipid peroxidation by lipoxygenase induction increased as the fermentation time progressed. However, the overall amount of hydroperoxide produced in the lipoxygenase-free dough tended not to increase. The rate of expansion during this time is shown in Fig. 2.

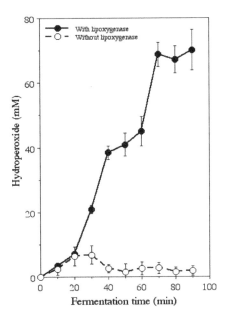

Figure 1. Changes in the amount of hydroperoxide produced with the fermentation of dough. Each value represents the mean ± standard error in triplicate.

For up to 40 min, dough with lipoxygenase expanded rapidly after the start of fermentation, but after this time the expansion tended to decrease abruptly. In contrast, lipoxygenase-free dough reached its maximum rate of expansion in 30 min from the start of fermentation, and this tended to decrease gradually afterwards. Further, in the dough without lipoxygenase, changes in the rate of expansion per unit time were smaller than in the dough with lipoxygenase. Changes brought about by this phenomenon are quite likely due to the effect that hydroperoxide, which is produced by lipid peroxidation, has on the fermentation stage.

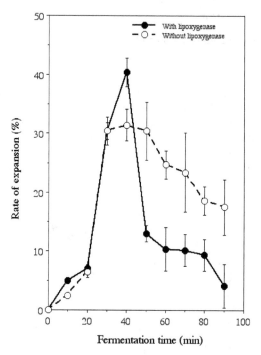

Figure 2. Changes in the rate of expansion with the fermentation of dough. Each value represents the mean ± standard error in triplicate.

Relationship between the rate of expansion and hydroperoxide

A model system of gluten and linoleic acid was created, and the involvement of the hydroperoxide that was produced in the fermentation of dough was examined. These results are shown in Fig. 3. The rate of dough expansion was affected by hydroperoxide concentration, and in this experiment the rate of expansion reached its maximum at a hydroperoxide concentration of 30-40 mM. Based on these results, the fermentation of dough was influenced by the concentrations of hydroperoxide that were produced.

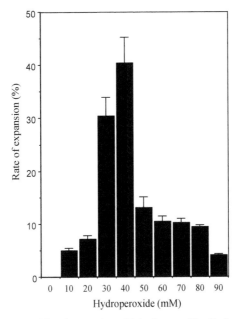

Figure 3. Changes in the rate of dough expansion with hydroperoxides. Each value represents the mean ± standard error in triplicate.

Effects of yeast and gluten during fermentation

The effects of the yeast on dough fermentation were also studied. The comparison of the dough with and without lipoxygenase is shown in Fig. 4. Both doughs with ≤ 2.0% yeast had similar rates of expansion that tended to increase with fermentation time. There were almost no changes in the rate of expansion with yeast concentrations of ≥ 2.5%; in fact, the rate of expansion tended to decrease. However, the rate of expansion of dough with lipoxygenase tended to increase more than the rate of the dough without lipoxygenase. Since a detailed study of the relationship between yeast and hydroperoxide produced was not done in this experiment, further study is needed.

Next, changes in the amount of hydroperoxide were studied; these results are shown in Fig. 5. For the dough with lipoxygenase, the amount of hydroperoxide reached its maximum when the yeast concentration was 1%; as the yeast concentration increased, the amount of hydroperoxide that was produced tended to decrease. Comparing these results with those in Fig. 4 indicates that there is a relationship between the amount of hydroperoxide produced and the yeast concentration; the specifics of this relationship need to be further investigated. Lipoxygenase-free dough produced almost no hydroperoxide. However, based on the results in Fig. 4, the production of hydroperoxide may not be the sole factor involved in the fermentation of dough. Thus, the hydroperoxide that is produced may be synergistically involved in the mechanism of yeast fermentation.

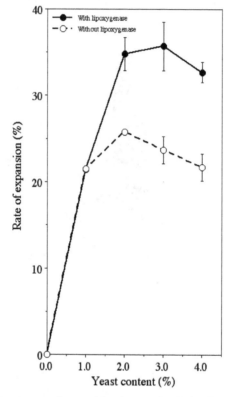

Figure 4. Effect of yeast contents on the rate of dough expansion. Each value represents the mean ± standard error in triplicate.

Next, the effect of differences in gluten content on dough fermentation was studied. The results are shown in Fig. 6. The expansion of dough began at a gluten content of 40%, and the rate of expansion reached its maximum at a gluten content of 60%. Beyond this concentration, the rate of expansion tended to gradually decrease. Dough with lipoxygenase had a rate of expansion of about 35% at a gluten content of 60%, while lipoxygenase-free dough had a rate of expansion of 15%. Thus, the presence of hydroperoxide had an effect on the rate of expansion; the hydroperoxide that was produced promoted fermentation.

The relationship between gluten and hydroperoxide

Hydroperoxide is involved in the fermentation of dough in a facilitatory manner, and, as a result, the rate of dough expansion is increased. Consequently, this phenomenon has a positive effect on dough. To study the effect of the hydroperoxide that is produced during dough fermentation on gluten, the gluten was separated and purified after the completion of fermentation using affinity chromatography, so that, ultimately, the gluten fraction was obtained. This gluten fraction was subjected to SDS–gel electrophoresis, and the relationship between gluten and hydroperoxide in the fermentation stage was studied; the results are shown in Fig. 7. In the dough without lipoxygenase, there were almost no changes in the molecular weight of gluten during 100 min of fermentation time. In contrast, in the dough with lipoxygenase, changes in the molecular weight of gluten were seen with fermentation, and formation of gluten polymers was noted with fermentation. This phenomenon was caused by hydroperoxide that was produced, which acted on the gluten and may have induced denaturation. A comparison of the results shown in Figure 4 and 5 shows that the gluten network was tightened, because the hydroperoxide that was produced by the addition of lipoxygenase denatured the gluten and, subsequently, increased dough expansion.

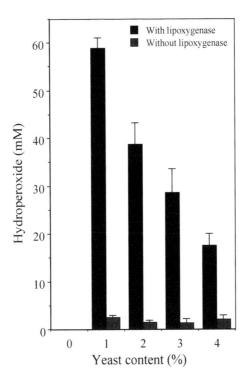

Figure 5. Effect of yeast contents on the amount of hydroperoxide produced in dough. Each value represents the mean ± standard error in triplicate.

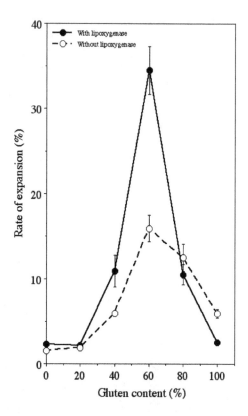

Figure 6. Effect of gluten content on the rate of dough expansion. Each value represents the mean ± standard error in triplicate.

Figure 7. Changes in the molecular weight of gluten when dough was fermented for 100 min.

The mechanism by which hydroperoxide acts to promote fermentation

The various experimental results that were obtained were comprehensively analyzed to determine the mechanism of action by which hydroperoxide acts to promote fermentation; this is shown in Fig. 8. During gluten formation, gluten is formed when gliadin and glutenin form a network structure. When gluten is crosslinked in the presence of hydroperoxide, the molecules themselves form macromolecules. As a result, expansion is promoted by the uptake of large amounts of carbon dioxide gas produced during dough fermentation. This phenomenon is ultimately advantageous when baking dough, and it improves the bread's texture. When very little lipid peroxidation is induced, the unoxidized linoleic acid has no interaction with gluten, and, as a result, gluten crosslinking does not occur, which results in baked bread with a poor texture.

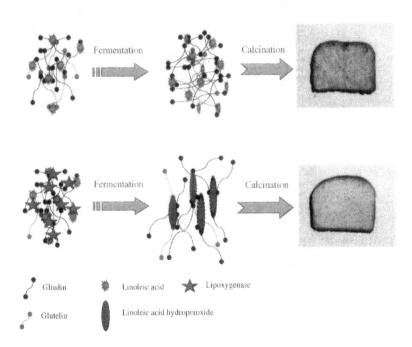

Figure 8. The mechanism by which hydroperoxides accelerate fermentation.

4. Conclusions

The current research demonstrated that well-fermented dough can be produced by the induction of lipid peroxidation when fermenting dough. The induction of lipid peroxidation was achieved in the current study by using lipoxygenase induction, but a similar phenomenon should also occur with lipid peroxidation induced by other methods. This phenomenon is advantageous when baking bread and can be used to enhance the quality of baked bread. Based on the results of these tests of physical properties, further detailed study is needed of the effect of lipid peroxidation on the flavor of baked bread.

Author details

Toshiyuki Toyosaki

Department of Foods and Nutrition, Koran Women's Junior College, Fukuoka, Japan

5. References

[1] Ames, B.M., Shigena, K., & Hagen, T.M. (1994). Oxidants, antioxidants and the degenerative diseases of aging. *Proceeding National Academy Science*, 90, 7915-7922.

[2] Cathcart, R., Schwiers, E., & Ames, N. (1984). Detection of picomole levels of lipid hydroperoxides using a dichloro-fluorescein fluorescent assay. In L. Packer (Ed), *Methods in Enzymology*, vol. 105 (p. 352). New York; Academic Press Inc.

[3] Gardnaer, H.W. (1996). Lipoxygenase as a versatile biocatalyst. *Journal of American Oil Chemical Society*, 73, 1347-1350.

[4] Grosch, W. (1987). Reactions of hydroperoxides-products of low molecular weight. In H.W.-S Chan (Ed), *Autoxidation of Unsaturated Lipids*. London; Academic Press Inc.

[5] Laemmli, U.K. (1979). Cleavage of structural proteins during the assembly of the head of bacteriophage T4. *Nature*, 227, 680-685.

[6] Lowry, O.H., Rosebrough, N.J., Farr A.L., & Randall, R.J. (1951). Protein measurement with the Folin-Phenol reagent. *Journal of Biological Chemistry*, 193, 265-275.

[7] Mercier, M., & Gelinas, P. (2001). Effect of lipid oxidation on dough bleaching. *Cereal Chemistry*, 78, 36-38.

[8] Pokorny, J. (1999). *Handbook of Food Preservation, Antioxidants in Food Preservation*. New York; Marcel Dekker.

[9] Shewfelt, R.L., & Del Rosario, B.A. (2000). The role of lipid peroxidation in storage disorders of fresh fruits and vegetables. *Horticulture Science*, 35, 575-579.

[10] Steel, R.G.D., & Torrie, J.H.(1980). *Principles and procedures of statistics: a biometrical approach* (2nd ed). New York: McGraw-Hill.

[11] Toyosaki, T., & Sakane, Y. (2002). Antioxidant effect of NaCl on the aqueous solution, emulsified and enzymic lipid peroxidation. *Bulletin of the Society of Sea Water Science, Japan*, 56, 10-16.

[12] Toyosaki, T., & Koketsu, M. (2004). Oxidative stability of silky fowl eggs. Comparison with hen eggs. *Journal of Agriculture and Food Chemistry*, 52, 1328-1330.

Chapter 10

The Effect of Plant Secondary Metabolites on Lipid Peroxidation and Eicosanoid Pathway

Neda Mimica-Dukić, Nataša Simin, Emilija Svirčev,
Dejan Orčić, Ivana Beara, Marija Lesjak and Biljana Božin

Additional information is available at the end of the chapter

1. Introduction

Inflammation, free radical damage and oxidative stress have become major health issues in recent years and the subject of plenty of research. These processes are implicated in cancer [1], cardiovascular diseases [2], multiple sclerosis [3], diabetes mellitus [4], Alzheimer's and Parkinson's diseases [5], rheumatoid arthritis [6], premature aging [7] and almost any other degenerative condition. Reactive oxygen species (ROS), which are involved in these physiological functional changes, are often either by-products of the normal cellular processes or are formed by action of exogenous factors - xenobiotics, ionizing radiation, stress, pathogens etc. Overproduction of ROS leads to oxidative stress, with biomolecules, including lipids, proteins and nucleic acids undergoing oxidative alterations.

1.1. Lipid peroxidation

Lipid peroxidation (LP) – oxidative degradation of polyunsaturated fatty acids caused by ROS – is responsible for degradation of membrane lipids resulting in cell damage and formation of many toxic products. LP is a free radical chain reaction where three major steps - the initiation, propagation and termination – can be recognised (Figure 1.).

In initiation phase, highly reactive hydroxyl radical, formed in Fenton reaction, abstracts hydrogen atom in α position relative to the polyunsaturated fatty acid double bond. This results in the formation of fatty acid radical, highly unstable, short-lived intermediate that stabilises by abstracting hydrogen from another chemical species, or reacts with triplet oxygen to generate different radical species, including fatty acid peroxyl radical. In the termination step peroxyl radicals transform into nonradical compounds – hydrocarbons, aldehydes, alcohols, volatile ketones and lipid polymers, some of which are harmful (Figure 1.) [8, 9].

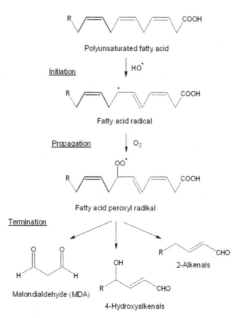

Figure 1. Steps in lipid peroxidation process

Organism uses a number of endogenous antioxidants, such as glutathione, α-lipoic acid, coenzyme Q10, bilirubin and antioxidant enzymes (glutathione peroxidase, catalase, superoxide dismutase), to protect itself from oxidative stress. When they are insufficient, it becomes necessary to introduce exogenous antioxidants. Most of these compounds are primarily taken into the body by food and are predominantly of herbal origin (phenolics, carotenoids, terpenoids and vitamins – ascorbic acid and tocopherol). Herbal antioxidants exhibit their activity through a wide variety of mechanisms, such as inhibition of oxidising enzymes, chelation of transition metals, transfer of hydrogen or single electron to radicals, singlet oxygen deactivation, or enzymatic detoxification of ROS [8]. They can stop LP process in either initiation or propagation step. Herbal antioxidants have become subjects of growing interest and targets of numerous scientific research. However, enormous number of plant species is still waiting to be investigated in this manner and explored as potential medical drugs or dietary supplements.

1.2. Eicosanoid pathway

In addition to direct detrimental effects on biomolecules and cellular structures, lipid peroxidation is also involved in biosynthesis of eicosanoids – arachidonic acid metabolites serving as inflammatory mediators. Arachidonic acid, released from phospholipids by action of phospholipase A, can be converted into these products by three different pathways: cyclooxygenase, leading to the formation of prostanoids (prostaglandins and thromboxanes), lipoxygenase, where leukotrienes and certain mono-, di- and trihydroxy acids are synthesised, and epoxygenase pathway, which includes cytochrome P-450 and gives epoxides as final products [10].

Cyclooxygenase (COX), key enzyme in cyclooxygenase pathway (Figure 2.), exists in three forms, COX-1, COX-2, and recently discovered COX-3. Despite the differences in structure, localisation and regulation, reactions catalysed by COX isoforms follow the same mechanism. All of them transform arachidonic acid into prostanoids: prostaglandins (PGH_2, PGE_2, $PGF_{2\alpha}$), thromboxanes (TXA_2, TXB_2), and 12(S)-hydroxyheptadeca-5Z,8E,10E-trienoic acid (12(S)-HHT) as a co-product.

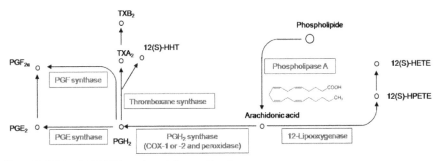

Figure 2. COX and 12-LOX branches of eicosanoid pathway

COX-catalysed transformation of arachidonic acid starts with tyrosyl radical generation through Tyr^{385} oxidation by heme in COX active site. Formed tyrosyl radical abstracts hydrogen from C-13 of arachidonic acid. In subsequent steps, free electron migration, reaction with oxygen yielding peroxyl radical, and cyclisation reactions give PGG_2 which is converted to PGH_2 through the action of peroxidase (Px) (Figure 3.) [11].

COX-1 is expressed constitutively in different tissues, blood monocytes and platelets, and is involved in normal cellular homeostasis. In contrast, COX-2 may be induced by a series of pro-inflammatory stimuli and its role in the progress of inflammation, fever and pain has been known [12]. Furthermore, COX-2 has been targeted in many cancers including: colon cancer, colorectal cancer, breast cancer, gliomas, prostate cancer, esophageal carcinoma, pancreatic cancer, lung carcinoma, gastric carcinoma, ovarian cancer, Kaposi's sarcoma and melanoma [13].

In lipoxygenase branch of arachidonic acid metabolism, there are three types of lipoxygenases, termed 5-, 12- and 15-lipoxygenase. In 12-lipooxigenase (12-LOX) pathway, the first reaction is abstraction of hydrogen from C-10 of arachidonic acid, which includes reduction of Fe^{3+} to Fe^{2+} in enzyme active site. This results in the formation of arachidonic acid radical, which than reacts with oxygen and generates 12-hydroperoxyeicosatetraenoic acid (12-HPETE). Finally, formation of 12(S)-hydroxy-(5Z,8Z,10E,14Z)-eicosatetraenoic acid (12-HETE) is catalysed by glutathione peroxidase (GPx), whereby glutathione (GSH) is oxidised to GS-SG (Figure 3.) [14, 15].

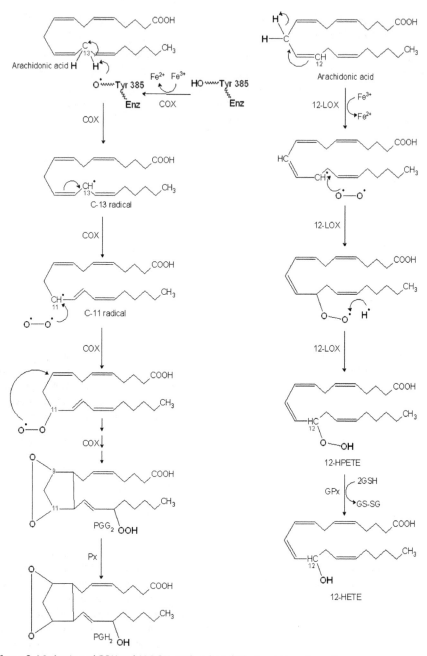

Figure 3. Mechanism of COX and 12-LOX-catalysed arachidonic acid transformation.

12-HETE is implicated in regulation of platelet aggregation, angiogenesis, as well as the progression of several human diseases like various cancers, rheumatoid arthritis and psoriasis [13, 16, 17]. Also, 12-HETE is known to take part in the metastatic cascade as a crucial intracellular signalling molecule which activates protein kinase C and mediates the biological functions of growth factors and cytokines.

Nowadays, there is a great need for a new anti-inflammatory compounds with minimal side effects. Natural products, especially phenolics frequently consumed in a diet of plant origin, are known to have great anti-inflammatory potential considering inhibition of COX and LOX enzymes [18, 19]. Thus, screening of plants for COX/LOX inhibitory activity, followed by effect-guided fractionation, can be a useful tool for discovering new secondary biomolecules with anti-inflammatory potential. Although plant species widely used in traditional medicine may be a good starting point, there is also a vast number of currently unexplored species of unknown composition and activity. Therefore, this study included the poorly investigated plant species classified into four families – Alliaceae, Cupressaceae, Plantaginaceae and Polygonaceae – wild growing in Serbia, that are part of our continuing research [20-29]. Most of these species are used in diet or used in folk medicine for healing various disorders.

Considering the fact that lipid peroxidation is involved in an inflammation process, the aim of this study was to compare antioxidant activity (more specifically, the ability to inhibit lipid peroxidation) of the selected plant extracts with their ability to inhibit production of particular arachidonic acid metabolites and to correlate these activities with the phenolic and flavonoid content.

2. Methods

2.1. Plant material and extract preparation

Plant material used in this study was collected from different locations in Serbia during the period between 2008–2010, during a flowering phase or as ripe specimens. Species belonging to four families: Alliaceae (genus Allium – *A. flavum* L., *A. carinatum* L., *A. melanantherum* Panč., *A. pallens* L., *A. rhodopeum* Velen., *A. paniculatum* L.), Cupressaceae (genus *Juniperus* L. – *J. communis* L., *J. sibirica* Burgsdorf., *J. foetidissima* Willd.), Plantaginaceae (genus *Plantago* L. – *P. major* L., *P. maritima* L., *P. media* L., *P. lanceolata* L., *P. altisima* L.) and Polygonaceae (genus *Rumex* L. – *R. patientia* L., *R. crispus* L., *R. obtusifolius* L.) were investigated. The voucher specimens were deposited in the Herbarium of the Department of Biology and Ecology (BUNS Herbarium), University of Novi Sad Faculty of Sciences.

Air-dried and grounded plant material (30 g) (whole plants of *Allium* sp., needles and cones of *Juniperus* sp., aerial parts of *Plantago* sp. and herbs and rhizomes of *Rumex* sp.) was extracted by maceration with 80% aqueous methanol (8 mL per 1 g of drug) during 72 h at room temperature. After filtration, the solvent was evaporated to dryness under reduced pressure. All raw extracts except those of *Allium* sp. and of *Rumex* sp. rhizomes were resuspended in hot distilled water (to a final concentration of approx. 1 g/mL), washed exhaustively with petroleum ether (fraction 40–60 °C) to remove nonpolar pigments, and evaporated to dryness under vacuum.

Dried extracts were dissolved in 80% aqueous methanol and DMSO for evaluation of the antioxidant and anti-inflammatory activity, respectively, to obtain 300 mg/mL stock solutions.

2.2. Determination of total phenolic content

Total phenolic content was determined according to method of Singleton et al. [30], modified for 96-well microplates. Gallic acid was used as a standard for calibration curve construction. Thirty microliters of each extract or standard solution was added to 150 μL of 0.1 mol/L Folin-Ciocalteu reagent and after 10 min mixed with 120 μL of sodium carbonate (7.5%). The same mixture, with solvent instead of extract, was used as a blank. Absorbance at 760 nm was read after 2 h. The phenolics concentration was determined by using the calibration curve of gallic acid. The total phenolics value was expressed as milligrams of gallic acid equivalents per gram of dry weight (dw).

2.3. Determination of total flavonoid content

The aluminium chloride spectrophotometric method described by Chang et al. [31] and modified for 96-well microplates, was used for determination of of the total flavonoid content. Quercetin was used as a standard to prepare a calibration curve. The reaction mixture was comprised of 30 μL of the extract or standard solution, 90 μL of methanol, 6 μL of 10% aluminium chloride (substituted with distilled water in blank probe), 6 μL of 1 mol/L potassium acetate and 170 μL of distilled water. Absorbance at 415 nm was measured after 30 min. Flavonoid content was calculated according to the standard calibration curve and are expressed in milligrams of quercetin equivalents per gram of dw.

2.4. Lipid peroxidation

There are several methods for measuring the ability of plant extracts to inhibit lipid peroxidation [8]. Some of the methods are based on monitoring of malondialdehyde (MDA), a degradation product of polyunsaturated fatty acids peroxidation. Possible ways for quantification of MDA are GC-FID after derivatisation, HPLC with DAD or fluorimetric detector and spectrophotometric method. The latter is based on the reaction of MDA with thiobarbituric acid (TBA) and it is commonly used both in *in vitro* and *in vivo* studies. Formation of the red-coloured MDA-TBA adduct is measured at 532 nm. Different substrates can be used in this test: lecithin liposomes, free fatty acids, LDL and body fluids [8, 32]. Also, a few different initiators of LP, such as ionizing radiation, chemical agents – metal ions, free radicals and metalloproteins may be used [8].

In this study, we used spectrophotometric TBA assay [28, 33] for evaluation the ability of extracts to inhibit LP. Linseed oil, used as a source of polyunsaturated fatty acids (69.7 % linolenic, 13.5 % linoleic acid, as determined by GC-MS), was obtained from linseed by

Soxhlet extraction. Oil was added to 0.067 mol/L phosphate buffer, pH 7.4, in the presence of 0.25% Tween-80 to obtain a 0.035% suspension and sonicated for 1 hour. This suspension (3.0 mL) was mixed with 20 µL of $FeSO_4$ (4.58 mmol/L), 20 µL of ascorbic acid (87 µmol/L), and 20 µL of extract (or solvent in control); 3.0 mL of phosphate buffer and 20 µL of extract were added in the blank probe. After incubation at 37 °C for 1 hour, 0.2 mL 3.72% EDTA was added to all samples followed by 2 mL of an aqueous mixture containing TBA (3.75 mg/mL), $HClO_4$ (1.3%), and trichloroacetic acid (0.15 g/mL). Reaction mixtures were heated at 100 °C for 15 min, cooled, centrifuged at 1600 g for 15 min, and absorbance was measured at 532 nm. All samples and control were made in triplicate. IC_{50} values were determined from inhibition vs. concentration plots.

2.5. Anti-inflammatory activity

There are a great number of different *in vitro* methods used for estimation of inhibitory activity of COX and LOX enzymes including a number of commercial kits. Some assays, as a source of enzymatic activity, include native or recombinant enzymes, animals or human origin, while others use different cell lines that express desirable activities. Arachidonic acid is often added in reaction medium and in some cases is radio labelled. Induction of inflammatory respond also differs, and in most cases is performed by bacterial lipopolysaccharide, various cytokines and tumour necrosis factor. Different techniques are used for detection of enzymatic activity, such as different chromatographic techniques (TLC, HPLC-UV), as well as EIA [19, 34].

In this study, COX-1 and 12-LOX inhibitory activity was investigated using *ex vivo* assay according to modified method of Safayhi *et al.* [29, 35]. Intact cells (human platelets) were used as a source of COX-1 and 12-LOX enzymes. Arachidonic acid metabolites (12-HHT and 12-HETE) were determined by use of LC–MS/MS technique [29].

An aliquot of human platelet concentrate (viable but outdated for medical use) containing $4 \cdot 10^8$ cells was suspended in buffer (0.137 mol/L NaCl, 2.7 mmol/L KCl, 2.0 mmol/L KH_2PO_4, 5.0 mmol/L Na_2HPO_4 and 5.0 mmol/L glucose, pH 7.2) to obtain final volume of 2 mL. This mixture was slowly stirred at 37 °C for 5 min. Subsequently, 0.1 mL of extracts or standard compounds solutions in DMSO (concentration ranging from 10.0 to 200.0, 0.156 to 5.0 and 0.01 to 0.6 mg/ml for extracts, quercetin and aspirin, respectively) and 0.1 mL of calcimycin (Calcium Ionophore A23187, 125 µmol/L in DMSO) were added and incubated for 2 min at 37 °C, with moderate shaking. The exact volume of extract in control and calcimycin in blank probe were substituted with solvent (DMSO). Thereafter, 0.3 mL of $CaCl_2$ aqueous solution (16.7 mmol/L), substituted with water in blank probe, was added and the mixture was incubated for further 5 min at 37 °C with shaking. Acidification with cold 1% aqueous formic acid (5.8 mL) to pH 3 terminated the reaction. If gel formation occurred, vortexing was applied before mixing with the acid. Prostaglandin B_2 (50 µL of 6 µg/mL solution in DMSO) was added as internal standard, and extraction of products was done with mixture of chloroform and methanol (1:1, 8.0 mL) with vigorous vortexing for 15 min. After centrifugation at 7012 × g for 15 min at 4 °C, organic layer was separated, evaporated to dryness, dissolved in methanol (0.5 mL), filtered and used for further LC–MS/MS analysis. All samples and control were made in triplicate.

Test for estimation of the anti-inflammatory activity, applied in our research, has a lot of advantages. Firstly, the advantage is avoidance of the undesirable *in vivo* tests on experimental animals. Even though the exact anti-inflammatory activity can be validated only through *in vivo* tests, creating *in vitro* assays, in which physiological conditions similar to *in vivo* assays are used, can provide valuable information about inhibitory potential of the compounds tested. Platelets are a suitable cell system for testing inhibitory activity, because they can provide physiological cell conditions and possibility to examine the inhibition of both enzymes at the same time. Secondly, for determination of the formed metabolites as a measure of level of inhibition of COX and LOX activity, LC-MS/MS technique was used. LC-MS/MS provided a highly sensitive and specific detection of desirable metabolites within a short analysis time [29].

2.6. Statistical analysis

Percent of lipid peroxidation inhibition achieved by different concentration of extracts was calculated by the following equation: $I(\%) = (A_0-A)/A_0 \times 100$, where A_0 was the absorbance of the control reaction and A was the absorbance of the examined samples, corrected for the value of blank probe. Percent of COX-1 and 12-LOX inhibition achieved by different concentrations of extracts was calculated by the following equation: $I(\%) = 100 \times (R_0-R)/R_0$, where R_0 and R were response ratios (metabolite peak area divided by internal standard peak area) in the control reaction and in the examined samples, respectively. Both R and R_0 were corrected for the value of blank probe. For both assays, corresponding inhibition-concentration curves were drawn using Origin software, version 8.0 (Origin Labs) and IC_{50} values (concentration of extract that inhibited lipid peroxidation and COX-1 and 12-LOX metabolites formation by 50%) were determined. All of the results were expressed as mean ± SD of three replicates. Correlation analyses were done using Statistica software version 6 (StatSoft). Concentrations of total phenolics and the total flavonoids were used as independent variables, while inhibitory activities towards LP and 12-HETE synthesis (expressed as $1/IC_{50}$) were used as dependent variables. Due to a wide range of values, log-log plots were applied. Pearson's correlation coefficients were calculated.

3. Results

In this study, we tested the effect of 17 taxa from four families (*Alliaceae, Cupressaceae, Plantaginaceae* and *Polygonaceae*) on lipid peroxidation and metabolism of arachidonic acid. The total phenolic and flavonoid contents were determined in the plant extracts, as well. All results are presented in Figures 4, 5, 6 and 9.

Among all tested samples, herb and rhizome extracts of *Rumex* species were the most potent inhibitors of LP (0.009-0.047 mg/mL). Slightly lower activity was shown by herb extracts of *Plantago* species (0.025-0.178 mg/mL), while extracts of cones and needles of *Juniperus* species, as well as the whole plant extracts of *Allium* species expressed a much lower activity (0.117-0.887 and 0.68-1.986 mg/mL, respectively).

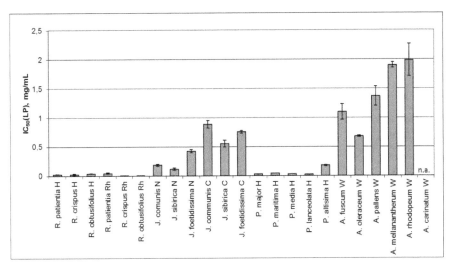

Legend: H – herb, Rh – rhizome, C – cones, N – needles, W – whole plant, n.a. - not achieved.

Figure 4. Results of lipid peroxidation inhibition assay.

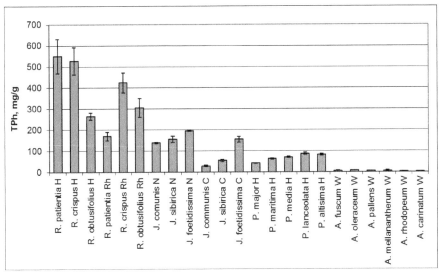

Legend: H – herb, Rh – rhizome, C – cones, N – needles, W – whole plant.

Figure 5. Results of total phenolic content assay (expressed as mg eq. gallic acid / 1 g dw).

Total phenolic content in examined genera decreased in following order: *Rumex*, *Juniperus*, *Plantago* and *Allium*, with *R. patentia* and *R. crispus* extracts being by far the richest in phenolics (550 and 527 mg eq. gallic acid per 1 g dw, respectively). Regarding the content of the total flavonoids, significant intrageneric variations were observed, hence it was not possible to point out a genus with the highest content. Flavonoids were the most abundant in *J. foetidissima* needles extract (60 mg quercetin eq per 1 g dw), and scarcest in *A. rhodopeum* (0.2 mg quercetin eq per 1 g dw).

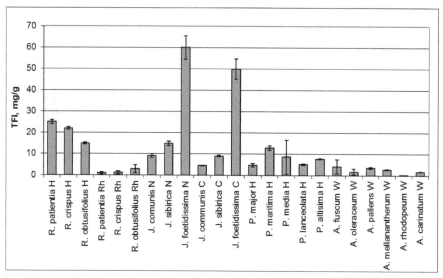

Legend: H – herb, Rh – rhizome, C – cones, N – needles, W – whole plant.

Figure 6. Results of total flavonoid content assay (expressed as mg eq. quercetin / 1 g dw).

Plant phenolics present in extracts can counteract lipid peroxidation in either initiation or propagation step. Possible mechanisms are chelation of transition metals (Fe^{2+}), reduction of Fe^{3+} and neutralisation of lipid peroxidation radical intermediates by transfer of hydrogen or single electron [8]. An attempt was made to correlate the observed lipid peroxidation inhibitory activity (given as $1/IC_{50}$) with the content of the total phenolics and total flavonoids. The corresponding plots are given in Figure 7. and Figure 8.

A high degree of correlation was observed between the lipid peroxidation inhibitory activity and the total phenolic content ($r = 0.7713$), while correlation with the total flavonoids content was not established ($r = 0.1410$). Thus, flavonoids do not contribute to the total observed activity to a significant extent. The lower activity of flavonoids, compared to other phenolics, can be explained through Porter's polar paradox. Namely, flavonoids present in the extracts investigated were predominantly in glycosylated form and thus highly polar. It is demonstrated that compounds of lower polarity are more effective in polar reaction media, such as oil-in-water emulsion used in our experiments, since they exhibit their activity at oil-water interface [36]. In addition, hydrogen atom donation ability of flavonoids decrease with glycosylation, especially if the most active hydroxyl groups (C-3, C-3' or C-4') are occupied by carbohydrate moiety [37, 38].

Regarding the eicosanoid pathway, *Rumex* and *Plantago* species showed, on average, very high 12-LOX inhibitory effect, with IC_{50} in range 0.75–3.59 mg/mL (Figure 9.). While

Juniperus and *Allium* also exhibited dose-dependent inhibition of 12-HETE production, their IC_{50} values were higher, ranging from 1.45 mg/mL to 11.04 mg/mL.

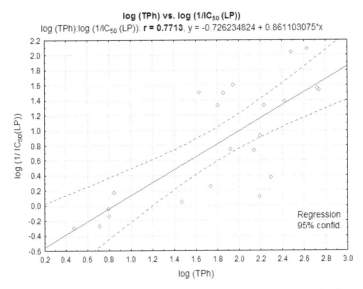

Figure 7. Correlation between total phenolic content and ability of extracts to inhibit lipid peroxidation

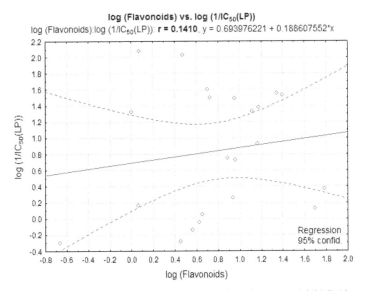

Figure 8. Correlation between total flavonoids content and ability of extracts to inhibit lipid peroxidation

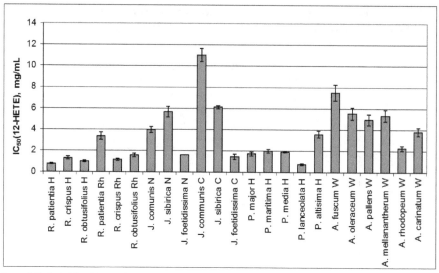

Legend: H – herb, Rh – rhizome, C – cones, N – needles, W – whole plant.

Figure 9. Results of 12-LOX inhibition assay.

Moreover, certain inhibitory activity of the examined extracts towards COX-1 was also confirmed in the anti-inflammatory assay applied in this research. Determined IC_{50} values ranged from 0.34 to 8.00 mg/mL, with no genus exhibiting significantly higher activity than others. However, the results are not shown due to lack of correlation with the total phenolics and flavonoid content. This is in agreement with our previous findings [27].

The exact mechanism of COX and LOX enzymes inhibition by natural products is still not fully elucidated. However, bearing in mind their structures and chemical properties, several mechanisms could be speculated. Both pathways of arachidonic acid metabolism include free radical reactions (Figure 3.). Due to their radical scavenging activity or reducing properties, many plant natural products can interfere with reactions catalysed by COX and LOX [39]. They can neutralise radical intermediates, thus terminating the reaction. In addition, they can reduce Fe^{3+} ion that is a part of active site of both enzymes and is necessary for initiation reaction. Some natural products, including acetylenes, can bind covalently to enzymes and inhibit them irreversibly [19]. Also, in inhibition of LOX, isoprenyl moiety of some phenolics and terpenoid backbone structure could play an important role. Prenylated phenolics are usually more hydrophobic than conventional ones. Terpenoids are also mostly non-polar compounds. These characteristics suggest an easy penetration through the cell membrane and their good 12-lipoxygenase inhibitory properties [19]. Finally, the COX or LOX activity can be decreased by suppressing their transcription by phenolics [18], although this effect is observable only in long-duration experiments.

To identify the compound class responsible for the observed 12-LOX-inhibitory activity, $1/IC_{50}$ was correlated with the content of the total phenolics and flavonoids. The corresponding plots are given in Figure 10. and Figure 11. As with the lipid peroxidation, the correlation between the content of the total phenolics and 12-LOX-inhibitory activity was also observed, although slightly weaker, with Pearson's correlation coefficient $r = 0.6037$. At the same time, no relationship was found between the total flavonoids content and 12-HETE production ($r = 0.2825$). Bearing in mind a good correlation of 12-LOX inhibition with the phenolic content, and the lack thereof with the flavonoid content, it is possible that only small phenolic molecules, but not the voluminous flavonoid glycosides, can enter 12-LOX active site and exhibit the inhibitory effect there. Thus, the observed differences in the total phenolics and the total flavonoids effects can at least partially be explained by steric hindrance.

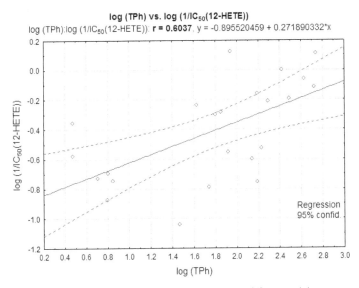

Figure 10. Correlation between total phenolic content and 12-LOX inhibitory activity

The differences in COX and LOX inhibition (LOX inhibition being correlated with phenolic content) can be attributed to the differences in reaction mechanism and active site three-dimensional structure. Namely, both mechanisms include abstraction of hydrogen from arachidonic acid leading to formation of radical species. However, hydrogen acceptor in COX is tyrosyl radical (formed through tyrosine oxidation by Fe^{3+}) while in LOX, electron is transferred directly to Fe^{3+}. Phenolics from plant extracts could reduce Fe^{3+} to Fe^{2+}, thus inactivating both enzymes. However, the presence of tyrosyl residue in COX active site could provide steric protection and prevent phenolics from approaching the Fe^{3+} ion.

Finally, the correlation between the ability of extracts to counteract lipid peroxidation and inhibit production of 12-HETE is shown in Figure 12. High correlation coefficient ($r = 0.6819$) suggests that extracts with a high inhibitory effect on LP, also represent potent inhibitors of 12-LOX pathway. Methods for examination of anti-inflammatory activity are labour-intensive, expensive and sometimes involve ethical issues due to the usage of laboratory animals. Therefore, *in vitro* measurement of lipid peroxidation inhibition and the total phenolic content could be a useful tool for screening of plant extracts with a potential anti-inflammatory activity.

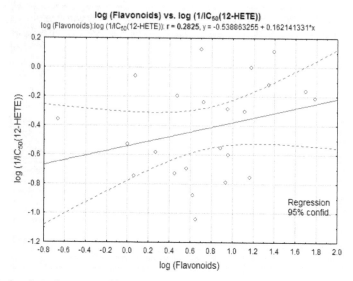

Figure 11. Correlation between total flavonoid content and 12-LOX inhibitory activity

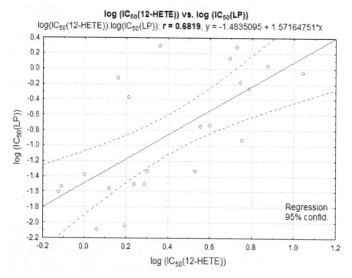

Figure 12. Correlation between the ability of extracts to inhibit 12-HETE production and lipid peroxidation

To summarise, the most of the examined species expressed high lipid peroxidation and anti-inflammatory activity, especially *Rumex* and *Plantago* species. High inhibitory activity of these species towards 12-LOX pathway makes them good candidates for further research taking into account role of this enzyme in cancer development. A good correlation was found between total phenolic content of 23 investigated plant extracts and their ability to inhibit lipid peroxidation and 12-LOX pathway. Consequently, LP inhibitory activity was also highly correlated with 12-LOX inhibition. Thus it can serve as indicator for preliminary selection of plant extracts further to be tested for anti-inflammatory activity by expensive and time-consuming methods.

Author details

Neda Mimica-Dukić, Nataša Simin, Emilija Svirčev, Dejan Orčić, Ivana Beara and Marija Lesjak
Department of Chemistry, Biochemistry and Environmental Protection, University of Novi Sad, Faculty of Sciences, Novi Sad, Republic of Serbia

Biljana Božin
Department of Pharmacy, Faculty of Medicine, University of Novi Sad, Novi Sad, Republic of Serbia

Acknowledgement

The Ministry of Education and Science, Republic of Serbia (Grant No. 172058) funded this research. We wish to thank to Goran Anačkov, PhD, for the providing voucher specimens, and Gordana Vlahović, MSc, for language corrections. Platelet concentrate was kindly provided by The Institute for Blood Transfusion of Vojvodina, Novi Sad, Republic of Serbia.

4. References

[1] Kawanishi S., Hiraku Y., Murata M., Oikawa S. The role of metals in site-specific DNA damage with reference to carcinogenesis. Free Radical Biology & Medicine 2002;32(9) 822-832.

[2] Manach C., Mazur A., Scalbert A. Polyphenols and prevention of cardiovascular diseases. Current Opinion in Lipidology 2005;16(1) 77-84.

[3] Gonsette RE. Review: Oxidative stress and excitotoxicity: a therapeutic issue in multiple sclerosis? Multiple Sclerosis 2008;14(1) 22-34.

[4] Maritim AC., Sanders RA., Watkins JB. Diabetes, Oxidative Stress, and Antioxidants: A Review. Journal of Biochemical and Molecular Toxicology 2003;17(1) 24-38.

[5] Ramassamy C. Emerging role of polyphenolic compounds in the treatment of neurodegenerative diseases: a review of their intracellular targets. European Journal of Pharmacology 2006;545(1) 51-64.

[6] Wruck CJ., Fragoulis A., Gurzynski A., Brandenburg L., Kan YW., Chan K., Hassenpflug J., Freitag-Wolf S., Varoga D., Lippross S., Pufe T. Role of oxidative stress

in rheumatoid arthritis: insights from the Nrf2-knockout mice. Annals of the Rheumatic Diseases 2010;doi:10.1136/ard.2010.132720

[7] Alvarado C., Alvarez P., Jimenez L., De la Fuente M. Oxidative stress in leukocytes from young prematurely aging mice is reversed by supplementation with biscuits rich in antioxidants. Developmental & Comparative Immunology 2006;30(12) 1168-1180.

[8] Laguerre M., Lecomte J., Villeneuve P. Evaluation of the ability of antioxidants to counteract lipid oxidation: Existing methods, new trends and challenges. Progress in Lipid Research 2007;46(5) 244-282.

[9] Gordon MH. The mechanism of antioxidant action in vitro. In: Hudson BJF. (ed.) Food antioxidants. Amsterdam: Elsevier; 1990. p1-18.

[10] Smith WL. The eicosanoids and their biochemical mechanisms of action. Biochemical Journal 1989;259(2) 315-324.

[11] Marnett LJ., Rowlinson SW., Goodwin DC., Kalgutkar AS., Lanzo CA. Arachidonic acid oxygenation by COX-1 and COX-2. Mechanisms of catalysis and inhibition. The Journal of Biological Chemistry 1999;274(33) 22903-22906.

[12] Hawkey CJ. COX-2 inhibitors. The Lancet 1999;353(9149) 307-314.

[13] Greene ER., Huang S., Serhan CN., Panigrahy D. Regulation of inflammation in cancer by eicosanoids. Prostaglandins and Other Lipid Mediators 2011;96(1-4): 27-36.

[14] Yamamoto S., Suzuki H., Ueda N., Takahashi Y., Zoshimoto T. Mammalian Lipoxygenases. In: Curtis-Prior PB. (ed.) The Eicosanoids. Cambridge: Wiley; 2004. p53-61.

[15] Müller-Decker K. Cyclooxygenases. In: Marks F., Fürstenberger G. (ed.) Prostaglandins, leukotrienes, and other eicosanoids: from biogenesis to clinical applications. Weinheim: Wiley-VCH; 1999. p65-83.

[16] Liagre B., Vergne P., Rigaud M., Beneytout JL. Expression of arachidonate platelet-type 12-lipoxygenase in human rheumatoid arthritis type B synoviocytes. FEBS Letters 1997;414(1) 159-164.

[17] Müller K. 5-lipoxygenase and 12-lipoxygenase: attractive targets for the development of novel antipsoriatic drugs. Archiv der Pharmazie. 1994;327(1) 3-19.

[18] Lee JH., Kim GH. Evaluation of Antioxidant and Inhibitory Activities for Different Subclasses Flavonoids on Enzymes for Rheumatoid Arthritis. Journal of Food Science. 2010;75(7) H212-H217.

[19] Schneider I., Bucar F. Lipoxygenase Inhibitors from Natural Plant Sources. Part 2: Medicinal Plants with Inhibitory Activity on Arachidonate 12-lipoxygenase, 15-lipoxygenase and Leukotriene Receptor Antagonists. Phytotherapy Research. 2005;19(4) 263-272.

[20] Beara IN., Lesjak MM., Orčić DZ., Simin NĐ., Četojević-Simin DD., Božin BN., Mimica-Dukić NM. Comparative analysis of phenolic profile, antioxidant, anti-inflammatory and cytotoxic activity of two closely-related Plantain species: *Plantago altissima* L. and *Plantago lanceolata* L. LWT - Food Science and Technology. 2012;47(1) 64-70.

[21] Lesjak MM., Beara IN., Orčić DZ., Anačkov GT., Balog KJ., Francišković MM., Mimica-Dukić NM. *Juniperus sibirica* Burgsdorf. as a novel source of antioxidant and anti-inflammatory agents. Food Chemistry. 2011;124(3) 850-856.

[22] Orčić DZ., Mimica-Dukić NM., Francišković MM., Petrović SS., Jovin ED. Antioxidant activity relationship of phenolic compounds in *Hypericum perforatum* L. Chemistry Central Journal. 2011;5(1) 34-41.

[23] Beara I. Phytochemical screening and evaluation of antioxidant and anti-inflammatory potential of secondary metaboliotes of *Plantago* L. species. PhD thesis. University of Novi Sad; 2010.

[24] Orčić D. Scandiceae tribe (Apiaceae Lindley 1836, subfam. Apioideae) species – potential resources of biologically and pharmacologically active secondary biomolecules. PhD thesis. University of Novi Sad; 2010.

[25] Mimica-Dukić NM., Simin NĐ., Cvejić JM., Jovin ED., Orčić DZ., Božin BN. Phenolic compounds in field horsetail (*Equisetum arvense* L.) as natural antioxidants. Molecules. 2008;13(7) 1455-1464.

[26] Božin BN., Mimica-Dukić NM., Simin NĐ., Anačkov GT. Characterization of the volatile composition of essential oils of some lamiaceae spices and the antimicrobial and antioxidant activities of the entire oils. Journal of Agricultural and Food Chemistry. 2006;54(5) 1822-1828.

[27] Lesjak M. Biopotential and chemical characterization of extracts and essential oils of species from *Juniperus* L. genus (Cupressaceae). PhD thesis. University of Novi Sad; 2011.

[28] Beara I., Lesjak M., Jovin E., Balog K., Anačkov G., Orčić D., Mimica-Dukić N. Plantain (*Plantago* L.) Species as novel sources of flavonoid antioxidants. Journal of Agricultural and Food Chemistry. 2009;57(19) 9268-9273.

[29] Beara I., Orčić D., Lesjak M., Mimica-Dukić N., Peković B., Popović M. Liquid chromatography/tandem mass spectrometry study of anti-inflammatory activity of Plantain (*Plantago* L.) species. Journal of Pharmaceutical and Biomedical Analysis. 2010;52(5):701-706.

[30] Singleton VL., Orthofer R., Ramuela-Raventos RM. Analysis of total phenols and other oxidation substrates and antioxidants by means of Folin-Ciocalteu reagent. Methods in Enzymology. 1999;299 152-178.

[31] Chang CC., Yang MH., Wen HM. Chern JC. Estimation of total flavonoid content in propolis by two complementary colorimetric methods. Journal of Food and Drug Analysis. 2002;10(3) 178-182.

[32] Mavi A., Lawrence GD., Kordalia A., Yildirim A. Inhibition of iron-fructose-phosphate-induced lipid peroxidation in lecithin liposome and linoleic acid emulsion systems by some edible plants. Journal of Food Biochemistry. 2011;35(3) 833-844.

[33] Aust SD. Lipid peroxidation. In: Greenwald RA. (ed.) Handbook of methods for oxygen radical research. Boca Raton: CRC Press, 1985. p203-207.

[34] Pairet M., Ryn van J. Experimental models used to investigate the differential inhibition of cyclooxygenase-1 and cyclooxygenase-2 by non-steroidal anti-inflammatory drugs. Inflammation Research. 1998;47(2) S93-101.

[35] Safayhi H., Mack T., Sabieraj J., Anazodo MI., Subramanian LR., Ammon HPT. Boswellic acids: novel, specific, nonredox inhibitors of 5-lipoxygenase. Journal of Pharmacology and Experimental Therapeutics. 1992;261(3) 1143-1146.

[36] Porter WL. Paradoxical behavior of antioxidants in food and biological systems. Toxicology and Industrial Health. 1993;9(1-2) 93-122.

[37] Rice-Evans CA., Miller NJ., Paganga G. Structure-antioxidant activity relationships of flavonoids and phenolic acids. Free Radical Biology and Medicine. 1996;20(7) 933-56.

[38] Rice-Evans CA., Miller NJ., Paganga G. Antioxidant properties of phenolic compounds. Trends in Plant Science. 1997;2(4) 152-159.

[39] Hostettmann K. Assays for immunomodulation and effects on mediators of inflammation. In: Dey PM., Harborne JB. (ed.) Methods in Plant Biochemistry vol. 6, London: Academic Press Limited; 1991. p207-211.

Chapter 11

Repeatedly Heated Vegetable Oils and Lipid Peroxidation

Kamsiah Jaarin and Yusof Kamisah

Additional information is available at the end of the chapter

1. Introduction

Deep frying is the most common and one of the oldest methods of food preparation worldwide. It involves heat and mass transfer. To reduce the expenses, the oils tend to be used repeatedly for frying. When heated repeatedly, changes in physical appearance of the oil will occur such as increased viscosity and darkening in colour [1], which may alter the fatty acid composition of the oil. Heating causes the oil to undergo a series of chemical reactions like oxidation, hydrolysis and polymerization [2]. During this process, many oxidative products such as hydroperoxide and aldehydes are produced, which can be absorbed into the fried food [3].

Palm and soy oil are the most commonly used vegetable oils in the household and industry in Malaysia for deep frying purposes. Both palm and soy oils are rich in tocopherols [4-5]. In addition to the tocopherols, palm oil also contains an abundant amount of tocotrienols. The latter form of vitamin E was consistently shown to possess better antioxidant activity than the former form [6]. The soy oil has bigger proportion of polyunsaturated fatty acid compared to the palm oil. Whereas in the palm oil, the major fatty acids present are the monounsaturated and saturated fatty acids [7].

A survey conducted in Kuala Lumpur recently had revealed that majority of the respondents admitted using repeatedly heated cooking oil [8]. The public level of awareness regarding such usage is influenced by the socioeconomic status. Respondents with higher income and education level had higher level of awareness [9]. Chronic consumption of repeatedly heated vegetable oils could be detrimental to health. It was shown to demonstrate genotoxic and preneoplastic change in the rat liver [10]. It also impaired fluid and glucose intestinal absorption in rats [11]. In rats given alcohol plus heated sunflower, an apparent liver damage as well as increased cholesterol level was observed [12]. Soriguer et al. [13] found an independent positive association between the risk of hypertension and intake of heated cooking oil. These accumulating data suggest chronic intake of heated cooking oils increases the risk of cancer and cardiovascular diseases.

The incidence of cardiovascular disease is higher in men compared to women of similar age [14]. However, after menopause, the risk increases in women due to deficiency of estrogen level. The estrogen was shown to be cardioprotective in ovariectomized female rats [15]. It also possesses antioxidant activity which can lower the risk of lipid peroxidation [16]. Ovariectomized female rats have been used widely as a postmenopausal model [17-18].

In this study, we used two models, normal male rats and ovariectomized female rats to compare the effects of prolonged consumption of repeatedly heated palm oil and soy oil on in vivo plasma lipid peroxidation content. The oxidative stability of these repeatedly heated palm oil and soy oil were also compared.

2. Materials and methods

2.1. Animals, materials and chemicals

Male and female Sprague Dawley rats (180-200 gram) were obtained from the Laboratory Animal Resource Unit of Universiti Kebangsaan Malaysia. They were kept in polyethylene cages in a well ventilated room at room temperature. Food and water were provided *ad libitum*. Palm oil (Lam Soon Edible Oil, Malaysia) and soy oil (Yee Lee Edible Oil, Malaysia) were used in this study. Sweet potatoes were purchased from the same source at a local market. All chemicals and enzymes were obtained from Sigma-Aldrich (St. Louis, MO, USA), unless otherwise stated.

Ethical approval regarding the experimental procedure and humane animal handling in the study was obtained from the Universiti Kebangsaan Malaysia Animal Ethical Commitee and Medical Research Ethics Committee.

2.2. Frying procedure

The oils were heated according to the method of Owu et al. [19]. A kilogram of sweet potato slices were fried in a stainless steel wok containing two and half litres of palm oil or soy oil for 10 minutes at 180°C. Upon completion of the frying process, once heated oil was obtained. The process was repeated four times to obtain five times heated oil with a cooling interval of at least five hours. The food quantity was proportionately adjusted with the amount of vegetable oil left. No fresh oil was added between the frying processes to make up for the loss due to uptake by the frying materials. After being heated, small quantity of the oils was extracted for the peroxide value, fatty acid composition and vitamin E content measurements.

2.3. Diet preparation

Diets enriched with heated palm or soy oils were prepared by mixing 850 g grinded standard mouse pellet with 150 g heated oil. While to prepare high cholesterol (2%) diets fortified with heated palm or soy oils, 150 g palm or soy oil were mixed with 20 g cholesterol

(MP Biomedical Inc., Australia) with 830 g grinded standard mouse pellet. The pellets were then remolded and dried in an oven at 80°C overnight.

2.4. Effects of heated oils in male rats

After one week of acclimatization period, forty-two male Sprague Dawley rats were randomly divided into seven groups. The first group was given standard mouse pellet (control). The second, third and fourth groups were given fresh, once and five times heated palm oil diets, while the fifth, sixth and the last groups were given fresh, once and five times heated soy oil diets, respectively for four months. Blood was sampled before and at the end of treatment duration.

2.5. Effects of heated oils in ovariectomized female rats fed high cholesterol diet

Forty-two female Sprague Dawley rats were allowed to acclimatize for a week before the treatment was started and were ovariectomized ahead of the study. They were randomly divided into seven groups. Group 1 was fed 2% cholesterol diet (control), while groups 2, 3 and 4 were respectively fed 2% cholesterol diet added with fresh, once and five times heated palm oil. Groups 5, 6 and 7 were respectively given 2% cholesterol diet added with fresh, once and five times heated soy oil. The treatment duration was four months, after which the rats were sacrificed and blood samples were taken. Blood sample was also taken prior to the treatment.

2.6. Peroxide content measurement

Measurement of peroxide values of the heated oils was done according to the American Oil Chemists' Society (AOCS) Official Methods Cd 8-53 [20]. Briefly, five grams of the oil sample were added with 30 ml of acetic acid-chloroform (3:2) in a flask. The flask was then swirled before the addition of 0.5 ml saturated potassium iodide. The solution was swirled again for a minute. An amount of 30 ml distilled water and a few drops of starch solution (10%) were added. The solution was titrated against 0.01 N sodium thiosulphate solution which was priorly standardised using potassium dichromate and potassium iodide, until blue colour disappeared. The peroxide value in the oils was calculated as the difference in the volume of sodium thiosulphate solution (ml) used for samples and blank, divided by its normality. The values were expressed in miliequivalents of peroxide per kilogram of the sample.

2.7. Fatty acid composition measurement

Fatty acid composition in the fresh and heated oils was analysed using gas chromatography (GC-17A, Shimadzu, Japan), which consisted of a flame ionisation detector, a BPX 70 capillary column (30 m x 0.25 mm x 0.25 μm), programmable injector temperature, set at 250°C and detector temperature, set at 280°C. The oil samples (100 μl) were first transesterified to fatty acid methyl ester using 1 ml of 1 M sodium methoxide in 1 ml hexane before injected into the gas chromatographic system. The injection volume was 1 μl. Nitrogen at a flow rate of 0.40 ml/min was used as carrier gas in the analysis. Identification of fatty acid methyl ester peaks was carried out by comparing their retention times with their authentic standards. The fatty acid composition was expressed as the percentage of the total fatty acids.

2.8. Vitamin E content measurement

The vitamin E content in the oil samples (20 μl sample) was analysed using an analytical high performance liquid chromatography (HPLC) using a programmable fluorescence detector at excitation 295 nm and emission 330 nm (Hewlett Packard HP1100, USA). The chromatographic system consisted of an isocratic pump and the stationary phase was a 150 mm silica normal phase column (YMC 5U) with an internal diameter 6 mm. The mobile phase was 0.5% isopropanol in hexane at a flow rate of 1 ml/min. The oil samples were injected directly into the HPLC system without any processing, after being cooled from the heating process. The vitamin E standard was obtained from the Malaysian Palm Oil Board (Bangi, Malaysia). Vitamin E measurement in the oils was done for six samples (n=6) for each of the three corresponding groups; fresh, once and five times heated. The vitamin E content in the oils was expressed as part per million (ppm). The estimated percentage of difference compared to the respective fresh oils was also calculated.

2.9. Plasma lipid peroxidation measurement

Lipid peroxidation content in the plasma measured as thiobarbituric acid reactive substance (TBARS) was determined following a method described by Ledwozyw et al. [21] with some modification. Briefly, 2.5 ml trichloroacetic acid (1.22 M in 0.6 M HCl) was used to acidify 0.5 ml plasma and incubated at room temperature for 15 minutes. Next, 1.5 ml of 0.67% thiobarbituric acid (in 0.05 M NaOH) was added. The samples were then incubated at 100°C for 30 minutes. After being cooled, the lipid peroxide content was extracted by the addition of 4 ml butanol using vigorous shaking. Later, the samples fluorescence unit was read at 515 nm (excitation wavelength) and 553 nm (emission wavelength) using a spectrofluorometer (Shimadzu RF500, Japan). 1,1,3,3-Tetraethoxypropane was used as the standard. The unit of the plasma lipid peroxidation was nmol malondialdehyde/mg protein. The results were shown as the difference percentage of post-treatment content compared to the pretreatment content.

The protein content in the plasma was carried out according to the method of Lowry et al. [22], using bovine serum albumin as the standard. A plasma sample (0.5 ml) was added with 5 ml mixture of sodium carbonate (2%), sodium potassium tartrate (2%) and copper sulphate (1%) solution at a ratio of 100 : 1 : 1, prior to incubation at room temperature for 15 minutes. Subsequently, 0.5 ml Folin-Ciocalteau phenol reagent (0.5 N) was added and then was left to stand at room temperature for 35 minutes. The absorbance of the samples was read at 700 nm with a spectrophotometer (Shimadzu UV-160A, Japan).

2.10. Statistical analysis

The results were expressed as the means ± standard error of mean (SEM). Normality of the data was analysed using Kolmogorov-Smirnov test. For normally distributed data, they were then analysed using one-way analysis of variance (ANOVA) followed by Tukey's HSD post-hoc test. While for not normally distributed data, the differences among the groups were determined using Kruskal-Wallis H and Mann-Whitney U test. Values of P<0.05 were considered statistically significant. All statistical analyses were performed using SPSS version 13.0 (SPSS Inc., Chicago, IL, USA).

3. Results

3.1. Peroxide value in the oils

The peroxide values measured in the oils are shown in Figure 1. The values in the once and five times heated palm and soy oils were significantly elevated compared to the fresh oil respectively. In five times heated oils, the values were also significantly higher than the once heated oils. Fresh and five times heated soy oils had bigger peroxide values compared to those of palm oils. According to the American Oil Chemists' Society (AOCS) [23], only five times heated soy oil had peroxide value exceeded the maximum allowable peroxide value for edible oils (> 10 meq/kg).

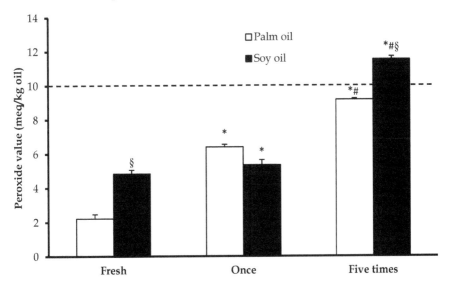

Figure 1. The peroxide values (meq/kg) in fresh and heated palm and soy oils. Bars represent mean ± SEM (n=6). *Significantly different from the fresh oils respectively (P<0.05). #Significantly different from the once heated oils respectively (P<0.05). §Significantly different from the palm oil groups respectively (P<0.05). Dashed horizontal line indicates maximum allowable peroxide value for edible oils according to the American Oil Chemists' Society (AOCS) [23].

3.2. Fatty acid composition of the oils

The fatty acid composition in both palm and soy oils is tabulated in Table 1. All the main components of fatty acid were present in the oils regardless of the frequency of heating. Once and five times heated palm oils had similar percentages of saturated, monounsaturated and polyunsaturated fatty acids composition compared to the fresh palm oil. In heated soy oil, generally the fatty acid composition was somewhat similar to the fresh oil. However, its polyunsaturated fatty acid percentage seemed to be lower than once heated and fresh oils. Overall, heating did not affect saturated and monounsaturated and polyunsaturated fatty acids components of palm oil. However, the repeated heating reduced the percentage of polyunsaturated and monounsaturated fatty acids, and increased saturated fatty acids components of the soy oil.

	Saturated	Monounsaturated	Polyunsaturated
Palm oil			
Fresh	42.87	48.94	8.18
Once heated	42.64	49.24	8.52
Five times heated	43.25	48.21	7.97
Soy oil			
Fresh	16.69	25.00	52.48
Once heated	17.14	26.10	51.78
Five times heated	18.10	24.21	41.72

Table 1. The percentage of saturated, monounsaturated and polyunsaturated fatty acids in the fresh, once and five times heated palm and soy oils.

3.3. Vitamin E content in the oils

The vitamin E isoforms, namely α-, γ- and δ-tocopherols, α-, γ- and δ-tocotrienols content in the oils is tabulated in Table 2. In palm oil, only α-tocopherol, α-, γ- and δ-tocotrienols were present. Whilst in soy oil, only tocopherol isoforms (α, γ and δ) were detected but none of tocotrienols. Fresh palm oil had larger content of total vitamin E compared to fresh soy oil. Once and five times heated palm oils had significantly lower content of all vitamin E isoforms than the fresh palm oil. In five times heated palm oil, all the isoforms content were also significantly reduced compared once heated palm oil. For soy oil, once and five times heating decreased the α-tocopherol content significantly compared to the fresh oil. The contents of γ- and δ-tocopherols were only reduced significantly in five times heated soy oil compared to once heated and fresh soy oils.

The difference in vitamin E content relative to fresh palm oil is diagrammatically shown in Figure 2. The relative reductions in α-tocopherol and all tocotrienol isoforms as well as total contents were greater in five times heated palm oil than once heated palm oil. The relative reduction was the least seen in the δ-tocotrienol content. In once heated soy oil, a big relative

	Tocopherols (T) (ppm)			Tocotrienols (T3) (ppm)			Total
	αT	γT	δT	αT3	γT3	δT3	(ppm)
Palm oil							
Fresh	178.3 ± 3.7	ND	ND	188.5 ± 16.3	260.8 ± 19.7	69.8 ± 2.2	697.4
Once heated	79.0 ± 8.1*	ND	ND	80.2 ± 14.6*	193.2 ± 21.4*	59.8 ± 10.4*	412.2
Five times heated	3.3 ± 3.3*#	ND	ND	3.8 ± 3.8*#	31.8 ± 20.8*#	35.8 ± 11.6*#	74.7
Soy oil							
Fresh	66.2 ± 1.4	247.2 ± 15.0	122.7 ± 0.7	ND	ND	ND	436.1
Once heated	25.3 ± 1.7*	254.7 ± 20.1	127.7 ± 0.7	ND	ND	ND	407.7
Five times heated	11.5 ± 2.0*	97.8 ± 18.0*#	80.5 ± 4.9*#	ND	ND	ND	189.8

Table 2. Composition of tocopherols and tocotrienols (ppm) in the fresh, once and five times heated palm and soy oils. Values are mean ± SEM (n=6). *Significantly different from the fresh oils respectively (P<0.05), #Significantly different from once heated oil respectively (P<0.05). ND, not detectable.

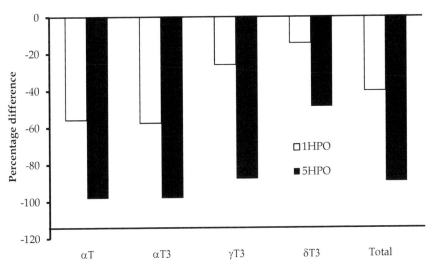

Figure 2. The percentage difference in α-tocopherol (αT), α-tocotrienol (αT3), γ-tocotrienol (γT3) and δ-tocotrienol (δT3) contents in heated palm oils (once and five times) in comparison to fresh palm oil.

decrease was seen in α-tocopherol content, whereas other isoforms content were not much affected. However, there was only a slight relative decrease in total vitamin E content. In five times heated soy oil, the relative reductions were seen in all tocopherol isoforms (α, γ and δ isoforms) (Figure 3). However, the least relative reduction was noted in the δ-tocopherol content. It thus appeared that δ-tocopherol and δ-tocotrienol were more resistant to heat compared to α and γ isoforms.

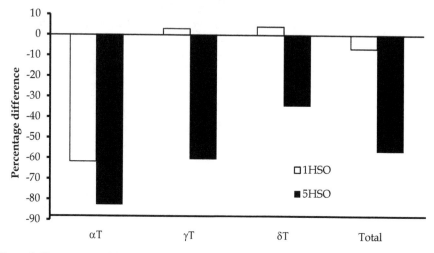

Figure 3. The percentage difference in α-tocopherol (αT), γ-tocopherol (γT) and δ-tocopherol (δT) contents in heated soy oils (once and five times) in comparison to the fresh soy oil.

3.4. Plasma lipid peroxidation in male rats

Relative plasma lipid peroxidation, measured as TBARS was increased significantly in male rats that were given diet containing 15% once and five times heated palm or soy oil for 4 months compared to the control and fresh oils, respectively (Figure 4). The five times heated groups also had significantly higher TBARS content than the once heated groups, respectively. In the once and five times heated groups, the TBARS content was significantly elevated in the soy oil-fed group compared to the palm oil-fed group, respectively. Both fresh palm and soy oil groups had significantly lower relative plasma TBARS content than the control.

3.5. Plasma lipid peroxidation in ovariectomized female rats

Ovariectomized female rats that were fed once and five times heated palm and soy oils in addition to 2% cholesterol for four months had significantly elevated relative plasma TBARS compared to both groups that were given either control diet or fresh oil respectively. The rats that ingested five times heated oils had significantly higher plasma TBARS than the

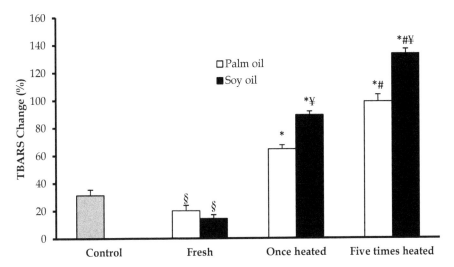

Figure 4. The change percentage of thiobarbituric acid reactive substance (TBARS), a lipid peroxidation product in male rats that were fed 15% once or five times heated (w/w) palm or soy oils for 4 months. Bars represent mean ± SEM (n=6). *Significantly different from the control and fresh oil respectively (P<0.05). #Significantly different from once heated soy oil (P<0.05). ¥Significantly different from the heated palm oil, respectively (P<0.05). §Significantly different from the control (P<0.05).

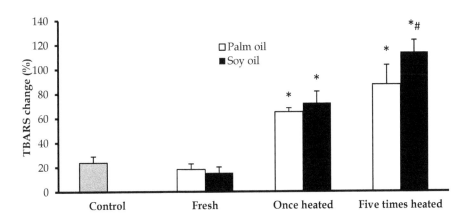

Figure 5. The percentage of change of lipid peroxidation product measured as thiobarbituric acid reactive substance (TBARS) in ovariectomised female rats that were fed 2% cholesterol together with 15% once or five times heated (w/w) palm or soy oils for 4 months. Bars represent mean ± SEM (n=6). *Significantly different from the control and fresh oil respectively (P<0.05). #Significantly different from once heated soy oil (P<0.05).

once heated groups respectively. The plasma TBARS in palm oil-fed groups were similar to the soy oil-fed groups except in five times heated groups which the palm oil group had a lower plasma TBARS (P<0.05). The plasma TBARS of the fresh oil groups was not different from that of the control.

4. Discussion

Repeatedly heated cooking oil is often used interchangeably with thermoxidized or recycled cooking oil. Repeated use of this oil has become a common practice due to low level of awareness among the public about the bad effect of this practice [9]. Nowadays, the consumption of deep-fried food has gained popularity which may cause increased risk of obesity [24].

During frying, food is immersed in hot oil at a high temperature of 150 °C to 190 °C. The heat and mass transfer of oil, food and air that occurs during deep frying produces the unique and desirable quality of fried foods [2]. It was shown in the present study that the peroxide values were increased with the increasing frequency of heating in both types of oil. Increased values indicate increased lipid peroxidation byproduct content, mainly the peroxides that were formed in the oil during heating process. The extent of oxidation in the oils was affected by the number of frying. Other than the peroxides, there are other oxidized components that are formed during oil heating such as oxidative dimers and oxidized triacylglycerols [25]. Only five times heated soy oil exceeded the upper limit of peroxide value set by the American Oil Chemists' Society (AOCS) that is 10 meq/kg oil [23]. However, if the Food Sanitation Law of Japan guideline (peroxide value ≤ 30 meq/kg oil) is used instead, all fresh and heated oils can be considered safe for human ingestion. However, a recent study done by Awney [26] has demonstrated that in male rats fed thermally oxidized soy oil with peroxide value of 14 meq/kg, had significantly increased lipid peroxidation content in various organs such as liver, kidney, testes and brain. Chronic intake of repeatedly heated cooking could be harmful to health and increase risk of many diseases including hypertension [13] and cancer [10].

During heating at high temperature, several complex chemical reactions take place in the oil. It is dependent on the temperature, duration of heating, type of frying materials, type of oils, presence of antioxidant and prooxidant as well as the amount of oxygen [2,27]. Repeatedly heated oils at high temperature will undergo chemical conversion of fatty acid configuration from cis to trans isomer [28]. Diets containing trans fatty acid could be detrimental to cardiovascular health because it was reported that this fatty acid isomer could induce inflammation of the blood vessels and decrease its nitric oxide production [29-30]. The polymer and polar compounds content are also increased more than 37% and 47% respectively when the oil is used to fry [31]. The repeatedly heating would reduce the quality of cooking oil by darkening its color and changing the smell as well as the taste [1].

We found that the fresh and repeatedly heated soy oil had greater peroxide value than palm oil. The peroxide value is often used as an indicator for oxidative stability or the extent of degradation for fats and oils [32]. Therefore, greater values found in soy oil suggest that the soy oil were more susceptible to oxidative modification than palm oil. Vegetable oils rich in polyunsaturated fatty acid are more prone to oxidation compared to the oils which are rich in monounsaturated fatty acids [33]. Fats are usually oxidized by prooxidants or free radicals at the unsaturated bonds of the fatty acids, which are abundantly present in the polyunsaturated fatty acids such as linoleic (C18:2) and linolenic (C18:3) acids. The bigger the numbers of unsaturated bonds, the more prone the fatty acids are to oxidation. Intense heating of oils decreases unsaturation of the fatty acids [34].

The fresh soy oil contained about five times more polyunsaturated fatty acid compared to the palm oil. It seemed that five times heating had reduced about 10% of the polyunsaturated fatty acid content in the soy oil. The content of monounsaturated fatty acid in the fresh palm oil was higher than that of the fresh soy oil. Palm oil had a quite balanced ratio of saturated and unsaturated fatty acids, whereas more than 70% of soy oil fatty acid was unsaturated (polyunsaturated and monounsaturated). This unique fatty acid composition of palm oil renders its stability against oxidative insult. The fatty acid composition of heated soy oil [26] and palm oil [35] was similarly obtained in previous studies.

The examples of the monounsaturated are oleic (C18:1) and palmitoleic (C16:1), whereas palmitic (C16:0) and stearic (C18:0) are saturated fatty acids [26]. Soy oil majorly contains linoleic (54%) and oleic (23%) acids. It also contains linolenic (8%), palmitic (11%) and stearic (4%) acids [26]. While palm oil has a bigger proportion of oleic (46%) and palmitic acids (36%), a smaller amount of linoleic (12%) and linolenic (< 1%) acids, and similar proportion of stearic acid (4.5%) compared to soy oil [36]. However in the present study, the individual fatty acid composition in the oil was not determined.

Most antioxidants are heat labile. Vitamin E is not an exception. When subjected to heat at high temperature, we found that all tocotrienol isoforms (α, γ and δ) as well as α-tocopherol in the palm oil were lost. It became more prominent when the oil was heated five times. The fresh palm oil contained almost 700 ppm of total vitamin E, but once and five times heating had further reduced it to barely 400 and 75 ppm, respectively. γ-Tocotrienol content was the highest in fresh and once heated palm oil, followed by α-tocotrienol, α-tocopherol and the least being the δ-tocotrienol. Once heating caused almost 50% reduction in α-tocopherol and α-tocotrienol contents in palm oil. While the loss of other isoforms (γ- and δ-tocotrienols) were at smaller percentages. In five times heated palm oil, the percentage loss of all vitamin E components were almost 100% except δ-tocotrienol (almost 50%), compared to the fresh palm oil. It seemed that amongst the tocotrienol isoforms, δ-tocotrienol had the highest resistance to heat.

Different from palm oil, the soy oil exclusively contained α-, γ- and δ-tocopherols. γ-Tocopherol was the largest constituent in the soy oil, followed by δ-tocopherol and lastly α-tocopherol. The total content of vitamin E in fresh soy oil was about 60% of that of fresh palm oil. Compared to tocopherols, tocotrienols are less widespread in plants [37]. Other than palm oil, tocotrienols are found abundantly in barley [38], rice bran [39], grape seed oil [40], rye, oat, maize and wheat germ [41]. Tocopherols on the other hand, are distributed at a wider range of food and vegetable oils such as soybean, corn, sunflower and cottonseed oils [40,42].

Once heating did not significantly affect γ- and δ-tocopherols contents in the soy oil, but the α-tocopherol content was apparently reduced. However in five times heated, all tocopherol isoforms content were significantly decreased. When the percentage difference relative to the fresh soy oil calculated, the most prominent reduction was seen in α-tocopherol content after once heating, followed by the γ-tocopherol and δ-tocopherol contents. Quite a similar finding was reported by Rennick & Warner [43] that the α-tocopherol content of soy oil was significantly decreased when was heated for 5 hours and was further decreased after heated for 10 hours. The loss of the tocopherol content was accompanied by an increasing appearance of α-tocopherolquinones.

From the findings obtained, it seemed that the loss of tocotrienol was more than the same isoform of tocopherol. The vitamin E chemical structure is consisted of a chromanol ring and a 15-carbon phytyl side chain. Tocotrienol differs from tocopherol by the presence of three unsaturated bonds in its phytyl chain [44]. The presence of unsaturations in the tocotrienol phytyl chain may explain the susceptibility of the compound to repeated heating. It was also noted that the α isoform of both tocopherol and tocotrienol was the mostly susceptible to oxidative loss in both palm and soy oils. The second sensitive isoform was the γ isoform (both γ-tocopherol and tocotrienol), and the least sensitive being the δ isoform. The α isoform has three methylated groups, the γ isoform has two while the δ isoform has only one methylated group on the chromanol ring [45]. However Miyagawa et al. [46] had reported that the decomposition rate of γ-tocopherol was the fastest, followed by δ- and α-tocopherol when a mixture of soy and rapeseed oil was used to deep fry potato. The discrepancy could be due to the different type of oils used in their study. It can be postulated that the degree and position of the methylation on the chromanol ring also determine the susceptibility of the vitamer to oxidative loss in addition to the degree of saturation of the phytyl chain.

The higher vitamin E content in the palm oil than soy oil might contribute to lesser peroxide value in the former oil. Both tocopherols and tocotrienols possess antioxidant property, with the latter having greater antioxidant property [6]. Vitamin E could effectively protect the fatty acids in the oil from oxidation. During oil heating, the vitamin E is consumed by scavenging the lipid free radicals which are derived from the oxidation of unsaturated fatty acids in the oils. α-Tocopherol addition to frying oil increased the stability and resistance of polyunsaturated fatty acid against oxidation [47]. Inclusion of antioxidants lemon seed extract and tert-butylhydroquinone was demonstrated to retard lipid oxidation and contribute to the α-tocopherol retention in the soy oil heated at high temperatures for several hours [48].

Male rats that were fed heated oils had elevated plasma lipid peroxidation content. The elevation was more prominent with the five times heated oils. The increased peroxide value of the heated oils may be associated with the significant increase in plasma lipid peroxidation. Chronic ingestion of heated oil was shown to cause an elevation of blood pressure [49] and necrotic cardiac changes [50]. This would disrupt the endogenous antioxidant defense in our body in order to overcome the overwhelming of oxidative stress. Linoleyl, peroxy, and alkoxy radicals were reported to be produced in heated oils. These radicals act on the fatty acids of the oil producing oxidized products via hemolytic β-scission [25]. Dietaries fresh palm and soy oil were shown to attenuate oxidative stress and augment antioxidant enzymes activities in rat models [51-53]. Their findings are in agreement with the present study where a significant reduction in plasma lipid peroxidation was observed in the fresh oil-fed male rats. This positive effect was attributed to the rich antioxidant content of the oils.

Heating diminished the vitamin E content and rendered the oils to lose their beneficial effects. Chronic ingestion of heated soy oil diet was shown to induce the increase in hepatic lipid peroxidation in rats [54]. They also observed a reduction in serum and hepatic vitamin E content in rats that were fed heated oil containing diet. An evidence of hepatic damage was also reported with a significant elevation of aspartate and alanine transaminases, the marker enzymes for liver function in rats that were given combination of heated soy and rapeseed oils [55]. The administration of dietary heated soy oil affected the activities of antioxidant enzymes in various rat organs. It increased superoxide dismutase in the liver and brain, while increment in glutathione reductase was seen in the liver and kidneys [26]. Heated palm oil also increased catalase, glutathione peroxidase and glutathione S-transferase in the rat liver [35]. It shows dietary heated oil contained prooxidant substances that would evoke the body primary defenses.

Plasma lipid peroxidation was increased significantly in the rats that were given heated soy oil (once and five times heated) when compared to the heated palm oil-fed groups. This finding was suggestive of the better effect of the palm oil in terms of oxidative stability when exposed to extreme heat. High composition of saturated fats in palm oil confers it to withstand thermal oxidative changes, in addition to its rich content of tocotrienols. The better effect of heated palm oil compared to heated soy oil was reported elsewhere [56].

Ovariectomized female rats were often used as an experimental postmenopausal model. Ovariectomy results in an estrogen deficiency state. The estrogen provides protection to the body against oxidative stress [16]. In the present study, dietaries once heated and five times heated palm and soy oil for four months had significantly increased the plasma lipid peroxidation in ovariectomized female rats. The result is in agreement with other studies which also reported that in ovariectomized-induced estrogen deficiency, an increase in oxidative stress accompanied by a reduction in antioxidant status was seen [57-58]. A study by Sánchez-Rodríguez [59] had recently found that menopause could be one of the main risk factors for oxidative stress. Therefore, this finding suggests that chronic ingestion of repeatedly heated vegetable oils is detrimental to health and could further aggravate the increase in oxidative stress in postmenopausal women. Shuid et al. [56] also had demonstrated that fresh and once heated palm and soy oil diets for six months reduced the risk of osteoporosis in ovariectomized rats, a positive effect that was not seen in rats that were fed five times heated oils.

The patterns of plasma lipid peroxidation increase in both male and ovariectomized female rats were somewhat similar. However, the plasma lipid peroxidation of the fresh oil-fed groups in the ovariectomized female rats that were similar to the control group, different from the one observed in the male rats. The addition of dietary 2% cholesterol and the ovariectomy status of the female rats made them more susceptible to oxidative insult. This may explain why there was no significant reduction in plasma TBARS seen in the female rats fed fresh oils compared to the control. The plasma lipid peroxidation was significantly elevated in the five times heated soy oil group compared to the same heating frequency of palm oil. The difference was also attributed to the higher component of polyunsaturated fatty acids present in the soy oil, which are prone to oxidation.

5. Conclusion

Our findings suggest that it is recommended not to heat cooking oil more than once in view of its possible deleterious effect on health. The use of palm oil possibly has better effect on health due to its stability against oxidative insult.

Author details

Kamsiah Jaarin and Yusof Kamisah
Department of Pharmacology, Faculty of Medicine,Universiti Kebangsaan Malaysia, Malaysia

Acknowledgement

The study was funded by grants from Universiti Kebangsaan Malaysia (UKM-GUP-SK-08-21-299) and Malaysian Ministry of Science, Technology and Innovations (IRPA 06-02-02-0050-EA242). The authors also would like to thank Prof. Dr Jumat Salimon, Ms Xin-Fang Leong and Mr. Chun-Yi Ng for their technical help as well as Ms Jurika Sharaton Abdul Wahed for editorial assistance.

6. References

[1] Rani AKS, Reddy SY, Chetana R (2010) Quality changes in trans and trans free fats/oils and products during frying. Eur. Food Res. Technol. 230(6): 803–811.

[2] Choe E, Min DB (2007) Chemistry of deep-fat frying oils. J. Food Sci. 72(5):R77-R86.

[3] Choe, E. & Min, D. B (2006) Chemistry and reactions of reactive oxygen species in foods. Crit. Rev. Food Sci. Nutr. 46(1): 1-22.

[4] Kamisah Y, Adam A, Wan Ngah WZ, Gapor MT, Azizah O, Marzuki A (2005) Chronic intake of red palm olein and palm olein produce beneficial effects on plasma lipid profile in rats. Pakistan J. Nutr. 4(2): 89-96

[5] Clemente TE, Cahoon EB (2009) Soybean oil: genetic approaches for modification of functionality and total content. Plant Physiol. 151(3): 1030-1040.

[6] Bardhan J, Chakraborty R, Raychaudhuri U (2011). The 21st century form of vitamin E - Tocotrienol. Curr. Pharm. Des. 17(21): 2196-2205.

[7] Edem DO (2002) Palm oil: biochemical, physiological, nutritional, hematological, and toxicological aspects: a review. Plant Foods Hum. Nutr. 57(3-4): 319-341.

[8] Azman, SM Shahrul, XS Chan, AP Noorhazliza, M Khairunnisak, MF Nur Azlina, HMS Qodriyah, Y Kamisah, K Jaarin (2012) Level of knowledge, attitude and practice of night market food outlet operators in Kuala Lumpur regarding the usage of repeatedly heated cooking oil. Med. J. Malays. 67(1): 97-107.

[9] Abdullah A, Shahrul S, Chan XS, Noorhazliza AP, Khairunnisak M, Qodriyah HMS, Kamisah Y, Nur Azlina MF, Kamsiah J (2010) Level of awareness amongst the general public regarding usage of repeatedly heated cooking oil in Kuala Lumpur, Malaysia. Inter. Med. J. 17(4): 310-311.

[10] Srivastava S, Singh M, George J, Bhui K, Murari Saxena A, Shukla Y (2010) Genotoxic and carcinogenic risks associated with the dietary consumption of repeatedly heated coconut oil. Br. J. Nutr. 104(9): 1343-1352.

[11] Obembe AO, Owu DU, Okwari OO, Antai AB, Osim EE (2011) Intestinal Fluid and Glucose Transport in Wistar Rats following Chronic Consumption of Fresh or Oxidised Palm Oil Diet. ISRN Gastroenterol. 2011: 972838.

[12] Latha P, Chaitanya D, Rukkumani R (2010) Protective effect of Phyllanthus niruri on alcohol and heated sunflower oil induced hyperlipidemia in Wistar rats. Toxicol Mech. Methods. 20(8): 498-503.

[13] Soriguer F, Rojo-Martínez G, Dobarganes MC, García Almeida JM, Esteva I, Beltrán M, Ruiz De Adana MS, Tinahones F, Gómez-Zumaquero JM, García-Fuentes E, González-Romero S (2003) Hypertension is related to the degradation of dietary frying oils. Am. J. Clin. Nutr. 78(6): 1092-1097.

[14] Rossouw JE (2002) Hormones, genetic factors and gender differences in cardiovascular disease. Cardiovasc. Res. 53: 550-557.

[15] Lu H, Meléndez GC, Levick SP, Janicki JS (2012) Prevention of adverse cardiac remodeling to volume overload in female rats is the result of an estrogen-altered mast cell phenotype. Am. J. Physiol. Heart Circ. Physiol. 302(3): H811-H817.

[16] Abbas AM, Elsamanoudy AZ (2011) Effects of 17β-estradiol and antioxidant administration on oxidative stress and insulin resistance in ovariectomized rats. Can. J. Physiol. Pharmacol. 89(7): 497-504.

[17] Shuid AN, Mohamad S, Muhammad N, Fadzilah FM, Mokhtar SA, Mohamed N, Soelaiman IN (2011) Effects of α-tocopherol on the early phase of osteoporotic fracture healing. J. Orthop. Res. 29(11): 1732-1738.

[18] Mvondo MA, Njamen D, Fomum ST, Wandji J, Vollmer G (2011) A postmenopause-like model of ovariectomized Wistar rats to identify active principles of Erythrina lysistemon (Fabaceae). Fitoterapia. 82(7): 939-949.

[19] Owu DU, Osim EE, Ebong PE (1998) Serum liver enzymes profile of Wistar rats following chronic consumption of fresh or oxidized palm oil diets. Acta Trop. 69(1): 65-73.

[20] AOCS (2003) Official methods and recommended practices of the American Oil Chemists' Society. 4th ed. AOCS Press, Champaign , Illinois.

[21] Ledwozyw A, Michalak J, Stepian A, Kadziolka A (1986) A relationship between plasma triglycerides, cholesterol, total lipids and lipid peroxidation products during human atherosclerosis. Clin. Chim. Acta. 155: 275-285.

[22] Lowry OH, Rosebrough NJ, Farr AL, Randall RJ (1951) Protein measurement with Folin Phenol reagent. J. Biol. Chem. 193: 265-275.

[23] Matthaus B (2006) Utilization of high-oleic rapeseed oil for deep-fat frying of French fries compared to other commonly used edible oils. Eur. J. Lipid Sci. Tech. 108(3): 200-211.

[24] Sayon-Orea C, Bes-Rastrollo M, Basterra-Gortari FJ, Beunza JJ, Guallar-Castillon P, de la Fuente-Arrillaga C, Martinez-Gonzalez MA (2012) Consumption of fried foods and weight gain in a Mediterranean cohort: The SUN project. Nutr. Metab. Cardiovasc. Dis. (in press)

[25] Picariello G, Paduano A, Sacchi R, Addeo F (2009) Maldi-tof mass spectrometry profiling of polar and nonpolar fractions in heated vegetable oils. J. Agric. Food Chem. 57(12): 5391-5400.

[26] Awney HA (2011) The effects of Bifidobacteria on the lipid profile and oxidative stress biomarkers of male rats fed thermally oxidized soybean oil. Biomarkers. 16(5): 445-452.

[27] Leskova E, Kubikova J, Kovacikova E, Kosicka M, Pobruska J, Holcikova K (2006) Vitamin losses: Retention during heat treatment and continual changes expressed by mathematical models. J. Food Comp. Anal. 19: 252–276.

[28] Velasco J, Marmesat S, Bordeaux O, Márquez-Ruiz G, Dobarganes C (2004) Formation and evolution of monoepoxy fatty acids in thermoxidized olive and sunflower oils and quantitation in used frying oils from restaurants and fried-food outlets. J. Agr. Food Chem. 52(14): 4438-4443.

[29] Iwata NG, Pham M, Rizzo NO, Cheng AM, Maloney E, Kim F (2011) Trans fatty acids induce vascular inflammation and reduce vascular nitric oxide production in endothelial cells. PLoS One. 6(12): e29600.

[30] Bryk D, Zapolska-Downar D, Malecki M, Hajdukiewicz K, Sitkiewicz D (2011) Trans fatty acids induce a proinflammatory response in endothelial cells through ROS-dependent nuclear factor-κB activation. J. Physiol. Pharmacol. 62(2): 229-238.

[31] Saka S, Aouacheri W, Abdennour C (2002) The capacity of glutathione reductase in cell protection from the toxic effect of heated oils. Biochimie 84(7): 661-665.

[32] Karoui IJ, Dhifi W, Jemia MB, Marzouk B (2011) Thermal stability of corn oil flavoured with Thymus capitatus under heating and deep-frying conditions. J. Sci. Food Agr. 91(5): 927-933.

[33] Kochhar SP, Henry CJ (2009) Oxidative stability and shelf-life evaluation of selected culinary oils. Inter. J. Food Sci. Nutr. 60(7 Suppl): 289-296.

[34] Moya Moreno MC, Mendoza Olivares D, Amézquita López FJ, Gimeno Adelantado JV, Bosch Reig F. Analytical evaluation of polyunsaturated fatty acids degradation during thermal oxidation of edible oils by Fourier transform infrared spectroscopy. Talanta. 1999 Sep 13;50(2):269-75.

[35] Purushothama S, Ramachandran HD, Narasimhamurthy K, Raina PL (2003) Long-term feeding effects of heated and fried oils on hepatic antioxidant enzymes, absorption and excretion of fat in rats. Mol. Cell Biochem. 247(1-2): 95-99.

[36] Vara Prasad SS, Jeya Kumar SS, Kumar PU, Qadri SS, Vajreswari A (2010) Dietary fatty acid composition alters 11β-hydroxysteroid dehydrogenase type 1 gene expression in rat retroperitoneal white adipose tissue. Lipids Health Dis. 9: 111.

[37] Horvath G, Wessjohann L, Bigirimana J, Jansen M, Guisez Y, Caubergs R, Horemans N (2006) Differential distribution of tocopherols and tocotrienols in photosynthetic and non-photosynthetic tissues. *Phytochemistry.* 67: 1185–1195.

[38] Shen Y, Lebold K, Lansky EP, Traber MG, Nevo E (2011) 'Tocol-omic' diversity in wild barley, short communication. Chem. Biodivers. 8(12): 2322-2330.

[39] Huang SH, Ng LT (2011) Quantification of tocopherols, tocotrienols, and γ-oryzanol contents and their distribution in some commercial rice varieties in Taiwan. J. Agric. Food Chem. 59(20): 11150-11159.

[40] Hassanein MM, Abedel-Razek AG (2009) Chromatographic quantitation of some bioactive minor components in oils of wheat germ and grape seeds produced as by-products. J. Oleo Sci. 58(5): 227-233.

[41] Milagros Delgado-Zamarreño M, Bustamante-Rangel M, Sierra-Manzano S, Verdugo-Jara M, Carabias-Martínez R (2009) Simultaneous extraction of tocotrienols and tocopherols from cereals using pressurized liquid extraction prior to LC determination. J. Sep. Sci. 32(9): 1430-1436.

[42] Traber MG (2007) Vitamin E regulatory mechanisms. *Annu. Rev. Nutr.* 27: 347–362.

[43] Rennick KA, Warner K (2006) Effect of elevated temperature on development of tocopherolquinones in oils. J. Agric. Food Chem. 54(6): 2188-2192.

[44] Sen CK, Rink C, Khanna S (2010) Palm oil-derived natural vitamin E alpha-tocotrienol in brain health and disease. J. Am. Coll. Nutr. 29(3 Suppl): 314S-323S.

[45] Aggarwal BB, Sundaram C, Prasad S, Kannappan R (2010) Tocotrienols, the vitamin E of the 21st century: its potential against cancer and other chronic diseases. Biochem. Pharmacol. 80(11): 1613-1631.

[46] Miyagawa K, Hirai K, Takezoe R (1991) Tocopherol and fluorescence levels in deepfrying oil and their measurement for oil assessment. J. Am. Oil Chem. Soc. 68: 163–166.

[47] Quiles JL, Ramírez-Tortosa MC, Ibáñez S, Alfonso González J, Duthie GG, Huertas JR, Mataix J (1999) Vitamin E supplementation increases the stability and the in vivo antioxidant capacity of refined olive oil. Free Radic. Res. 31 Suppl: S129-S135.

[48] Luzia DM, Jorge N (2009) Oxidative stability and alpha-tocopherol retention in soybean oil with lemon seed extract (Citrus limon) under thermoxidation. Nat. Prod. Commun. 4(11): 1553-1556.

[49] Leong X, Najib MNM, Das S, Mustafa MR, Jaarin K (2009) Intake of repeatedly heated palm oil causes elevation in blood pressure with impaired vasorelaxation in rats. Tohoku J. Exp. Med. 219: 71-78.

[50] Leong XF, Aishah A, Aini UN, Das S, Jaarin K (2008) Heated palm oil causes rise in blood pressure and cardiac changes in heart muscle in experimental rats. Arch. Med. Res. 39: 567-572.

[51] Bayorh MA, Abukhalaf IK, Ganafa AA (2005) Effect of palm oil on blood pressure, endothelial function and oxidative stress. Asia Pac. J. Clin. Nutr. 14(4): 325-339.

[52] Narang D, Sood S, Thomas M, Dinda AK, Maulik SK (2005) Dietary palm olein oil augments cardiac antioxidant enzymes and protects against isoproterenol-induced myocardial necrosis in rats. J. Pharm. Pharmacol. 57(11): 1445-1451.

[53] Hassan HA, Abdel-Wahhab MA (2012) Effect of soybean oil on atherogenic metabolic risks associated with estrogen deficiency in ovariectomized rats : Dietary soybean oil modulate atherogenic risks in overiectomized rats. J. Physiol. Biochem. (in press).

[54] Corcos-Benedetti P, Di Felice M, Gentili V, Tagliamonte B, Tomassi G (1990) Influence of dietary thermally oxidized soybean oil on the oxidative status of rats of different ages. Ann. Nutr. Metab. 34(4): 221-231.

[55] Totani N, Burenjargal M, Yawata M, Ojiri Y (2008) Chemical properties and cytotoxicity of thermally oxidized oil. J. Oleo. Sci. 57(3): 153-160.

[56] Shuid AN, Chuan LH, Mohamed N, Jaarin K, Fong YS, Soelaiman IN (2007) Recycled palm oil is better than soy oil in maintaining bone properties in a menopausal syndrome model of ovariectomized rat. Asia Pac J Clin Nutr. 16(3): 393-402.

[57] Huang YH, Zhang QH (2010) Genistein reduced the neural apoptosis in the brain of ovariectomised rats by modulating mitochondrial oxidative stress. Br. J. Nutr. 104(9): 1297-1303.

[58] Topçuoglu A, Uzun H, Balci H, Karakus M, Coban I, Altug T, Aydin S, Topçuoglu D, Cakatay U (2009) Effects of estrogens on oxidative protein damage in plasma and tissues in ovariectomised rats. Clin. Invest. Med. 32(2): E133-143.

[59] Sánchez-Rodríguez MA, Zacarías-Flores M, Arronte-Rosales A, Correa-Muñoz E, Mendoza-Núñez VM (2012) Menopause as risk factor for oxidative stress. Menopause. 19(3): 361-367.

Chapter 12

Modification of Fatty Acid Composition in Meat Through Diet: Effect on Lipid Peroxidation and Relationship to Nutritional Quality – A Review

Gema Nieto and Gaspar Ros

Additional information is available at the end of the chapter

1. Introduction

The use of nutritional strategies to improve quality of food products from livestock is a new approach that emerges at the interface of food science and animal science. These strategies have emphasized in the alteration of nutritional profile, for example increasing the content of polyunsaturated fatty acid (PUFA), and in the improvement of the oxidative stability, such as supplementation of animal with natural antioxidants to minimize pigment and lipid oxidation in meat.

The interest in the modification of fatty acid of meat is due to that fatty acid composition plays an important role in the definition of meat quality because it is related to differences in sensory attributes and in the nutritional value for human consumption [1]. Meat is a major source of fat in the diet, especially of saturated fatty acids (SFA), which have been implicated in diseases, especially in developed countries, such as cardiovascular diseases and some types of cancer.

One of the key goals of nutritional research focuses on establishing clear relationships between components of diet and chronic diseases, considering that nutrients could provide beneficial health results. The incidence of these diseases in humans is associated with the amount and the type of fat consumed in the diet. Diets high in SFA contribute to increase LDL-cholesterol level, which is positively related to the occurrence of heart diseases. However, some monounsaturated fatty acids (MUFA) and PUFA, in particular long-chain n-3 PUFA have favourable effects on human health.

In recent years, consumers' pressure to reduce the composition and quality of fat in meat has led to attempts to modify meat by dietary strategies. Where as in recent years consumers have been advised to limit their intake of saturated fats and to reach a ratio of *PUFA:SFA* greater than 4 and the type of polyunsaturated fatty acid is now being emphasized and a higher ratio of n-3: n-6 fatty acids is advocated [1]. There is also now concern about the consumption of unsaturated fatty acids that are formed during high-temperature hydrogenation of oils for use in food products: the *trans*-unsaturated fatty acids in which the double bonds are in the *trans*-stereometric position.

Nutritional approaches to improve the oxidative stability of muscle foods are often more effective than direct addition of food ingredients since the antioxidants are preferentially deposited where it is most needed. In addition, diet often represents the only technology available to alter the oxidative stability of intact muscle foods, where utilization of exogenous antioxidants additives is difficult if not impossible. Since product composition is altered biologically, nutritional alteration of muscle composition is more label-friendly since no additive declarations are required.

Among the strategies used, meat and meat products can be modified by adding ingredients considered beneficial for health where the ingredients are able to eliminate or reduce components that are considered harmful. In this sense, several studies have shown that animal diet can strongly influence the fatty acid composition of meat. Scerra et al. [2] showed that feeding ewes with pasture increases the *PUFA* content of intramuscular fat of the lamb infant compared with diets consisting of concentrate. Nieto et al. [3] showed that feeding Segureña ewes with thyme increases the *PUFA* content of intramuscular fat of the lamb meat compared with control diets. Similarly, Elmore et al. [4] showed that feeding lambs with diets rich in fish oil can modify the fatty acid profile of meat (increasing the level of *PUFA*). Moreover Bas et al. [5] used linseed diet and Ponnampalam et al. [6] used fish oil, in order to increase the content of long-chain n-3 fatty acids in lamb meat.

The variation of fatty acid compositions has profound effects on meat quality, because fatty acid composition determines the firmness/oiliness of adipose tissue and the oxidative stability of muscle, which in turn affects flavour and muscle colour. It is well known that high *PUFA* levels may produce alterations in meat flavour due to their susceptibility to oxidation and the production of unpleasant volatile components during cooking [7]. Therefore, it's important to study the implications of the modification of fatty acid in the quality of the meat and the lipid stability, for that it would be interesting the use of liposomes to study the lipid oxidation.

Since liposomes mimic cellular structures [8], the feasibility to protect lipid membranes in the presence of natural antioxidants can be investigated in model systems prior to administration trough feeding. Such previous experiments are particularly interesting for meat industry as they furnish preliminary insights with respect to lipid oxidation at relatively short timescales [9].

2. Lipid digestion in ruminants and non-ruminants

It is well known that lipid digestion is different in ruminant and non-ruminant and that the nature of lipid digestion by the animal has an important effect on the transfer of fatty acids from the diet into the animal product. In case of non-ruminant, the principal site of digestion of dietary lipid is the small intestine, where the pancreatic lipase breaks the triacylglycerols down to mainly 2-monoacylglycerols and free fatty acids and the formation of micelles aids absorption, with lipid uptake mediated by the lipoprotein lipase enzyme, which is widely distributed throughout the body. Therefore dietary fatty acids in the non-ruminant are absorbed unchanged before incorporation into the tissue lipids. Dietary lipid sources have a direct and generally predictable effect on the fatty acid composition of pig and poultry products and the supply of unsaturated fatty acids (UFA) to tissues may be simply increased by increasing their proportion in the diet [10].

However, digestion and metabolism of ingested lipids in the rumen results in the exit of mainly long-chain, saturated fatty acids from the rumen. The rumen microorganisms in the ruminant digestive system have a major impact on the composition of fatty acids leaving the rumen for absorption in the small intestine. Microbial enzymes are responsible for the isomerisation and hydrolysis of dietary lipid and the conversion of UFA to various partially and fully saturated derivatives, including stearic acid ($C_{18:0}$). Although linoleic ($C_{18:2}$ n-6) and linolenic ($C_{18:3}$ n-3) acids are the main UFA in the diet of ruminants, the processes within the rumen ensure that the major fatty acid leaving the rumen is $C_{18:0}$. The intestinal absorption coefficient of individual fatty acids is higher in ruminants than nonruminants, ranging from 80% for SFA to 92% for PUFA in conventional low fat diets. Therefore, the higher absorption efficiency of SFA by ruminants has been attributed to the greater capacity of the bile salt and lysophospholipid micellar system to solubilise fatty acids, as well as the acid conditions within the duodenum and jejunum (pH 3.0–6.0).

3. Fatty acid in meat

Taking into accounts that fat is currently an unpopular constituent of meat and however contributes to meat quality and is important to the nutritional value of meat. This section considers the fatty acid composition in different species and the roles of the fat in meat quality.

Doing a brief introduction of the importance of fatty acids, firstly we will highlight the essential unsaturated fatty acids, linoleic ($C_{18:2}$), linolenic ($C_{18:3}$) and arachidonic ($C_{20:4}$). They are necessary constituents of mitochondria and cell walls. These fatty acids are specials, because contrary to the production from saturated sources, the body can not produce any of the fatty acid mentioned above, unless one of them is available in the diet. Oleic, linoleic and linolenic acids each belong to a different family of compounds in which unsaturation occurs at the n–9, the n–6 and n–3 carbon atoms, respectively, in the hydrocarbon chain numbering from the methyl carbon (n). They are thus referred to as the ω –9, ω –6 and ω –3 series. Linoleic acid is abundant in vegetable oils and at about 20 times the concentration found in meat; and linolenic acid is present in leafy plant tissues [11].

Doing a comparative data between the content of PUFA in the muscular tissue of the beef, lamb and pork (Table 1), it is clear that linoleic acid ($C_{18:2}$) is markedly greater in the lean meat of pigs than in that of either the beef or lamb.

	C18:2	C18:3	C20:4	C22:5	C22:6
Beef	2.0	1.3	1.0	Tr.	-
Lamb	2.5	2.5	-	Tr.	-
Pork	7.4	0.9	Tr.	Tr.	1.0

Table 1. Polyunsaturated fatty acids and cholesterol in lean meat (as % total fatty acids)

In addition, the Table 2 shows the study of Enser at al. [12], who obtained 50 samples of beef sirloin steaks, pork chops, and lamb chops and determined the fatty acid profile of the muscle portions of these retail meat cuts. In the same way that Table 1, the most notable difference among the ruminant species and pork was the fivefold greater concentration of linoleic in pork and significantly greater proportions of $C_{20:3}$, $C_{20:4}$, and $C_{22:6}$, and $C_{14:0}$. For example, pork have a proportions of linoleic acid ($C_{18:2}$ n-6): 302 mg/100g of loin muscle, while beef and lamb contains 89 and 25 mg/100g, respectively. The reason of this is because linoleic acid is derived entirely from the diet. It passes through the pig's stomach unchanged and is then absorbed into the blood stream in the small intestine and incorporated from there into tissues. When linoleic acid is ingested, they are metabolized by animal liver to produce two families of long chain polyunsaturated fatty acids which are specific to animals, respectively, the n-6 and n-3 series.

Fatty acid	Pork	Beef	Lamb
$C_{12:0}$ (lauric)	2.6	2.9	13.8
$C_{14:0}$ (myristic)	30	103	155
$C_{16:0}$ (palmitic)	526	962	1101
$C_{18:0}$ (stearic)	278	507	898
$C_{18:1}$ (trans)	-	104	231
$C_{18:1}$ (oleic)	759	1395	1625
$C_{18:2}$ n-6 (linoleic)	302	89	125
$C_{18:3}$ n-3 (α-linolenic)	21	26	66
$C_{20:3}$ n-6 (lauric)	7	7	2
$C_{20:4}$ n-6 (arachidonic)	46	22	29
$C_{20:5}$ n-3 (eicosopentaenoic)	6	10	21
$C_{22:5}$ n-3 (docosopentaenoic)	13	16	24
$C_{22:6}$ n-3 (docosohexaenoic)	8	2	7
Total	2255	3835	4934
P:S	0.58	0.11	0.15
n-6:n-3	7.22	2.11	1.32

Source: Enser et al. [12]

Table 2. Fatty acid Content (mg/100g) of loin muscle in steaks or chops.

However, in ruminants, linoleic acid ($C_{18:2}$ n–6) and α-linolenic acid ($C_{18:3}$ n-3) which are at present in many concentrate feed ingredients, are degraded into monounsaturated (*MUFA*) and saturated fatty acids (*SFA*) in the rumen by microbial biohydrogenation (70–95% and 85-100%, respectively) and only a small proportion, around 10% of dietary consumption, is available for incorporation into tissue lipids. By that reason, beef and lamb contain lower content of linoleic acid, compared with pork meat. Muscle also contains significant proportions of long chain (C20-22) *PUFAS* which are formed from $C_{18:2}$ n–6 and $C_{18:3}$ n–3 by the action of Δ5 and Δ 6 desaturase and elongase enzymes. Important products are arachidonic acid ($C_{20:4}$ n-6) and eicosapentaenoic acid (EPA, $C_{20:5}$ n-3).

Taking into accounts that in ruminants, rumen microorganisms hydrogenate a substantial proportion of *PUFA* diet, resulting in high levels of *SFA* for deposition in muscle tissue, lamb or beef meat contain a low relationship between fatty acids *PUFA* and *SFA* (ratio P/S), which increases the risk of cardiovascular problems and other diseases.

The consequences of a greater incorporation of $C_{18:2}$ n-6 into pig muscle fatty acids compared with ruminants produces higher levels of $C_{20:4}$ n-6 by synthesis and the net result is a higher ratio of n-6:n-3 *PUFA* compared with the ruminants. If the nutritional advice is for ratios <4.0, the present value of 7 on pig muscle is unbalanced relative to that of the ruminants (1.32 in lamb and 2.11 in beef). In addition, another ratio is the ratio of all *PUFA* to *SFA* (P:S). The ratio P/S in a normal diet is 0.4 [13] and in lamb meat is 0.15 and in beef 0.11, while in pork is 0.58.

For all these reasons, there is an increase interesting in research intended to modify the fatty acid composition in meat, especially reducing the concentration of *SFA* and increasing *PUFA*.

4. Dietary modification of fatty acid in meat

Doing the comparison between ruminants and non-ruminants, the fatty acid composition of stored lipids of the ruminant is relatively unwilling to changes in the fatty acid profile of ingested lipids. This logic has been the basis for expression of the concept that ruminant fats are more saturated than those of non-ruminants. Although effects of ruminal biohydrogenation on ruminant tissue fatty acid profiles are generalized, numerous researchers have demonstrated that ruminant lipids can be manipulated by dietary means to contain a higher proportion of unsaturated fatty acids. The next paragraphs will show the different strategies to modify the fatty acids of ruminants and nonruminants.

4.1. Altering quality of muscle from monogastric

Monogastric farm animals are worldwide a main source of high-quality products with a high content of highly available protein, minerals, and vitamins. Pigs and chickens are the main monogastric farm animals [1].

To altering the quality of muscle of monogastric, it's necessary to know the digestion of nutrients in these animals. Anaerobic microorganisms are able to hydrogenate unsaturated fatty acids (UFA) preferably polyunsaturated fatty acids (PUFA). During these processes they build trans-fatty acid as well as conjugated fatty. While in ruminants these fatty acids are absorbed to a great extent, monogastric animals excrete most of them with the feces as they are produced in the lower parts of the digestive tract.

4.2. Altering quality of muscle from ruminants

Meat from ruminants is a major source of essential nutrients (amino acids, iron, zinc and vitamins from the B group). Meat from ruminants (huge diversity of breeding systems and pieces) is characterised by great variations in fats, quantitatively and qualitatively. Some saturated ($C_{14:0}$ and $C_{16:0}$) and monounsaturated trans fatty acids are not recommended for human consumption and it is possible to reduce their concentrations in meats by increasing the proportions of polyunsaturated fatty acids absorbed by the animals from their diets. To achieve this goal, fatty acids must be protected against hydrogenation in the rumen. Dietary intake of PUFA from the n-3 series and especially from the n-6 series by the animals favour the production of conjugated linoleic acid by the rumen bacteria. Some of these fats, such as CLA (conjugated linoleic acid), could be beneficial to human health. CLA is important in the prevention of specific cancers and in the treatment of obesity that has been demonstrated in animal models and, at least partly, in humans.

Alteration of quality in food products from ruminants requires knowledge of the nutritional and metabolic principles that influence product composition. Ruminants are unique among mammals due to their pregastric fermentation. Microflora and microfauna present in the ruminant forestomach dramatically modify the ingested nutrients and consequently have a large impact on the metabolism and composition of the muscle and milk [1].

5. Fatty acid sources

It's important to take into accounts several factors to choice the ingredient and the form by which it is included in the feed: (a) the cost and availability; (b) the impact of the ingredients and its fatty acid composition on feed digestibility; (3) the influence of consumers and retailers regarding the introduction of ingredients into the food chain and (4) animal feed regulations regarding permitted supplements.

Recognizing that fatty acids are readily absorbed from the diet and incorporated into tissue fat, producers have attempted to improve the nutritional quality of meat by incorporating various sources of n-3-PUFA [7, 14]. The main dietary sources of n-3 fatty acids fed to pig are vegetable oils [15-20], fish oil/fish meal [21,22] and forage [23]. Novel oil sources such as chia seed, marine algae, lupin and camelina have been investigated as lipid sources in animal feeds.

5.1. Lipid sources for ruminants

5.1.1. Forages

Several studies have shown that ruminants consuming fresh pasture have higher content of UFA in their meat that those receiving a cereal-based concentrate diet.

Grass is a good source of n–3 *PUFA* although there can be variation due to maturity and variety. Grass lipids contain high proportions of the unsaturated linolenic acid ($C_{18:3}$ *n–3*). Other studies have suggested that the (n–6)/(n–3) ratio in phospolipids may be useful to discriminate grass-fed from grain-fed lambs [1, 24].

Therefore, pasture-raised animals have higher proportions of linolenic acid in their fat than stallfed animals [25].

French et al. [15] compared the effect of offering grazed grass, grass silage and concentrates on the fatty acid composition of intramuscular fat in steers. Similar low intramuscular fat contents (<4.5 g/100 g muscle) were determined in meat from all diets offered, hence a possible confounding effect due to differences in the amount of fat deposited was avoided. Decreasing the proportion of concentrate in the ration effectively increased the proportion of grass intake and resulted in a linear increase in *PUFA:SFA* ratio ($P<0.01$) and a linear decrease in the concentration of SFA ($P<0.001$). The highest concentration of *PUFA* in the intramuscular fat was found in those animals that had consumed grass only (22 kg grazed grass). Grass and grass silage had a much greater proportion of α-linolenic acid than the concentrates, although levels of linoleic acid were similar. The content of linoleic acid in the intramuscular fat was not significantly different between treatments but concentrations of α-linolenic acid and total conjugated linoleic acid were significantly higher for grass-fed steers than for steers offered grass silage and/or concentrates.

Moreover, Nuernberg et al. [26] showed that the concentration of lauric acid was higher in subcutaneous fat and muscle of lambs fed on pasture compared to lambs fed concentrate. Similar results were found by Demirel et al. [27], who studied the fatty acids of lamb meat from two breeds fed different forage: concentrate ratio. And Scerra et al. [2], who showed that lamb meat derived from pasture-fed ewes had a lower levels of lauric and palmitic acid (compared with diets with concentrate) that are though to be a public health risk

Sañudo et al. [28] studied British lambs compared with lambs fed grass fed grain, the result showed that a higher percentage of linolenic acid in the meat of grass-fed lambs, as result of the introduction of this natural antioxidant.

Realinia et al. [29] studied thirty Hereford steers that were finished either on pasture or concentrate to determine dietary and antioxidant treatment effects on fatty acid composition and quality of beef. These authors reported that the percentages of $C_{14:0}$, $C_{16:0}$, and $C_{18:1}$ fatty acids were higher ($P<0.01$) in the intramuscular fat of concentrate-fed steers, whereas pasture-fed cattle showed greater ($P<0.01$) proportions of $C_{18:0}$, $C_{18:2}$, $C_{18:3}$, $C_{20:4}$, $C_{20:5}$, and $C_{22:5}$. Total conjugated linoleic acid (CLA) was higher ($P<0.01$) for pasture- than concentrate-fed cattle. Therefore, these authors reported that finishing cattle on pasture enhanced the unsaturated fatty acid profile of intramuscular fat in beef including CLA and omega-3 fatty acids. Results from this study suggest that the negative image of beef attributed to its highly saturated nature may be overcome by enhancing the fatty acid profile of intramuscular fat in beef through pasture feeding from a human health perspective.

5.1.2. Oilseeds

Many studies to manipulate the fatty acid composition of meat using whole oilseeds have been conducted. For example, the effect of the physical form of linseed offered on the fatty acid composition of meat has been reported by several workers: Raes et al. [30] reported that the replacement of whole soyabean with extruded linseed or crushed linseed in the finishing diet of Belgian Blue young bulls increased α-linolenic acid. Mach et al. [31] reported that whole canola seed (α-linolenic acid content 10.6 g/100 g total FA) or whole linseed (α-linolenic acid content 54.2 g/100 g total FA), at three lipid levels (50, 80 and 110 g/kg DM) to 54 Holstein bulls increased linearly with lipid level the concentration of n–3 PUFA in the longissimus dorsi muscle.

Elmore et al. [4] reported that the feeding of lamb with diets rich in fat and oils (fish oils, kelp and flax seed) increased the level of polyunsaturated fatty acids. Similarly, Nute et al. (2007) studied the oxidative stability and quality of fresh meat from lambs fed different levels of n-3 PUFA from linseed oil, fish oil, a supplement produced from flax seed (PLS), seed sunflower and soybean meal, seaweed, and combinations of these different oils. They reported that the fatty acid composition of semimembranosus muscle phospholipids was affected by diet.

The rabbit meat was also used in several studies with the objective of fatty acid modification. As the study of Kouba et al. [32], who studied rabbits fed with a diet containing 30 g of extruded linseed/kg. Feeding the linseed diet increased (P < 0.005) the content of 18:2n-3 in muscles, perirenal fat, and raw and cooked meat. The long chain n-3 polyunsaturated fatty acid (PUFA) contents were also increased (P < 0.01) in the meat. The linseed diet produced a decrease in the n-6/n-3 ratio. These authors highlights that the inclusion of linseed in rabbit diets is a valid method of improving the nutritional value of rabbit meat.

5.1.3. Marine algae

Marine algae are an alternative to fish oil as a dietary source of n-3 long chain PUFA (LCPUFA), eicosapentaenoic acid (EPA) and docosahexaenoic acid (DHA).

In the study of Cooper [33], the marine algae were included in the sheep diet not only as a source of DHA (fish oil/algae diet supplied 15 g/100 g total FA as DHA) but also because it had been previously shown to undergo a lower level of biohydrogenation than fish oil [33]. In another study of the same author [34] studied the manipulation of the n-3 PUFA fatty acid content of muscle and adipose tissue lamb was studied. For that fifty lambs, with an initial live weight of 29 ± ⊙2.1 kg, were allocated to one of five concentrate-based diets formulated to have a similar fatty acid content (60 g/kg DM), but containing either linseed oil (high in 18:3n⊙3); fish oil (high in 20:5n⊙3 and 22:6n⊙3); protected linseed and soybean (PLS; high in 18:2n⊙6 and 18:3n⊙3); fish oil and marine algae (fish/algae; high in 20:5n⊙3 and 22:6n⊙3); or PLS and algae (PLS/algae; high in 18:3n⊙3 and 22:6n⊙3). Lambs fed either diet containing marine algae contained the highest ($P < 0.05$) percentage of 22:6n⊙3 in the phospholipid (mean of 5.2%), 2.8-fold higher than in sheep fed the fish oil diet.

A more limited number of studies have looked into the effects of dietary supplementation with DHA-rich marine algae on the fatty acid composition of muscle tissue of rabbits [35], lambs [4] and pigs [36, 37].

5.2. Lipid sources for non-ruminants

The cereal-based diet commonly offered to poultry and pigs supplies mainly n–6 *PUFA* and a small amount of n–3 *PUFA*. This is reflected in the fatty acid composition of the animal product. Dietary modification of poultry meat, eggs or pork to increase the n–3 *PUFA* content requires a supply of n–3 *PUFA* from the diet.

The actual strategies to non-ruminants are focused on assessing the effect of offering terrestrial versus marine sources of n–3 PUFA and the subsequent implications for product quality.

5.2.1. Vegetable oils

Enrichment of poultry diets with plant oils has been shown to have different impacts on abdominal fat and the site of fatty acid deposition depending on the *SFA*, *MUFA* and *PUFA* content of the oil [38].

Crespo & Esteve-García [38] studied broiler chickens fed with a basal diet supplemented for 20 days before slaughter with 10% inclusion of linseed oil, sunflower oil and olive oil. As expected with non-ruminant, the fatty acid profile of the deposited fat in the broiler carcase reflected the dietary fat source. The supplementation with olive oil resulting in the highest proportion of C$_{18:1}$, sunflower oil supplementation resulting in the highest proportion of linoleic acid (51.1 g/100 g total body FA), while linseed oil contributed the highest amount of n–3 *PUFA* and most favourable n–6:n–3 ratio in the carcase fat.

In addition, Lu et al. [39] investigated the effects of soybean oil and linseed oil on the fatty acid compositions of pork. The three dietary treatments were: (a) no oil supplement; (b) 3% soybean oil supplement; (c) 3% linseed oil supplement. Dietary linseed oil and soybean oil significantly increased the contents of C$_{18:3}$ and C$_{18:2}$ in the neutral lipids and phospholipids in both longissimus muscle and biceps brachii muscle, respectively.

5.2.2. Linseed and fish oil

The n–3 long chain polyunsaturated fatty acids can be incorporated into non-ruminant products from dietary fish oil. The transfer of these fatty acids was found to be influenced by time and duration of feeding and the presence of other oil supplements. Haak et al. [40] offered to pigs a basal diet composed of barley, wheat and soyabean meal ad libitum alone or supplemented with 1.2% linseed or fish oil during: the whole fattening period; the first fattening phase (weeks 1–8) only; or the second fattening phase (6 weeks or 9 weeks, until slaughter at 100 kg). Haak et al. [40] reported that incorporation of α-linolenic acid into the longissimus thoracis muscle was similar (1.24 g/100 g total FA) when linseed was offered throughout the fattening period or only during the second phase. When fish oil was offered during either of the fattening phases, only the proportion of DHA incorporated was affected, being greater when fish oil was offered during the second fattening phase ($P<0.05$). Incorporation of EPA (Eicosapentaenoic acid) and DHA (Docosahexaenoic acid) following fish oil supplementation for the whole fattening period was 1.37 and 1.02 g/100 g total FA in the longissimus thoracis muscle, representing a six-fold increase compared to the basal diet and a three- and five-fold increase respectively, compared to the linseed diet. In agreement with other animal work [34], Haak et al. [40] concluded that a direct dietary source of DHA was required to increase DHA in animal muscle and that levels in pork could not be substantially influenced by dietary supply of precursors.

5.2.3. Marine algae

It's well known that microalgae are the original source of DHA in the marine food chain [41], dried marine algae have also been included in animal feeds to improve the DHA level of foods of animal origin. Studies has mainly focused on the quality of eggs [42, 43] and chicken meat [44]. A more limited number of studies have looked into the effects of dietary supplementation with DHA-rich marine algae on the fatty acid composition of muscle tissue of pigs [36, 37].

6. Effects of fatty acid modification on the nutritional value of meat

There is a growing consumers resistance to the incorporation of additives into foods, especially where the additives are of synthetic origin, even when they have a nutritional or health advantage. Dietary supplementation of the growing animal provides a unique method of manipulating the content of some micronutrients and other nonnutrient bioactive compounds in meat, with a view to improving the nutrient intake of consumers or improving their overall health.

Research on heart disease in humans has tended to implicate high intakes of saturated fat and cholesterol as contributory factors with a possible protective effect of polyunsaturated fat and a neutral effect of monounsaturated fat [45]. Overall, the advice to consumers has been to control the level of energy consumed as fat to under 35% and in particular, to limit saturated fats to 10%of energy intake [13]. It is also recommended that the proportion of short- and medium-chain saturated fatty acids be reduced and that intake of n-6 fatty acids be reduced relative to n-3 [45].

The nutritional properties of meat are largely related to its fat content and its fatty acid composition. In this sense, long-chain n-3 fatty acids, such as C20:5 n-3 and C22:6 n-3 have beneficial health effects, such as reduction in the thrombotic tendency of blood, associated with lower coronary heart disease in humans [46]. In addition, the role of dietary fat in human health is further complicated by the differing biological activity of some fatty acids when present at different stereospecific positions in triacylglycerols [47].

To avoid possible health dangers from the consumption of the meat of ruminants, a greater degree of unsaturation could be introduced into their fats. One example of the modification of fatty acid in meat resulting in an improvement of human health is the study of Diaz et al. [48]. These authors studied the fatty acid content and sensory characteristics of meat from light lambs fed three diets supplemented with different sources of n-3 fatty acids (fish oil, linseed and linseed plus microalgae) and a control diet during refrigerated storage. The meat from lambs fed linseed diets had the highest levels of C18:3 n-3,while animals fed fish oil, had the highest long-chain n-3 polyunsaturated fatty acids (*PUFA*). Thus, 100 g of meat from lamb fed the fish oil diet provided 183 mg of long-chain n-3 *PUFA*, representing 40% of the daily recommended intake. The levels of n-3, n-6 and long-chain n-3 *PUFA* decreased during a 7-day storage period. These authors reported that consumption of 100 g of lamb muscle from lambs fed control diet would provide about 5% of the daily recommended intake for long chain n-3 fatty acids (500 mg per day, according to EFSA [49]. In case of linseed plus microalgae and linseed diets, would provide nearly 10% of the daily recommended intake for long-chain n-3 fatty acids. The greatest supply of n-3 *PUFA* and long-chain n-3 fatty acids would come from lambs fed fish oil diet, which would provide about 34% of the daily recommended intake for long-chain n-3 fatty acids. Moreover, the highest *PUFA/SFA* ratio was found in lambs fed linseed and fish oil diets, which was close to the recommended value (0.35). The lowest value was observed in lambs fed control diet. During storage, the total content of *PUFA*, including n-3 *PUFA*, n-6 *PUFA* and the long-chain n-3 *PUFA*, decreased. Thus, meat from lambs fed fish oil could supply close to 40% of the daily recommended intake for long-chain n-3 fatty acids on day 0. On day 7 this meat supplies almost 31% and, therefore, this could be considered a reduction in the nutritional value of the meat. This decrease could be a consequence of oxidation changes, since *PUFA* are more prone to oxidation than *MUFA* or *SFA*; the meat from the supplemented groups and especially from animals fed fish diet, are more prone to oxidation than the control diet (with lower content in *PUFA* and long-chain n-3 *PUFA*). Therefore, the importance of the influence of the modification of fatty acid profile on the lipid peroxidation should be studied.

7. Quality of *PUFA* enriched animal products and relation with lipid peroxidation

One of the main factors limiting the quality of meat and meat products is lipid oxidation. Lipid oxidation results in rancid odour and flavour, sometimes referred to as warmed-over flavour. Fatty acids are oxidised into aldehydes, alkanes, alcohols and ketones by chemical (auto-oxidation) or enzymatic (β-oxidation) reactions. In this sense, rancid aroma is apparently due to the dominance of alkanal (hexanal, nonanal) or certain alcohols (1-penten-3-ol, 1-octen-3-ol). The reason is due to the first step of lipid oxidation, which involves the removal of hydrogen from a methylene carbon in the fatty acid. This becomes easier as the number of double bonds in the fatty acid increases, which is why polyunsaturated fatty acids are particularly susceptible to oxidation. Therefore, increasing the degree of unsaturation of muscle membranes reduces the oxidative stability of the muscle. In addition, the relative oxidation rates of fatty acids containing 1, 2, 3, 4, 5 or 6 double bonds are 0.025, 1, 2, 4, 6 and 8, respectively [50].

It is very interesting to correlate the fatty acid profile of the meat with the development of off-odour and off-flavour in order to understand the susceptibility of oxidative damage in the meat. For example, If the ratio of PUFA to SFA is higher in the meat, this softer fat is more susceptible to oxidative damage, and this may cause difficulties for the retailers who are increasingly turning toward centralized butchery and modified atmosphere packaging, both of which lead to meats being exposed to higher levels of oxygen for a longer period of time prior to retail.

There are few studies that examine the effect of an enrichment of the diet in n-3 PUFA and the oxidation potential of muscle. Some of these studies are made in rabbit meat [51-52, 32] and lamb meat [53]. While Kouba et al. [32], reported that the enriched Longissimus dorsi did not exhibit a lower oxidative stability, Castellini et al. [51] and Dal Bosco et al. [52] found that feeding a n-3 PUFA enriched diet lowered significantly TBARS level in, raw meat and Longissimus dorsi. One plausible explanation could be that these authors only supplemented the experimental diet with a high amount of vitamin E, which led to an increase of vitamin E level in tissues of rabbits fed this diet, as already described by Oriani et al. [54], and it is well known that the susceptibility of lipids to oxidation can be reduced by vitamin E, as described by Lin et al. [55] in poultry and Monahan et al. [56] in pigs.

The oxidative processes in living animals are dependent on the endocrine and enzymatic activities in the tissues. There is some evidence that differences between species and breeds of animals exists. However, a high individual variation has also to be assumed. Oxidative processes can occur at many different stages in animal nutrition. During digestion the nutrients are soluble and therefore can be easily oxidized. Extrinsic influences on the oxidative processes mainly derive from the composition of the feedstuffs and feed additives. Therefore, feed should be protected against oxidative damage already during storage.

With increased content of polyunsaturated fatty acids (PUFA) a higher oxidation rate in the feed, in the digest as well as in the intermediate metabolism, occurs. But feedstuffs can also contain antioxidants like vitamins, carotenoids, or phenols, or prooxidative compounds like some trace elements. To improve the oxidative stability of the feed, antioxidative additives are often used as supplements to the diets.

Notwithstanding the beneficial attributes of polyunsaturated fatty acids, it should be noted that lipid oxidation products are believed to adversely affect the health of cells. Fortunately muscular tissue contains several enzymes that protect cells against such change, the most important of which is glutathione peroxidase [57].

To avoid the lipid oxidation tendency shown in meat rich-PUFA, Díaz et al. [48], recommended the inclusion of antioxidants in the diet of lambs, in order to avoid the negative impact on the flavour and to prevent fatty acids from oxidation of these on lamb meat enriched in n-3 fatty acids. Therefore, the inclusion of antioxidants with the incorporation of the ingredients responsible of the fatty acid modification through the feed could be an interesting strategy to prevent oxidation of the meat. Similarly, it has been shown that some fatty acids (such as conjugated linoleic acid) can exert antioxidant activity in meat by reducing lipid oxidation [58]. In a previous study made by our group, the effectiveness of thyme leaves diet (during pregnancy and lactation of ewes) to improving the lamb meat lipid stability was attributed to the antioxidant effect of the phenolic compounds present in the thyme leaf. These bioactive compounds in the leaves may interfere with the propagation reaction of lipid oxidation, besides inhibiting the enzymatic systems involved in initiation reactions [59]. It has been shown that diet with natural antioxidants interferes with the metabolism of fatty acids in ruminants [60]. Taking into accounts another studies using plants of the family Labiatae in the diet, Youdim and Deans [61] showed that a dietary supply of thyme oil or thymol to ageing rats showed a beneficial effect on the antioxidative enzymes superoxide dismutase and glutathione peroxidase, as well as on the polyunsaturated fatty acid composition in various tissues. Animals receiving these supplements had higher concentrations of polyunsaturated fatty acids in phospholipids of the brain compared to the untreated controls. Similarly, Lee et al. [62] showed that the pattern of fatty acids of the abdominal fat of chicken was also altered by oregano oil and dietary carvacrol lowered plasma triglycerides. In animals for food productions, such effects are of importance for product quality: these supplement may improve the dietary value and lead to a better oxidative stability and longer shelf-life of fat, and meat [63].

7.1. Liposomes

Oxidative stress leads to oxidation of low-density lipoproteins (LDL), which plays a key role in the pathogenesis of atherosclerosis, which is the primary cause of coronary heart disease [64]. The nutritional manipulations of the fatty acid composition of meats increase the susceptibility of their lipids to peroxidation; because as have been explained in the previous sections, PUFA are known to act as substrates initiating the oxidative process in meat. In this sense, much attention has been paid to the use of the natural antioxidants, since potentially these components may reduce the level of oxidative stress in the feed.

Several methods have been described in the literature for assessing antioxidant activity. These include radical scavenging assays, ferric reducing assay, or inhibition of the oxidation of oils, emulsions, low-density lipoproteins (LDL), or liposomes. The use of LDL is an interesting method of assessing antioxidant properties relevant to human nutrition, since these systems allow investigation of the protection of a substrate by an antioxidant in a model biological membrane or a lipoprotein. Assessment of the activity of mixtures of lipid-soluble and water-soluble antioxidants in liposomes has clear advantages over other commonly used methods. The liposome system allows the lipid-soluble components to be present in the lipid phase without the presence of a cosolvent, while the water soluble antioxidants can be added to the aqueous phase of the liposome [8].

The liposome system also allows study the synergy between different antioxidants ingredients used in the manufacture of feed, as tocopherols or other water-soluble antioxidants to be demonstrated [65- 66], whereas synergy is not normally observed if these components are present in homogeneous solution.

Therefore, the use of a liposome system is an interesting strategy for a preliminary assessment of the antioxidant activity of ingredients used in the manufacture of feedstuffs.

This was the objective of a previous study [9] where the use of liposomes as biological membrane models to evaluate the potential of natural antioxidants as inhibitors of lipid peroxidation was described. For that, the antioxidative effects of by-products from manufacturing of essential oils, i.e., distilled rosemary leaf residues (DRL), distilled thyme leaf residues (DTL), and the combined antioxidative effects of DRL or DTL with α-tocopherol (TOH), ascorbic acid (AA), and quercetin (QC) on peroxidation of L-α-phosphatidylcholine liposomes as initiated by hydrophilic azo-initiators, were investigated. The results showed that the extracts from DRL and DTL all had an obvious antioxidative effect as evidenced by a lag phase for the formation of phosphatidylcholine-derived conjugated dienes. Combination of TOH or QC with DRL and DTL, respectively, showed synergism in prolonging of the lag phase. Distilled leaves of rosemary and thyme were found to be a rich source of antioxidants as shown by the inhibition of the formation of conjugated dienes in a liposome system. Based on this study, it can be concluded that rosemary and thyme residues, as by-products from distillation of essential oils, are a readily accessible source of natural antioxidants, which possibly provides a good alternative to using synthetic antioxidants in the protection of foods and meat products in particular.

After this study, it was reported that both distilled leaves (rosemary and thyme) were readily accessible source of natural antioxidants in animal feedstuffs, these by-products were added to the feed of pregnant ewes [67- 71]. As shown previously with the liposomes model system study, the meat of lambs from ewes fed with distilled rosemary and thyme leaf had lower levels of lipid oxidation and these additives were considered a good alternative to using synthetic antioxidant in animal diets.

8. Conclusions

This review suggest that the negative image of meat attributed to its highly saturated nature may be overcome by enhancing the fatty acid profile of intramuscular fat through feeding from a human health perspective. Increasing the n-3 PUFA content of animal feedstuffs can be a promising and sustainable way to improve the nutritional value of meat, without forcing consumers to change their eating habits.

It's well known that although dietary PUFA improves meat nutritional qualities, such meats are more susceptible to lipid oxidation during processing. Therefore, there is a need to study the differences in oxidative stability of the muscles in order to understand the effect of dietary on lipid peroxidation. For that the use of liposomes is an interesting strategy to study the lipid peroxidation in model system as preliminary studies (prior the administration of fatty acid sources through feeding).

When all of these considerations are taken into account, the possibility of preserving the nutritional qualities of processed meat rich in *PUFA* by an original dietary antioxidant strategy is recommended, in order to prevent the lipid peroxidation and the decrease of overall liking of meat.

Author details

Gema Nieto and Gaspar Ros
Department of Food Technology, Nutrition and Food Science,
Faculty of Veterinary Sciences, University of Murcia, Murcia, Spain

Acknowledgement

We thank the University of Murcia for the postdoctoral contract of Gema Nieto.

9. References

[1] Wood JD, Richardson RI, Nute GR, Fisher AV, Campo MM, Kasapidou E, Sheard PR, Enser, M.(2003). Effects of fatty acids on meat quality: A review. Meat Sci.66: 21–32.

[2] Scerra M, Caparra P, Foti F, GalofaroV, Sinatra MC, & Scerra V (2007) Influence of ewe feeding systems on fatty acid composition of suckling lambs. Meat Sci. 76: 390-394.

[3] Nieto G, Bañon S, Garrido MD (2012a) Incorporation of thyme leaves in the diet of pregnant and lactating ewes: Effect on the fatty acid profile of lamb. Small Ruminant Res. 105: 140-147.

[4] Elmore JS, Cooper SL, Enser M, Mottram DS, Sinclair LA, Wilkinson, RG, Wood JD (2005) Dietary manipulation of fatty acid composition in lamb meat and its effect on the volatile aroma compounds of grilled lamb. Meat Sci. 69: 233-242.

[5] Bas, P.; Berthelot, E.; Pottier, E. Normand, J. (2007) Effect of level of linseed on fatty acid composition of muscles and adipose tissues of lambs with emphasis on trans fatty acids. Meat Sci. 77: 678-688.

[6] Ponnampalam EN, Trout GR, Sinclair AJ, Egan AR, Leury BJ (2001) Comparison of the color stability and lipid oxidative stability of fresh and vacuum packaged lamb muscle containing elevated omega-3 and omega-6 fatty acid levels from dietary manipulation. Meat Sci. 58: 151–161.

[7] Wood JD, Enser M, Fisher AV, Nute GR, Richardson RI, Sheard PR. (1999) Manipulating meat quality and composition. Proceedings of the Nutrition Society. 58: 363– 370.

[8] Roberts WG, Gordon MH (2003) Determination of the total antioxidant activity of fruits and vegetables by a liposome assay. J. Agr. Food Chem. 51 (5): 1486–1493

[9] Nieto G, Huvaere K, Skibsted LH (2011a) Antioxidant activity of rosemary and thyme by-products and synergism with added antioxidant in a liposome system. Eur. Food Res.Technol. 233: 11-18.

[10] Woods VB, Fearon AM (2009) Dietary sources of unsaturated fatty acids for animals and their transfer into meat, milk and eggs: A review. Livest. Sci. 126:1– 20.

[11] Decker EA, Faustman C, Lope-Bote C.J (2000) Antioxidants in muscle foods. Nutritional strategies to improve quality. Wiley-Interscience. 465 p.

[12] Enser MS, Hallet K, Hewitt B, Fursey GAJ, Wood JD (1996) Fatty acid content and composition of English beef, lamb and pork retail. Meat Sci. 42: 443-456.

[13] Department of Health (1994) Nutritional aspects of cardiovascular disease (Report on health and social subjects no. 46). H.M. Stationery Office, London.

[14] Raes K, Haak L, Balcaen A, Claeys E, Demeyer D, De Smet S (2004) Effect of linseed feeding at similar linoleic acid levels on the fatty acid composition of double-muscled Belgian Blue young bulls. Meat Sci. 66 (2): 307–315.

[15] French P, Stanton C, Lawless F, O'Riordan EG, Monahan FJ, Caffrey PJ, Moloney AP (2000) Fatty acid composition, including conjugated linoleic acid, of intramuscular fat from steers offered grazed grass, grass silage or concentrate-based diets. J. Anim. Sci.78: (11): 2849–2855.

[16] Loor JJ, Hoover WH, Miller-Webster TK, Herbein JH, Polan CE (2003) Biohydrogenation of unsaturated fatty acids in continuous culture fermenters during digestion of orchard grass or red clover with three levels of ground corn supplementation. J. Anim. Sci. 81: 1611–1627.

[17] Romans JR, Johnson RC, Wulf DM, Libal GW, Costello WJ (1995) Effects of ground flaxseed in swine diets on pig performance and on physical and sensory characteristics and omega-3 fatty acids content of pork: I. Dietary level of flaxseed. J. Anim. Sci. 73: 1982– 1986.

[18] Leskanich CO, Matthews KR, Warkup CC, Noble RC, Hazzledine M (1997) The effect of dietary oil containing (n-3) fatty acids on the fatty acid, physicochemical and organoleptic characteristics of pig meat and fat. J. Anim. Sci. 75: 673– 683.

[19] Enser M, Richardson RI, Wood JD, Gill BP, Shaeard PR (2000) Feeding linseed to increase the n-3 PUFA of pork: fatty acid composition of muscle, adipose tissue, liver and sausage. Meat Sci. 55: 201–212.

[20] Kouba M, Enser M, Whittington FM, Nue GR, Wood JD (2003) Effect of a high & linolenic acid diet on lipogenic enzyme activities, fatty acid composition, and meat quality in the growing pig. J. Anim. Sci. 81: 1967– 1979.

[21] Øverland M, Taugbl O, Haug A, Sundstl E (1996) Effect of fish oil on growth performance, carcass characteristics, sensory parameters, and fatty acid composition in pigs. Acta Agric. Scand., A. Anim. Sci. 46: 11 –17.

[22] Kjos NP, Skrede A, Øverland M (1999) Effect of dietary fish silage and fish fat on growth performance and sensory quality of growing-finishing pigs. Can. J. Anim. Sci. 79: 139–147.

[23] Nilzen V, Babol J, Dutta PC, Lundeheim N, Enfalt AC, Lundstrom K (2001) Free-range rearing of pigs with access to pasture grazing-effect on fatty acid composition and lipid oxidation products. Meat Sci. 58: 267–275.

[24] Aurousseau B, Bauchart D, Calichon E, Micol D, Priolo A (2004) Effect of grass or concentrate feeding systems and rate of growth on triglyceride and phospholipid and their fatty acids in the M. *Longissimus thoracis* of lambs. Meat Sci. 66: 531–541.

[25] Wood JD, Enser M (1997) Factors influencing fatty acids in meat and the role of antioxidants in improving meat quality. British J. Nutrition. 78: S49–S60.

[26] Nuernberg K, Fischer A, Nuernberg G., Ender, K., & Dannenberger, D. (2008). Meat quality and fatty acid composition of lipids in muscle and fatty tissue of Skudde lambs fed grass versus concentrate. Small Ruminant Res. 74: 279-283.

[27] Demirel G, Ozpinar H, Nazli B, Keser O (2006) Fatty acids of lamb meat from two breeds fed different forage: concentrate ratio. Meat Sci. 72: 229–235.

[28] Sañudo C, Enser M, Campo MM, Nute GR, Maria G, Sierra I, Wood J D (2000) Fatty acid composition and sensory characteristics of lamb carcasses from Britain and Spain. Meat Sci. 54: 339–346.

[29] Realinia CE, Ducketta SK, Britob GW, Dalla Rizzab M, De Mattos D (2004) Effect of pasture vs. concentrate feeding with or without antioxidants on carcass characteristics, fatty acid composition, and quality of Uruguayan beef. Meat Sci. 66: 567–577.

[30] Raes, K., De Smet, S., & Demeyer, D. (2004). Effect of dietary fatty acids on incorporation of long chain polyunsaturated fatty acids and conjugated linoleic acid in lamb, beef and pork meat: a review. Anim. Feed Sci. Tech. 113: 199–221.

[31] Mach N, Devant M, Díaz I, Font-Furnols M, Oliver MA, García JA, Bach A (2006) Increasing the amount of n–3 fatty acids in meat from young Holstein bulls through nutrition. J. Anim. Sci. 84: 3039–3048.

[32] Kouba M, Benatmane F, Blochet JE, Mourot J (2008) Effect of a linseed diet on lipid oxidation, fatty acid composition of muscle, perirenal fat, and raw and cooked rabbit meat. Meat Sci. 77 (80): 829-834.

[33] Cooper SL (2002). Dietary manipulation of the fatty acid Composition of sheep Meat. Ph.D. Diss. Open University, Milton Keynes, UK.

[34] Cooper SL, Sinclair LA, Wilkinson RG, Hallett KG, Enser M, Wood, JD (2004) Manipulation of the n–3 polyunsaturated fatty acid content of muscle and adipose tissue in lambs. J. Anim. Sci. 82: 1461–1470.

[35] Tassinari M, Mordenti AL, Testi S, Zotti A (2002) Esperienze sulla possibilita` di arricchire con la dieta la carne di coniglio di LCPUFA n-3. Prog. Nutr. 4: 119– 124.

[36] Marriott NG, Garret JE, Sims MD, Abrul JR (2002) Composition of pigs fed a diet with docosahexaenoic acid. J. Muscle Foods. 13: 265– 277.

[37] Sardi L, Martelli G, Lambertini L, Parisini P, Mordenti A (2006) Effects of a dietary supplement of DHA-rich marine algae on Italian heavy pig production parameters. Livest. Sci. 103: 95–103.

[38] Crespo N, Esteve-García E (2002) Nutrient and fatty acid deposition in broilers fed different dietary fatty acid profiles. Poultry Sci. 81: 1533–1542.

[39] Lu P, Zhang LY, D Yin JD, Everts AKR, Li DF (2008) Effects of soybean oil and linseed oil on fatty acid compositions of muscle lipids and cooked pork flavour. Meat Sci. 80: 910–918.

[40] Haak L, De Smeet A, Fremaut D, Van Walleghem K, Raes K (2008) Fatty acid profile and oxidative stability of pork influence by duration and time of dietary linseed or fish oil supplementation. J. Anim. Sci. 86:1418–1425.

[41] Abril R, Garret J, Zeller SG, Sande WJ, Mast RW (2003) Safety assessment of DHA-rich microalgae from Schizochytrium sp.: Part V. Target animal safety/toxicity in growing swine. Regulatory Toxicology and Pharmacology. 37:73– 82.

[42] Van Elswyk, M.E., (1997). Comparison of n−3 fatty acid sources in laying hen rations for improvement of whole egg nutritional quality. A review. British Journal and Nutrition, Vol. 78 (Suppl. 1), 61–69.

[43] Meluzzi A, Sirri F, Tallarico N, Franchini A (2001) Effect of different vegetable lipid sources on the fatty acid composition of egg yolk and on hen performance. Arch. Geflugelkd. 65: 207– 213.

[44] Sirri F, Minelli G, Iaffaldano N, Tallarico N, Franchini A (2003) Oxidative stability and quality traits of n-3 PUFA enriched chicken meat. Italian J. Anim. Sci. 2 (1): 450–452.

[45] Gibney MJ (1993) Fat in animal products: Facts and perceptions. In Safety and Quality of Food from Animals, J.D. Wood and T.L.J. Lawrence (eds.), p.p.57-61. British Society of Animal Production Occasional Publication No. 17, BSAP, Edinburgh.

[46] EFSA (European Food Safety Authority) (2010a). Scientific opinion on dietary reference values for fats, including saturated fatty acids, polyunsaturated fatty acids, monounsaturated fatty acids, trans fatty acids, and cholesterol. EFSA Journal, 8(1461), 1–107.

[47] Decker EA (1996) The role of stereospecific saturated fatty acid positions on lipid nutrition. Nutrition Review. 54: 108–110.

[48] Díaz MT, Cañeque V, Sánchez S, Lauzurica C, Pérez C, Fernández C, Àlvarez I, De la Fuente J (2011) Nutritional and sensory aspects of light lamb meat enriched in n-3 fatty acids during refrigerated storage. Food Chem.124: 147–155.

[49] EFSA (European Food Safety Authority) (2010b). Outcome of the public consultation on the draft opinion of the scientific panel on dietetic products, nutrition, and allergies (NDA) on dietary reference values for fats, including saturated fatty acids, polyunsaturated fatty acids, monounsaturated fatty acids, trans fatty acids, and cholesterol. EFSA ON–1507, pp. 1–23.

[50] Horwitt MK (1986) Interpretations of requirements of thiamine, riboflavin, niacin-tryptophan and vitamin E plus comments on balance studies and vitamin B-6. The American Journal of Clinical Nutrition, 44: 5–973.

[51] Castellini C, Dal Bosco A, Bernardini M, Cyril HW (1998) Effect of dietary vitamin E on the oxidative stability of raw and cooked rabbit meat. Meat Sci. 50: 153–161.

[52] Dal Bosco A, Castellini C, Bianchi L, Mugnai C (2004) Effect of dietary alinolenic acid and vitamin E on the fatty acid composition, storage stability and sensory traits of rabbit meat. Meat Sci. 66: 407–413.

[53] Berthelot, V., Bas, P., Pottier, E., & Normand, J. (2012). The effect of maternal linseed supplementation and/or lamb linseed supplementation on muscle and subcutaneous adipose tissue fatty acid composition of indoor lambs. Meat Science, 90, 548–557.

[54] Oriani G, Salvatori G, Pastorelli G, Pantaelo L, Ritieni A, Corino C (2001) Oxidative status of plasma and muscle in rabbits supplemented with dietary vitamin E. J. Nutritional Biochem. 12: 138–143.

[55] Lin, C. F., Asghar, A., Gray, J. I., Buckley, D. J., Booren, A. M., Crackel, R. L., et al. (1989). Effects of oxidised dietary oil and antioxidant supplementation on broiler growth and meat stability. *British Poultry Science*, Vol. 30, p.p. 855–864.

[56] Monahan FJ, Buckley DJ, Morrissey PA, Lynch PB, Gray JI (1992) Influence of dietary fat and a-tocopherol supplementation on lipid oxidation in pork. Meat Sci. 31: 229–241.

[57] Halliwell B, Aeschbach R, Loeliger J, Auroma OI (1995) The characterization of antioxidants. Food Chem. Toxicol. 33: 601–617.

[58] Joo ST, Lee JI, Ha YL, Park GB (2002) Effects of dietary conjugated linoleic acid on fatty acid composition, lipid oxidation, color, and water-holding capacity of pork loin. J. Anim. Sci. 80: 108–112.

[59] You KM, Jong HG, Kim HP (1999) Inhibition of cyclooxygenase/ lipoxygenase from human platelets by polyhydroxylated/ methoxylated flavonoids isolated from medicinal plants. Arch. Pharm. Res. 22 (1): 18–24.

[60] Vasta V, Pennisi M, Lanza D, Barbagallo M, Bella A, Priolo A (2007) Intramuscular fatty acid composition of lambs given a tanniniferous diet with or without polyethylene glycol supplementation. Meat Sci. 76: 739–745.

[61] Youdim KA, Deans SG (2000) Effect of thyme oil and thymol dietary supplementation on the antioxidant status and fatty acid composition of the ageing rat brain. British J. Nutrition. 83: 878.

[62] Lee, K.W., Everts, H., Kappert, H.J., Frehner, M., Losa, R., & Beynen, A.C. (2003). Effects of dietary essential oil components on growth performance, digestive enzymes and lipid metabolism in female broiler chickens. *British Poultry Science*, Vol. 44, p.p. 450.

[63] Govaris A, Botsoglou E, Florou-Paneri P, Moulas A, Papageorgiou G (2005) Dietary supplementation of oregano essential oil and α-tocopheryl acetate on microbial growth and lipid oxidation of turkey breast fillets during storage. Int. J. Poultry Sci. 4: 969.

[64] Reaven PD, Witztum JL (1996) Oxidized low-density lipoproteins in atherogenesis: Role of dietary modification. Annu. Rev. Nutr. 16: 51-71.

[65] Niki, E., Noguchi, N., Tsuchihashi, H., & Gotoh, N. (1995). Interaction among vitamin C, vitamin E, and beta-carotene. Am. J. Clin. Nutr. 62: S1322-S1326.

[66] Livrea MA, Tesoriere L, D'Arpa D, Morreale M (1997) Reaction of melatonin with lipoperoxyl radicals in phospholipid bilayers. Free Radic. Biol. Med. 23: 706-711.

[67] Nieto G, Bañon S, Garrido MD (2010a) Effect on lamb meat quality of including thyme (*Thymus zygis* ssp. gracilis) leaves in ewes' diet. Meat Sci. 85: 82-88.

[68] Nieto G, Díaz P, Bañon S, Garrido MD (2010b) Dietary administration of ewes diets with a distillate from rosemary leaves (*Rosmarinus officinalis*): influence on lamb meat quality. Meat Sci. 84: 23-29.

[69] Nieto G, Bañon S, Garrido MD (2011b) Effect of supplementing ewes' diet with thyme (*Thymus zygis* ssp. gracilis) leaves on the lipid oxidation of cooked lamb meat. Food Chem. 125(4): 1147-1152.

[70] Nieto G, Estrada M, Jordán MJ, Garrido MD, Bañon S (2011c) Effects in ewe diet of rosemary by-product on lipid oxidation and the eating quality of cooked lamb under retail display. Food Chem. 124(4): 1423-1429.

[71] Nieto G, Bañon S, Garrido MD (2012b) Administration of distillate thyme leaves into the diet of Segureña ewes: effect on lamb meat quality. *Animal*, In press.

Permissions

The contributors of this book come from diverse backgrounds, making this book a truly international effort. This book will bring forth new frontiers with its revolutionizing research information and detailed analysis of the nascent developments around the world.

We would like to thank Angel Catala, for lending his expertise to make the book truly unique. He has played a crucial role in the development of this book. Without his invaluable contribution this book wouldn't have been possible. He has made vital efforts to compile up to date information on the varied aspects of this subject to make this book a valuable addition to the collection of many professionals and students.

This book was conceptualized with the vision of imparting up-to-date information and advanced data in this field. To ensure the same, a matchless editorial board was set up. Every individual on the board went through rigorous rounds of assessment to prove their worth. After which they invested a large part of their time researching and compiling the most relevant data for our readers. Conferences and sessions were held from time to time between the editorial board and the contributing authors to present the data in the most comprehensible form. The editorial team has worked tirelessly to provide valuable and valid information to help people across the globe.

Every chapter published in this book has been scrutinized by our experts. Their significance has been extensively debated. The topics covered herein carry significant findings which will fuel the growth of the discipline. They may even be implemented as practical applications or may be referred to as a beginning point for another development. Chapters in this book were first published by InTech; hereby published with permission under the Creative Commons Attribution License or equivalent.

The editorial board has been involved in producing this book since its inception. They have spent rigorous hours researching and exploring the diverse topics which have resulted in the successful publishing of this book. They have passed on their knowledge of decades through this book. To expedite this challenging task, the publisher supported the team at every step. A small team of assistant editors was also appointed to further simplify the editing procedure and attain best results for the readers.

Our editorial team has been hand-picked from every corner of the world. Their multi-ethnicity adds dynamic inputs to the discussions which result in innovative outcomes. These outcomes are then further discussed with the researchers and contributors who give their valuable feedback and opinion regarding the same. The feedback is then collaborated with the researches and they are edited in a comprehensive manner to aid the understanding of the subject.

Apart from the editorial board, the designing team has also invested a significant amount of their time in understanding the subject and creating the most relevant covers. They scrutinized every image to scout for the most suitable representation of the subject and create an appropriate cover for the book.

The publishing team has been involved in this book since its early stages. They were actively engaged in every process, be it collecting the data, connecting with the contributors or procuring relevant information. The team has been an ardent support to the editorial, designing and production team. Their endless efforts to recruit the best for this project, has resulted in the accomplishment of this book. They are a veteran in the field of academics and their pool of knowledge is as vast as their experience in printing. Their expertise and guidance has proved useful at every step. Their uncompromising quality standards have made this book an exceptional effort. Their encouragement from time to time has been an inspiration for everyone.

The publisher and the editorial board hope that this book will prove to be a valuable piece of knowledge for researchers, students, practitioners and scholars across the globe.

List of Contributors

Marisa Repetto, Jimena Semprine and Alberto Boveris
University of Buenos Aires, School of Pharmacy and Biochemistry, General and Inorganic Chemistry, Institute of Biochemistry and Molecular Medicine (IBIMOL-UBA-CONICET), Argentina

Hanaa Ali Hassan Mostafa Abd El-Aal
Zoology Department, Faculty of Science, Mansoura University, Mansoura, Egypt

Vessela D. Kancheva
Lipid Chemistry Department, Institute of Organic Chemistry with Centre of Phytochemistry, Bulgarian Academy of Sciences, Sofia, Bulgaria

Olga T. Kasaikina
N.N. Semenov Institute of Chemical Physics, Russian Academy of Sciences, Moscow, Russia

Paula M. González, Natacha E. Piloni and Susana Puntarulo
Physical Chemistry-PRALIB, School of Pharmacy and Biochemistry, University of Buenos Aires, Junín, Buenos Aires, Argentina

Jiri Sochor, Branislav Ruttkay-Nedecky, Vojtech Adam, Jaromir Hubalek and Rene Kizek
Department of Chemistry and Biochemistry, Faculty of Agronomy, Mendel University in Brno, Brno, Czech Republic, European Union
Department of Microelectronics, Faculty of Electrical Engineering and Communication, Brno University of Technology, Brno, Czech Republic, European Union

Petr Babula
Department of Natural Drugs, Faculty of Pharmacy, University of Veterinary and Pharmaceutical Sciences Brno, Brno, Czech Republic, European Union

Biljana Kaurinovic and Mira Popovic
Department of Chemistry, Biochemistry and Environmental Protection, University of Novi Sad, Novi Sad, Republic of Serbia

Natalia Fagali and Angel Catalá
Instituto de Investigaciones Fisicoquímicas Teóricas y Aplicadas, (INIFTA-CCT La Plata-CONICET), Facultad de Ciencias Exactas, Universidad Nacional de La Plata. Casilla de Correo 16, La Plata, Argentina

Toshiyuki Toyosaki
Department of Foods and Nutrition, Koran Women's Junior College, Fukuoka, Japan

Neda Mimica-Dukić, Nataša Simin, Emilija Svirčev, Dejan Orčić, Ivana Beara and Marija Lesjak
Department of Chemistry, Biochemistry and Environmental Protection, University of Novi Sad, Faculty of Sciences, Novi Sad, Republic of Serbia

Biljana Božin
Department of Pharmacy, Faculty of Medicine, University of Novi Sad, Novi Sad, Republic of Serbia

Kamsiah Jaarin and Yusof Kamisah
Department of Pharmacology, Faculty of Medicine, Universiti Kebangsaan Malaysia, Malaysia

Gema Nieto and Gaspar Ros
Department of Food Technology, Nutrition and Food Science, Faculty of Veterinary Sciences, University of Murcia, Murcia, Spain

Printed in the USA
CPSIA information can be obtained
at www.ICGtesting.com
JSHW011445221024
72173JS00004B/951

9 781632 394491